Introduction to the Theory of X-Ray and Electronic Spectra of Free Atoms

PHYSICS OF ATOMS AND MOLECULES

Recent volumes in this series:

ATOMIC PHOTOEFFECT
M. Ya. Amusia

ATOMIC SPECTRA AND COLLISIONS IN EXTERNAL FIELDS
Edited by K. T. Taylor, M. H. Nayfeh, and C. W. Clark

ATOMS AND LIGHT: INTERACTIONS
John N. Dodd

ELECTRON COLLISIONS WITH MOLECULES, CLUSTERS, AND SURFACES
Edited by H. Ehrhardt and L. A. Morgan

ELECTRON–MOLECULE SCATTERING AND PHOTOIONIZATION
Edited by P. G. Burke and J. B. West

THE HANLE EFFECT AND LEVEL-CROSSING SPECTROSCOPY
Edited by Giovanni Moruzzi and Franco Strumia

INTRODUCTION TO THE THEORY OF LASER–ATOM INTERACTIONS, Second Edition
Marvin H. Mittleman

INTRODUCTION TO THE THEORY OF X-RAY AND ELECTRONIC SPECTRA OF FREE ATOMS
Romas Karazija

MOLECULAR PROCESSES IN SPACE
Edited by Tsutomu Watanabe, Isao Shimamura, Mikio Shimizu, and Yukikazu Itikawa

POLARIZATION BREMSSTRAHLUNG
Edited by V. N. Tsytovich and I. M. Ojringel

POLARIZED ELECTRON/POLARIZED PHOTON PHYSICS
Edited by H. Kleinpoppen and W. R. Newell

THEORY OF ELECTRON–ATOM COLLISIONS, Part 1: Potential Scattering
Philip G. Burke and Charles J. Joachain

VUV AND SOFT X-RAY PHOTOIONIZATION
Edited by Uwe Becker and David A. Shirley

A Chronological Listing of Volumes in this series appears at the back of this volume.

A Continuation Order Plan is available for this series. A continuation order will bring delivery of each new volume immediately upon publication. Volumes are billed only upon actual shipment. For further information please contact the publisher.

Introduction to the Theory of X-Ray and Electronic Spectra of Free Atoms

Romas Karazija
Institute of Theoretical Physics and Astronomy
Lithuanian Academy of Sciences
Vilnius, Lithuania

Translated from Russian by
W. Robert Welsh

Springer Science+Business Media, LLC

Library of Congress Cataloging-in-Publication Data

Karazija, Romas.
[Vvedenie v teoriiu rentgenovskikh i ėlektronnykh spektrov svobodnykh atomov. English]
Introduction to the theory of x-ray and electronic spectra of free atoms / Romas Karazija ; translated from Russian by W. Robert Welsh.
p. cm. -- (Physics of atoms and molecules)
Translation of: Vvedenie v teoriiu rentgenovskikh i ėlektronykh spektrov svobodnykh atomov.
Includes bibliographical references and index.

1. Atomic spectra. 2. Molecular spectra. 3. X-ray spectroscopy. I. Title. II. Series.
QC454.A8K3713 1996
539.7--dc20 92-6731
CIP

ISBN 978-1-4899-1536-8 ISBN 978-1-4899-1534-4 (eBook)
DOI 10.1007/978-1-4899-1534-4

Originally published by Plenum Press, New York in 1996.
Softcover reprint of the hardcover 1st edition 1996

10 9 8 7 6 5 4 3 2 1

NOTATION

α	fine-structure constant, equal to $e^2/\hbar c \approx 0.00729735$;
ζ_{nl}	spin–orbit coupling constant, Eq. (1.225);
δ_l	scattering phase;
σ_{nl}	(outer) screening constant, Eq. (3.11);
σ_{nl}	total screening constant;
$\rho(r)$	spherically averaged probability density of finding electrons in an atom, Eq. (1.158);
$a_\nu^\dagger$, a_ν	creation and annihilation operators for an electron in a state ν;
$A_a(E)$	spectral function of an a^{-1} vacancy, Eq. (3.14);
c	speed of light;
$C^{(k)}$	spherical function operator, Eq. (1.100);
$D^{(1)} \equiv O^{(1)}$	dipole transition operator [Eq. (1.259) when $t = 1$];
e	absolute value of electron charge;
$\hat{\varepsilon}$	photon polarization unit vector;
F^k, G^k	electrostatic interaction integrals, Eqs. (1.108) and (1.109);
h	Planck's constant, $\hbar = h/2\pi$
H^e	electrostatic interaction operator, Eq. (1.96);
H^{so}	spin–orbit interaction operator, Eq. (1.132);
$I_{nl}(I_{nlj})$	bonding energy of an $nl(nlj)$ electron;
$\mathbf{k}$	photon wave vector;
m	rest mass of an electron;
$\eta(\gamma J)$	population density of γJ level;
$P_{nl}(r)$	radial wave function for an nl electron;
$P_{nlj}(r)$, $Q_{nlj}(r)$	the large and small components, respectively, of a relativistic wave function;
R	Rydberg's constant, equal to $me^4/4\pi\hbar^3 c$;
Ry	Rydberg, equal to 1/2 atomic units of energy;
$s_{n_1 l_1, n_2 l_2}^{(t)}$	one-electron submatrix element of the electric multipole transition operator, Eq. (1.264);

$U^{(k)}$, $V^{(kk')}$	operators composed from the unit tensors given by Eqs. (1.71) and (1.72);
Z	atomic number or nuclear charge in units of e; Z^* is the effective nuclear charge;
$\mathbf{q}$	electron wave vector;
${}_eO^{(t)}$, ${}_mO^{(t)}$	relativistic electric and magnetic multipole transition operators given, respectively, by Eqs. (1.277) and (1.278);
$Y_{lm}(\theta, \varphi)$	spherical harmonic;
$[j_1, j_2, j_3, \ldots]$	$(2j_1 + 1)(2j_2 + 1)(2j_3 + 1), \ldots$;
$\{l_1 l_2 l_3\}$	triangular condition, Eq. (1.11);
$(l_1 l_2 l_3)$	triangular condition with an even sum of parameters;
$\langle \mid \quad \mid \rangle$	matrix element for nonrelativistic wave functions;
$(\mid \quad \mid)$	matrix element for relativistic wave functions;
$(\Vert)$	coefficient of fractional parentage;
$\begin{pmatrix} j_1 & j_2 & j_3 \\ m_1 & m_2 & m_3 \end{pmatrix}$	Wigner symbol, Eq. (1.17);
$\begin{Bmatrix} j_1 & j_2 & j_3 \\ l_1 & l_2 & l_3 \end{Bmatrix}$	6j-symbol, Eq. (1.22);
$\begin{Bmatrix} j_1 & j_2 & j_3 \\ l_1 & l_2 & l_3 \\ k_1 & k_2 & k_3 \end{Bmatrix}$	9j-symbol Eq. (1.26).

PREFACE

Research into the regularities of x-ray radiation and absorption spectra, the formulation of Moseley's law, and the discovery of the Compton effect have all played a major role in shaping quantum mechanics. Modern atomic theory has in the main developed from the results of more accurate optical spectroscopy, whereas x-ray spectroscopy has been more closely associated with solid-state and molecular physics. It was in approximately the 1960s that x-ray, x-ray electron, and Auger spectroscopy (in short, electron spectroscopy) were developed significantly. The awarding of the 1981 Nobel Prize in Physics to K. Siegbahn ". . . for his contribution to the development of electron spectroscopy" bears witness to the extraordinary renaissance of x-ray physics and the importance of these methods in science and technology.

Methods from the many-electron theory of a free atom are being increasingly used and developed in x-ray and electron spectroscopy. Interest in the spectra of the gaseous and vapor states is increasing because these spectra contain valuable information on the properties of highly excited atoms, on the various processes that occur when the inner and outer electron shells participate, and on many-electron effects. Comparing the spectra of free atoms with those of the same atoms in molecules and solids makes it possible for us to distinguish the effects of the environment and the free atom.

A large number of works have examined the spectra of inert gas atoms, which can be quite easily studied both experimentally and theoretically, thus making these spectra a convenient means for testing the various models, approximations, and hypotheses used.

Compounds of the rare-earth elements exhibit some interesting physical and chemical properties that are widely used in science and technology. Their atoms contain an inner open *f*-shell. The model of a free atom as a zero approximation is useful for describing the multiple spectra of transition elements having an open *d*-shell. The effects of a crystal field are then considered in first-order perturbation theory.

If an x-ray or an Auger transition occurs when only the inner shells of the atom participate, then the approximation of a free atom or ion can be used to interpret the spectra of the other elements — the effects of the environment usually manifest themselves as merely an overall shift in the spectrum.

High-precision x-ray and electron spectra have a very complex structure that contains information on the various processes taking place concurrently in the inner shells of an atom, on many-electron and relativistic effects, and on the comparatively unstudied properties of the inner, especially the intermediate, shells. This has led to increased interest in x-ray and electron spectroscopy.

A number of monographs [1 – 10, etc.] have discussed the theory and interpretation of x-ray and electron spectra, although it has been mostly the spectra of solids and molecules that have been examined. The development and use of methods from the theory of a free atom in different areas of x-ray and electron spectroscopy have appeared in a number of handbooks [11 – 25] and review articles [26 – 31, etc.]. This book provides a general introduction to the physics of the inner shells of an atom and a discussion of the theoretical foundations for different x-ray and electron spectra. The author has attempted to consistently use the mathematical technique of irreducible tensorial operators and to extend the examination to expressions that can be used to calculate the spectra. In doing this, the question arises as to what are the relative role and range of applicability of relativistic and nonrelativistic methods, as well as one-configuration and correlation methods.

Relativistic effects increase rapidly as the nuclear charge increases and as the shells approach the nucleus. Unfortunately, relativistic methods are much more complex than nonrelativistic methods. Also, when open shells are present, their electrons can be distributed among the subshells by various means, and these relativistic configurations, or subconfigurations, intermix among themselves. For this reason a nonrelativistic one-configuration approximation corresponds to the superposition of a number of subconfigurations, which makes the relativistic description of an atom even more complex. In many cases, first-order perturbation theory can effectively take relativistic effects into account, especially for the intermediate shells. For this reason the nonrelativistic approximation is widely used when interpreting x-ray and electron spectra, and we will pay particular attention to this method here. Relativistic expressions will, basically, reduce to a one- or two-electron approximation that is useful for describing the transitions in the deep inner shells.

Methods that take the correlation between electrons into account are being increasingly used at present in x-ray and electron spectroscopy (we will call these correlation methods rather than many-electron methods, since many-electron effects are accounted for even in a one-configuration approximation, e.g., when using the wave functions of coupled angular momenta). For highly excited autoionization states, besides correlation effects, the interaction between the configurations and the continuous spectrum is also significant. Correlation between electrons increases rapidly for some intermediate and subvalent shells. Nevertheless, in many cases a one-configuration approximation allows us to describe the features of the different spectra and clarify their relationships not only qualitatively, but quantitatively as well. This one-configuration approximation is the clearest and most universal and allows us to derive a basis for more accurate calculations. On the other hand, many correlation methods — especially those from perturbation theory — have thus far been realized in practice for only the simplest configurations that have filled shells or a small number of electrons in the atom. Therefore, it is always necessary to carefully consider those cases in which the rather large amount of extra work involved in calculating correlation and relativistic effects in x-ray and electron spectra, especially in the zero approximation, is really necessary.

Interpreting specific spectra will not be considered here. The bibliography lists only a small number of works that deal with the wide range of topics that are examined. These are primarily the reviews and theoretical articles which, in the author's opinion, are the more important.

The theoretical methods that will be examined here are not limited to x rays only and can, with few exceptions, be used to describe the spectra of vacuum ultraviolet light.

Chapters 1 and 2 briefly discuss those topics in atomic theory which will be used in the remainder of the book. Chapter 3 discusses the peculiarities of configurations with vacancies and the structure of the energy spectrum. Chapter 4 acquaints the reader with the basic methods for calculating the populations of the excited states of the atom. Subsequent chapters discuss the theory of the various x-ray and electron spectra — the emission and absorption of x rays, photoelectrons, and Auger electrons, bremmstrahlung, and x-ray scattering.

The quantities e, m, $\hbar$ are used in the expressions, which are set to unity in the atomic system of units. The exceptions are those formulas in which the use of the atomic system of units is generally accepted; these are designated by an asterisk.

This book was made possible only through the profound influence of the author's teacher, Professor A. Jucys (Yutsis). It is mainly because of his initiative almost 15 years ago at the Institute of Physics and Mathematics that work on the theory of x-ray spectra began. The author is grateful to Professor Z. Rudzikas for his interest in this work and for valuable advice, and to his co-workers A. Savukynas, J. Kaniauskas, E. Naslines, S. Kuces, and A. Karssiene, and his colleagues from Rostov-on-Don I. D. Petrov and V. L. Sukhorukov, for their valuable comments while reading the manuscript. The author is especially grateful to the reviewers, Docent D. Grabauskas and Professor E. S. Parilis, for their constructive criticism of the manuscript.

CONTENTS

Chapter 1

FUNDAMENTALS OF ATOMIC THEORY. THE ONE-CONFIGURATION APPROXIMATION

The wave function of an atom containing N electrons depends on the $4N$ spatial and spin variables and can be defined exactly by solving Schrödinger's equation only when $N = 1$. A one-configuration model based on the following assumptions is used to approximately describe a many-electron atom.

1. There are one-electron states in an atom which are described by a one-electron wave function.

2. The field in an atom is spherically symmetric (the potential is a function of only the radial coordinate).

3. The one-electron states of electrons having the same quantum numbers nl (in a nonrelativistic theory) or nlj (in relativistic theory) are described by the same radial wave function.

The first assumption allows us to construct an atom's wave function as the antisymmetrization product of one-electron wave functions.

Because the field is spherically symmetric, the one-electron wave function can be divided into orbital, spin, and radial functions. The angular momentum of an electron in such a field is conserved; therefore, its orbital wave function must be an eigenfunction of the angular momentum and its projection onto the z axis; i.e., it has the standard form, determined only by the symmetry of the field. Because the one-electron nonrelativistic Hamiltonian also commutes with the electron's spin operator, the one-electron spin function is also standard. Only the radial wave function is dependent on the electron configuration and the peculiarities of the field in the atom. Additionally, in a spherically symmetric field the wave function for the entire atom can be conveniently constructed so that it will be the eigenfunction of the total angular momenta of the electron system, and the operators for the physical quantities will be irreducible tensorial operators. Then, when we calculate the mean values of these quantities, we can use the effective mathematical technique of irreducible tensorial operators, the foundation for which was laid in [32, 33]. We will describe this technique at the beginning of this chapter and then use it to find the matrix elements for the relativistic and nonrelativistic Hamiltonians and radiative transition operators. Section 1.5 discusses the Hartree–Fock method for calculating radial wave functions. The one-configuration approximation is examined only briefly and without exhaustive proofs, being primarily oriented to the applications in other chapters. The reader can find a systematic discussion of these methods in [34–39]. Topics in group theory [40, 41] are virtually untouched in this chapter, and the method of second quantization [40, 49] is discussed only briefly.

1.1. Angular Momentum and the 3*nj* Symbols

The concept of angular momentum and its properties forms the basis of the theoretical description of a many-electron atom [36, 42, 44].

The quantum mechanics definition of angular momentum is derived from the definition used in classical mechanics,

$$\mathbf{l} = [\mathbf{r} \times \mathbf{p}] \tag{1.1}$$

by replacing the radius vector $\mathbf{r}$ and momentum $\mathbf{p}$ with their respective operators. Expressions for the commutators of the angular momentum components j_i,

$$[j_x, j_y] = i\hbar j_z, \quad [j_y, j_z] = i\hbar j_x, \quad [j_z, j_x] = i\hbar j_y. \tag{1.2}$$

follow from the rules of commutation for these operators. Here, the angular momentum $\mathbf{l}$ is replaced by $\mathbf{j}$ since Eq. (1.2) is satisfied for not only the orbital, but the spin s and total $\mathbf{j} = \mathbf{l} + \mathbf{s}$ angular momenta. Thus, Eq. (1.2) is a more general definition of this quantity than is Eq. (1.1).

The square of the angular momentum commutes with any of its components; therefore, $\mathbf{j}^2$ and, for example, j_z form a set of commutating operators that have the same eigenfunctions $|\,jm\rangle$. From Eq. (1.2) for the commutation conditions, we obtain

$$\mathbf{j}^2 |jm\rangle = \hbar^2 j(j+1) |jm\rangle, \tag{1.3}$$

$$j_z |jm\rangle = \hbar m |jm\rangle. \tag{1.4}$$

The eigenvalues j and m are both integers or half integers, and $-j \leq m \leq j$.

Using the components j_x and j_y, we can compose the operators $j_\pm = j_x \pm ij_y$, which, by acting on the eigenfunction $|jm\rangle$, transform it into another function having a projection that is greater or less than m by one:

$$j_\pm |jm\rangle = \hbar [j(j+1) - m(m \pm 1)]^{1/2} |jm \pm 1\rangle. \tag{1.5}$$

The eigenfunctions of the square of the orbital angular momentum $\mathbf{l}^2$ and the component l_z are the spherical harmonics $Y_{lm}(\theta, \varphi)$ [44].

The nonrelativistic one-electron Hamiltonian commutes with $\mathbf{l}^2$, l_z and $\mathbf{s}^2$, s_z; therefore, the one-electron wave function (one-electron orbital) can be written

$$|lsm_l\mu\rangle = Y_{lm_l}(\theta, \varphi)\, \chi_\mu(\sigma)\, \frac{P_{nl}(r)}{r}. \tag{1.6}$$

Here, $P_{nl}(r)$ is the radial wave function and $\chi_\mu(\sigma) = \delta(\mu, \sigma)$ is a spin function, where μ is the projection of the electron's spin onto the z axis in units of $\hbar$; σ is the spin and can assume two possible values: $-1/2$ and $1/2$.

Angular momentum is closely related to rotation group. The rotation operator is R, which, when the coordinate system is rotated, transforms the wave function of the atomic electron ψ into a new wave function

$$\psi' = R\psi, \tag{1.7}$$

that is given in terms of the angular momentum operator [50]. Therefore, the functions $\psi = |jm\rangle$ form the basis for the irreducible representation of the rotation group and are transformed through one another by the operator R:

$$R|jm\rangle = \sum_{m'} D^{(j)}_{m'm} |jm'\rangle, \tag{1.8}$$

where $D_{m'm}^{(j)}$ is a matrix element of the operator R.

The nonrelativistic Hamiltonian for the electron system does not commute with the one-electron operators $\mathbf{l}_i^2$, $\mathbf{s}_i^2$, l_{iz}, or s_{iz}, although it does commute with the operators for the squares and z-projections of the resultant angular momenta

$$\mathbf{L}=\sum_i \mathbf{l}_i,\quad \mathbf{S}=\sum_i \mathbf{s}_i,\quad \mathbf{J}=\mathbf{L}+\mathbf{S}, \tag{1.9}$$

Therefore, the state of an atom with many electrons can be conveniently described by the eigenfunctions and eigenvalues of the commutating operators $\mathbf{L}^2$, $\mathbf{S}^2$, L_z, and S_z or $\mathbf{L}^2$, $\mathbf{S}^2$, $\mathbf{J}^2$, and J_z.

We will examine the construction of wave functions of the resultant angular momenta for the vector addition of two commuting angular momenta $\mathbf{j}_1 + \mathbf{j}_2 = \vec{\mathrm{j}}$ (this may be the angular momenta of the individual electrons or their groups, or the orbital and spin angular momenta of the same electron).

We can readily see that $\mathbf{j}^2$ commutes with $\mathbf{j}_1{}^2$, $\mathbf{j}_2{}^2$, and j_z, but does not commute with j_{1z} and j_{2z}. The transition from the basis of eigenfunctions for the operators $\mathbf{j}_1{}^2$, j_{1z}, $\mathbf{j}_2{}^2$, and j_{2z} to the basis of eigenfunctions $\mathbf{j}_1{}^2$, $\mathbf{j}_2{}^2$, $\mathbf{j}^2$, and j_z is made via the linear unitary transformation:

$$|j_1 j_2 jm\rangle = \sum_{m_1 m_2} \begin{bmatrix} j_1 & j_2 & j \\ m_1 & m_2 & m \end{bmatrix} |j_1 m_1\rangle\, |j_2 m_2\rangle . \tag{1.10}$$

The quantity in square brackets is the vector coupling (Clebsch–Gordan) coefficient, which has a rather complex expression [42, 44] although it can be greatly simplified when one of the parameters is 0, 1/2, or 1 [39, 44]. We will point out the basic features of the vector coupling coefficient. It is defined as a real quantity. The sum of the angular momenta and its projections in one column as well as the sum of all three angular momenta must be an integer. In addition, the coefficient does not disappear even when

$$|j_1 - j_2| \leqslant j \leqslant j_1 + j_2, \tag{1.11}$$

$$m = m_1 + m_2 . \tag{1.12}$$

The first equation is the triangular condition, designated $\{j_1, j_2, j_3\}$: the numbers j_1, j_2, j_3 form a triad (corresponds to the sides of triangle) if any one of them is greater than the absolute value of the difference between the other two and less than the sum of the other two.

The conditions for the vector coupling coefficients

$$\sum_{m_1 m_2} \begin{bmatrix} j_1 & j_2 & j \\ m_1 & m_2 & m \end{bmatrix} \begin{bmatrix} j_1 & j_2 & j' \\ m_1 & m_2 & m' \end{bmatrix} = \delta(j, j')\,\delta(m, m')\,\{j_1 j_2 j\}, \tag{1.13}$$

$$\sum_{jm} \begin{bmatrix} j_1 & j_2 & j \\ m_1 & m_2 & m \end{bmatrix} \begin{bmatrix} j_1 & j_2 & j \\ m_1' & m_2' & m \end{bmatrix} = \delta(m_1, m_1')\,\delta(m_2, m_2'). \qquad (1.14)$$

follow from the condition that the wave functions be orthonormal. They are satisfied by the permutation relations

$$\begin{aligned} \begin{bmatrix} j_1 & j_2 & j \\ m_1 & m_2 & m \end{bmatrix} &= (-1)^{j_1+j_2-j} \begin{bmatrix} j_2 & j_1 & j \\ m_2 & m_1 & m \end{bmatrix} = (-1)^{j_1-m_1} \left(\frac{2j+1}{2j_2+1} \right)^{1/2} \\ &\times \begin{bmatrix} j_1 & j & j_2 \\ m_1 & -m & -m_2 \end{bmatrix} = (-1)^{j_2+m_2} \left(\frac{2j+1}{2j_1+1} \right)^{1/2} \begin{bmatrix} j & j_2 & j_1 \\ -m & m_2 & -m_1 \end{bmatrix} \\ &= (-1)^{j-j_2+m_1} \left(\frac{2j+1}{2j_2+1} \right)^{1/2} \begin{bmatrix} j & j_1 & j_2 \\ -m & m_1 & -m_2 \end{bmatrix} = (-1)^{j-j_1-m_2} \left(\frac{2j+1}{2j_1+1} \right)^{1/2} \\ &\times \begin{bmatrix} j_2 & j & j_1 \\ m_2 & -m & -m_1 \end{bmatrix} = (-1)^{j_1+j_2-j} \begin{bmatrix} j_1 & j_2 & j \\ -m_1 & -m_2 & -m \end{bmatrix}. \end{aligned} \qquad (1.15)$$

The same vector coupling coefficient makes a transformation that is the inverse of the transformation given by Eq. (1.10):

$$j_1 j_2 m_1 m_2\rangle \equiv |j_1 m_1\rangle\,|j_2 m_2\rangle = \sum_{jm} \begin{bmatrix} j_1 & j_2 & j \\ m_1 & m_2 & m \end{bmatrix} |j_1 j_2 jm\rangle. \qquad (1.16)$$

A more symmetric quantity, the Wigner symbol, or the $3j$-symbol

$$\begin{pmatrix} j_1 & j_2 & j_3 \\ m_1 & m_2 & m_3 \end{pmatrix} = (-1)^{-j_1+j_2+m_3} [j_3]^{-1/2} \begin{bmatrix} j_1 & j_2 & j_3 \\ m_1 & m_2 & -m_3 \end{bmatrix}. \qquad (1.17)$$

is often used instead of the vector coupling coefficient. From this point on, we will use the notation $[j_1, j_2, \ldots] = (2j_1 + 1) \times (2j_2 + 1) \ldots$

The Wigner symbol has the following fundamental properties of symmetry:

$$\begin{pmatrix} j_1 & j_2 & j_3 \\ m_1 & m_2 & m_3 \end{pmatrix} = (-1)^{j_1+j_2+j_3} \begin{pmatrix} j_1 & j_2 & j_3 \\ -m_1 & -m_2 & -m_3 \end{pmatrix}, \qquad (1.18)$$

$$\begin{pmatrix} j_1 & j_2 & j_3 \\ m_1 & m_2 & m_3 \end{pmatrix} = \varepsilon \begin{pmatrix} j_i & j_k & j_p \\ m_i & m_k & m_p \end{pmatrix}, \tag{1.19}$$

where $\varepsilon = 1$, if the permutation $\begin{pmatrix} 1 & 2 & 3 \\ i & k & p \end{pmatrix}$ is even and $(-1)^{j_1+j_2+j_3}$ if it is odd.

A Wigner symbol having one angular momentum equal to zero has the simple expression

$$\begin{pmatrix} j_1 & j_2 & 0 \\ m_1 & m_2 & 0 \end{pmatrix} = \delta(j_1, j_2)\,\delta(m_1, -m_2)(-1)^{j_1-m_1}[j_1]^{-1/2}. \tag{1.20}$$

Specifically, it follows from Eqs. (1.10), (1.17), and (1.19) that the order in which the angular momenta are coupled is important:

$$|j_2 j_1 jm\rangle = (-1)^{j_1+j_2-j}\,|j_1 j_2 jm\rangle. \tag{1.21}$$

When three or more angular momenta are coupled, the eigenfunctions of the resultant angular momentum can be found by repeatedly applying Eq. (1.10), which is equivalent to using the so-called generalized vector coupling coefficients [36]. Functions corresponding to different momentum-coupling schemes are distinguished by not only a phase factor as in Eq. (1.21), but by intermediate angular momenta and by another dependence on the various angular momenta.

Transformations between wave functions in which the same angular momenta are coupled into the same resultant angular momentum by different means are accomplished by transformation matrices (Appendix 3). These are given in terms of $3nj$ symbols that are independent of the projections of the quantities that are sums of the products of the vector coupling coefficients.

The matrices for transformations between wave functions having a different coupling order for three angular momenta are given in terms of a $6j$-symbol or a Racah coefficient, which is the sum of the Wigner symbols:

$$\begin{Bmatrix} j_1 & j_2 & j_3 \\ l_1 & l_2 & l_3 \end{Bmatrix} = \sum_{m_i\, n_i} (-1)^{\varphi} \begin{pmatrix} j_1 & j_2 & j_3 \\ m_1 & m_2 & -m_3 \end{pmatrix} \begin{pmatrix} j_1 & l_2 & l_3 \\ -m_1 & n_2 & n_3 \end{pmatrix} \begin{pmatrix} l_1 & l_2 & j_3 \\ n_1 & -n_2 & m_3 \end{pmatrix} \times$$

$$\times\begin{pmatrix} l_1 & j_2 & l_3 \\ -n_1 & -m_2 & -n_3 \end{pmatrix},$$

$$\varphi=j_1-m_1+j_2-m_2+j_3-m_3+l_1-n_1+l_2-n_2+l_3-n_3. \tag{1.22}$$

Consequently, the parameters of the $6j$-symbol must satisfy four triangular conditions. The coefficient has 24 fundamental properties of symmetry:

$$\begin{Bmatrix} j_1 & j_2 & j_3 \\ l_1 & l_2 & l_3 \end{Bmatrix}=\begin{Bmatrix} j_a & j_b & j_c \\ l_a & l_b & l_c \end{Bmatrix}$$

$$=\begin{Bmatrix} j_a & l_b & l_c \\ l_a & j_b & j_c \end{Bmatrix} \quad (a,\ b,\ c=1,\ 2,\ 3). \tag{1.23}$$

Formulas for the $6j$-symbol with one parameter equal to zero are often used,

$$\begin{Bmatrix} j_1 & j_2 & j_3 \\ l_1 & l_2 & 0 \end{Bmatrix}=\delta(j_1,\ l_2)\,\delta(j_2,\ l_1)\{j_1j_2j_3\}(-1)^{j_1+j_2+j_3}[j_1,\ j_2]^{-1/2} \tag{1.24}$$

or unity when the other parameters are pairwise equal to.

$$\begin{Bmatrix} j_1 & j_2 & j_3 \\ j_2 & j_1 & 1 \end{Bmatrix}=(-1)^{j_1+j_2+j_3}\frac{2[j_3(j_3+1)-j_1(j_1+1)-j_2(j_2+1)]}{[2j_1(2j_1+1)(2j_1+2)\,2j_2(2j_2+1)(2j_2+2)]^{1/2}}. \tag{1.25}$$

The sum of the products of three $6j$-symbols yields a $3nj$-symbol of higher order – a $9j$-symbol:

$$\begin{Bmatrix} j_1 & j_2 & j_3 \\ l_1 & l_2 & l_3 \\ k_1 & k_2 & k_3 \end{Bmatrix}=\sum_x [x](-1)^{2x}\begin{Bmatrix} j_1 & j_2 & j_3 \\ l_3 & k_3 & x \end{Bmatrix}\begin{Bmatrix} l_1 & l_2 & l_3 \\ j_2 & x & k_2 \end{Bmatrix}\begin{Bmatrix} k_1 & k_2 & k_3 \\ x & j_1 & l_1 \end{Bmatrix}. \tag{1.26}$$

Rows and columns can be replaced in this symbol (if the replacement is odd, a phase factor will appear), and the parameters can also be transposed relative to both diagonals:

$$\begin{Bmatrix} j_1 & j_2 & j_3 \\ l_1 & l_2 & l_3 \\ k_1 & k_2 & k_3 \end{Bmatrix} = \begin{Bmatrix} j_1 & l_1 & k_1 \\ j_2 & l_2 & k_2 \\ j_3 & l_3 & k_3 \end{Bmatrix} = (-1)^{\varphi} \begin{Bmatrix} j_2 & j_1 & j_3 \\ l_2 & l_1 & l_3 \\ k_2 & k_1 & k_3 \end{Bmatrix} = (-1)^{\varphi} \begin{Bmatrix} l_1 & l_2 & l_3 \\ j_1 & j_2 & j_3 \\ k_1 & k_2 & k_3 \end{Bmatrix},$$

$$\varphi = \sum_{i=1}^{3} (j_i + l_i + k_i). \tag{1.27}$$

The condition for a $9j$-symbol with two equal rows or columns to become zero follows from the latter property:

$$\begin{Bmatrix} j_1 & j_2 & j_3 \\ j_1 & j_2 & j_3 \\ k_1 & k_2 & k_3 \end{Bmatrix} = 0, \tag{1.28}$$

if $k_1 + k_2 + k_3$ is odd. A $9j$-symbol with one zero is equal to

$$\begin{Bmatrix} j_1 & j_2 & j_3 \\ l_1 & l_2 & l_3 \\ k_1 & k_2 & 0 \end{Bmatrix} = \begin{Bmatrix} l_1 & l_2 & l_3 \\ k_1 & k_2 & 0 \\ j_1 & j_2 & j_3 \end{Bmatrix} = \begin{Bmatrix} k_1 & k_2 & 0 \\ j_1 & j_2 & j_3 \\ l_1 & l_2 & l_3 \end{Bmatrix} = \begin{Bmatrix} j_2 & j_3 & j_1 \\ l_2 & l_3 & l_1 \\ k_2 & 0 & k_1 \end{Bmatrix} = \begin{Bmatrix} l_2 & l_3 & l_1 \\ k_2 & 0 & k_1 \\ j_2 & j_3 & j_1 \end{Bmatrix} =$$

$$= \begin{Bmatrix} k_2 & 0 & k_1 \\ j_2 & j_3 & j_1 \\ l_2 & l_3 & l_1 \end{Bmatrix} = \begin{Bmatrix} j_3 & j_1 & j_2 \\ l_3 & l_1 & l_2 \\ 0 & k_1 & k_2 \end{Bmatrix} = \begin{Bmatrix} l_3 & l_1 & l_2 \\ 0 & k_1 & k_2 \\ j_3 & j_1 & j_2 \end{Bmatrix} = \begin{Bmatrix} 0 & k_1 & k_2 \\ j_3 & j_1 & j_2 \\ l_3 & l_1 & l_2 \end{Bmatrix} =$$

$$= \delta(k_1, k_2)\,\delta(j_3, l_3)\,(-1)^{k_1+l_1+j_2+j_3}\,[j_3, k_1]^{-1/2} \begin{Bmatrix} j_1 & j_2 & j_3 \\ l_2 & l_1 & k_1 \end{Bmatrix}. \tag{1.29}$$

If the three parameters are equal to zero, the $9j$-symbol is written in terms of simple multipliers:

$$\begin{Bmatrix} j_1 & j_1 & 0 \\ l_1 & l_1 & 0 \\ k_1 & k_1 & 0 \end{Bmatrix} = [j_1, l_1, k_1]^{-1/2} \tag{1.30}$$

Higher-order $3nj$-symbols appear in the expressions for the transformation matrices and matrix elements for the operators much less often and can be given in terms of the $6j$- and $9j$-symbols [36].

A graphical method [36, 42, 47] is more convenient for constructing and transforming the wave functions of the coupling angular momenta and finding the matrix elements in their basis along with the formulas for summing the $3nj$-symbols. The more important of these formulas are given in Appendix 2.

The method described above allows us to couple the angular momenta of nonequivalent electrons and open shells, but is inadequate for constructing wave functions of coupled angular momenta for the shells of equivalent electrons because it does not take the Pauli exclusion principle, which prohibits electrons from being in the same one-electron states, into consideration.

1.2. Antisymmetric Wave Functions for a System of Many Electrons

The Pauli exclusion principle is automatically satisfied if the electron wave function is a determinant of a one-electron wave function. Therefore, we can construct an antisymmetric wave function of coupled angular momenta as a linear combination of the determinants, which will be an eigenfunction of the resultant angular momenta.

We will examine the case of two nonequivalent electrons. The wave function for uncoupled angular momenta is

$$|n_1 l_1 m_1 \mu_1 \; n_2 l_2 m_2 \mu_2\rangle = \frac{1}{\sqrt{2}} \begin{vmatrix} \psi(n_1 l_1 m_1 \mu_1 | x_1) & \psi(n_1 l_1 m_1 \mu_1 | x_2) \\ \psi(n_2 l_2 m_2 \mu_2 | x_1) & \psi(n_2 l_2 m_2 \mu_2 | x_2) \end{vmatrix}, \quad (1.31)$$

where x_i denotes a set of space and spin coordinates; μ is the projection of the electron spin, and $\psi(nlm\mu | x) \equiv^{\cdot} | nlm\mu\rangle$.

Expanding the determinant and using the vector coupling coefficients to couple the individual orbital and spin angular momenta of the electrons, we obtain the antisymmetric wave function in the LS-coupling:

$$|n_1 l_1 \; n_2 l_2 \; LSM_L M_S\rangle^a = \frac{1}{\sqrt{2}} \sum_{m_1 \mu_1 m_2 \mu_2} \begin{bmatrix} l_1 & l_2 & L \\ m_1 & m_2 & M_L \end{bmatrix} \begin{bmatrix} 1/2 & 1/2 & S \\ \mu_1 & \mu_2 & M_S \end{bmatrix} \times$$
$$\times \{ \psi(n_1 l_1 m_1 \mu_1 | x_1) \psi(n_2 l_2 m_2 \mu_2 | x_2) -$$
$$- \psi(n_2 l_2 m_2 \mu_2 | x_1) \psi(n_1 l_1 m_1 \mu_1 | x_2) \}. \quad (1.32)$$

Using the expression for the coupled angular momenta (1.10), and the symmetry property, Eq. (1.21), we can write the antisymmetric wave function in terms of nonantisymmetric wave functions (in these functions the electron indicated by the first term has the coordinate x_1):

$$|n_1 l_1\, n_2 l_2\, LSM_L M_S\rangle^a = \frac{1}{\sqrt{2}}\,[\,|n_1 l_1\, n_2 l_2\, LSM_L M_S\rangle +$$

$$+(-1)^{l_1+l_2-L-S}|n_2 l_2\, n_1 l_1\, LSM_L M_S\rangle]. \qquad (1.33)$$

If we initially couple the orbital and spin angular momenta of each electron and then couple the total angular momenta of the electrons into the resultant J, we obtain a two-electron wave function in the jj-coupling:

$$|n_1 l_1 j_1\, n_2 l_2 j_2\, JM\rangle^a =$$

$$= \frac{1}{\sqrt{2}}\,[|n_1 l_1 j_1\, n_2 l_2 j_2\, JM\rangle - (-1)^{j_1+j_2-J}|n_2 l_2 j_2\, n_1 l_1 j_1\, JM\rangle]. \qquad (1.34)$$

Hereafter, when using antisymmetric wave functions, we can omit the symbol a when it is clear from the context.

It follows from Eqs. (1.33) and (1.34) that for two equivalent electrons the wave function does not go to zero if $L + S$ or J, respectively, is equal to an even number ($2j = 2l \pm 1$ is an odd number). In addition, the normalization factor $2^{-1/2}$ must be replaced by 2^{-1}.

If this determinant method is used to construct the wave function for a many-electron system, the expressions obtained will be very complicated and cumbersome. A more efficient method is based on using coefficients of fractional parentage. Before we look into this method, we will pause briefly to examine the classification of the states of an open shell.

The coupling between equivalent electrons is usually more nearly LS-coupling than jj-coupling. It is evident from Eq. (1.32) that the quantum numbers LSM_LM_S are sufficient to denote the states of the two nl^2 electrons, but that we need additional quantum numbers to designate the states of the three d- or f-electrons. Unfortunately, exact quantum numbers that might be eigenvalues of the operators that commute with the Hamiltonian do not exist. It turns out that, for the d-shell, we need only include a quantum number v, which is equal to the smallest number of electrons in the shell for which the given term — the quantum numbers L and S — first appeared. The irreducible representations of the continuous groups R_7 and G_2 are used to classify the states of the

f-shell [41, 40]. These groups, as well as the three-dimensional rotation groups, are described by a unitary transformation of the one-electron orbital wave functions:

$$|lm\rangle \rightarrow \sum_{m'} u_{m'm} |lm'\rangle, \tag{1.35}$$

as well as by electron-shell functions that were constructed from them. The quantity R_7 is the group of rotations in $(2l + 1)$-dimensional space of the $|\, lm\rangle$ functions with fixed l, and G_2 is a special subgroup of the shell. Their irreducible representations are respectively denoted by sets of integer-valued parameters $W \equiv (w_1 w_2 w_3)$ and $U \equiv (u_1 u_2)$ which are equal to

$$2 \geqslant w_1 \geqslant w_2 \geqslant w_3 \geqslant 0, \quad 4 \geqslant u_1 \geqslant u_2 \geqslant 0, \quad 4 \geqslant u_1 + u_2. \tag{1.36}$$

when $l = 3$.

The number of possible states for an l^N shell is called its statistical weight, equal to the binomial coefficient

$$g(l^N) = \binom{4l+2}{N} = \frac{(4l+2)!}{(4l+2-N)!\,N!}. \tag{1.37}$$

The statistical weight increases rapidly as N approaches $2l + 1$ and then falls off symmetrically for an almost filled shell.

According to the method of coefficients of fractional parentage, recursive procedures can be used to construct the antisymmetric wave function of the l^N-electrons from the antisymmetric functions of an l^{N-1} electrons that have been coupled (but not antisymmetrized) with the wave function for a single l-electron [33, 39]:

$$|l^{N-1}\bar{\gamma}\bar{L}\bar{S}lLSM_L M_S\rangle =$$

$$= \sum_{M_{\bar{L}} M_{\bar{S}} m \mu} \begin{bmatrix} \bar{L} & l & L \\ M_{\bar{L}} & m & M_L \end{bmatrix} \begin{bmatrix} \bar{S} & s & S \\ M_{\bar{S}} & \mu & M_S \end{bmatrix} |l^{N-1}\bar{\gamma}\bar{L}\bar{S}M_{\bar{L}} M_{\bar{S}}\rangle\, |lm\mu\rangle, \tag{1.38}$$

where $\bar{\gamma}$ is a set of quantum numbers which, along with the explicitly indicated quantum numbers, describe the state of the shell. The attached electron in Eq. (1.38) has an Nth coordinate.

In order to exclude terms prohibited by the Pauli exclusion principle from the LS terms obtained by vector addition, we compose linear combinations of the functions given by Eq. (1.38) for an l^N shell:

$$|l^N\gamma LSM_L M_S\rangle = \sum_{\bar{\gamma}\bar{L}\bar{S}} (l^{N-1}\bar{\gamma}\bar{L}\bar{S}l \| l^N\gamma LS)|l^{N-1}\bar{\gamma}\bar{L}\bar{S}lLSM_L M_S\rangle. \quad (1.39)$$

The coefficients from this expansion are called coefficients of fractional parentage and they satisfy the following orthogonality conditions:

$$\sum_{\bar{\gamma}\bar{L}\bar{S}} (l^N\gamma LS \| l^{N-1}\bar{\gamma}\bar{L}\bar{S}l)(l^{N-1}\bar{\gamma}\bar{L}\bar{S}l \| l^N\gamma' LS) = \delta(\gamma, \gamma'),$$

$$\sum_{\gamma LS} [L, S](l^{N-1}\bar{\gamma}\bar{L}\bar{S}l \| l^N\gamma LS)(l^N\gamma LS \| l^{N-1}\bar{\gamma}'\bar{L}\bar{S}l) = \quad (1.40)$$

$$= \delta(\bar{\gamma}, \bar{\gamma}') \frac{4l+3-N}{N} [\bar{L}, \bar{S}]. \quad (1.41)$$

Their phase may be chosen so that they will become real; then, the equation

$$(l^N\gamma LS \| l^{N-1}\bar{\gamma}\bar{L}\bar{S}l) = (l^{N-1}\bar{\gamma}\bar{L}\bar{S}l \| l^N\gamma LS). \quad (1.42)$$

is satisfied. In the simplest cases,

$$(l \| l^0 l) = 1, \quad (1.43)$$

$$(l^2 LS \| ll) = \frac{1}{2} [1 + (-1)^{L+S}]. \quad (1.44)$$

Equation (1.44) is a direct consequence of Eq. (1.33).

Algebraic expressions for the coefficients of fractional parentage, especially for the f^N-shell, are rather complex. Tables of these functions have been presented in [51, 52].

The coefficients of fractional parentage for an almost filled shell ($N > 2l + 1$) are given in terms of the coefficients for a partially filled shell ($N < 2l + 1$) [53]:

$$(l^{4l+2-N}\gamma LS \| l^{4l+1-N}\bar{\gamma}\bar{L}\bar{S}l) = (-1)^{L-S+\bar{L}+\bar{S}+l+s+\frac{v-\bar{v}-1}{2}} \times$$

$$\times \left[\frac{(N+1)(2\bar{L}+1)(2\bar{S}+1)}{(4l+2-N)(2L+1)(2S+1)}\right]^{1/2} (l^{N+1}\bar{\gamma}\bar{L}\bar{S} \| l^N\gamma LSl). \quad (1.45)$$

The phase factor used in Eq. (1.45) is completely compatible for coefficients of fractional parentage and wave functions [53] and does not produce the ambiguities, which are possible when other factors, proposed in [33, 43], are used.

In special cases, and keeping Eqs. (1.43) and (1.44) in mind, we have

$$(l^{4l+2}\,{}^1S\,||\,l^{4l+1}\,l) = 1, \tag{1.46}$$

$$(l^{4l+1}\,||\,l^{4l}\,\upsilon LSl) = (-1)^{\frac{\upsilon}{2}}\,\frac{1}{2}\,[1+(-1)^{L+S}]\left[\frac{(2L+1)(2S+1)}{(2l+1)(4l+1)}\right]^{1/2}. \tag{1.47}$$

More general coefficients of fractional parentage that make it possible to attach several electrons, most often two electrons, to the antisymmetric wave functions are also used [33, 43]. These are given in terms of ordinary coefficients of fractional parentage:

$$\begin{aligned}(l^N\gamma LS\,||\,l^{N-2}\gamma' L' S' l^2 L'' S'') = \sum_{\bar\gamma\bar L\bar S} (-1)^{L+S+L'+S'+1}[\bar L,\ \bar S,\ L'',\ S'']^{1/2}\times \\ \times\begin{Bmatrix} l & l & L'' \\ L & L' & \bar L\end{Bmatrix}\begin{Bmatrix} s & s & S'' \\ S & S' & \bar S\end{Bmatrix}(l^N\gamma LS\,||\,l^{N-1}\bar\gamma\bar L\bar S l)\times \\ \times(l^{N-1}\bar\gamma\bar L\bar S\,||\,l^{N-2}\gamma' L' S' l).\end{aligned} \tag{1.48}$$

When $N = 4l + 2$, Eq. (1.48) can be simplified:

$$\begin{aligned}(l^{4l+2}\,||\,l^{4l}\upsilon LSl^2LS) = (-1)^{\upsilon/2}\,\frac{1}{2}[1+(-1)^{L+S}] \\ \times\left[\frac{(2L+1)(2S+1)}{(2l+1)(4l+1)}\right]^{1/2}.\end{aligned} \tag{1.49}$$

If there are several open shells in the configuration the atom's wave function is constructed from antisymmetric functions of the individual shells by coupling their angular momenta. When the matrix elements are calculated the antisymmetry between the electrons of different shells is important only in the following situations [39]:

a) in the diagonal with respect to configurations, matrix element for the operator of two-electron type, if this operator acts on the electrons in the different shells (it ensures the use of a two-electron matrix element for the antisymmetric wave functions).

b) in a matrix element which is nondiagonal relative to the configurations (see Sec. 2.2).

1.3. Irreducible Tensorial Operators and Their Matrix Elements

The matrix elements of atomic operators take on their simplest form when these operators are used as irreducible tensors which, when the coordinate system is rotated, are transformed in the same manner as are the spherical harmonics – the eigenfunctions of the angular momentum when the coordinate system is rotated.

As we all know, the transformation $\psi \rightarrow R\psi$ corresponds to the transformation $T \rightarrow RTR^{-1}$. Thus, an irreducible tensor operator is the set of $(2k + 1)$ quantities $T_q^{(k)}$ $(q = -k, -k + 1, \ldots, k)$ that is transformed according to the irreducible representations of a rotation group

$$RT_q^{(k)}R^{-1} = \sum_{q'} D_{q'q}^{(k)} T_{q'}^{(k)}, \tag{1.50}$$

when the coordinate system is rotated, where k is the operator rank and q is the projection of the rank. The other notations are the same as in Eq. (1.8).

Because the rotation operator R is given in terms of the angular momentum, we can give another definition, equivalent to Eq. (1.50), of an irreducible tensorial operator by means of the rules of commutation with the components of the angular momentum [34]:

$$[j_z, T_q^{(k)}] = \hbar q T_q^{(k)}, \tag{1.51}$$

$$[j_\pm, T_q^{(k)}] = \hbar [k(k+1) - q(q \pm 1)]^{1/2} T_{q\pm 1}^{(k)}. \tag{1.52}$$

In comparing Eqs. (1.51) and (1.52) with similar equations for the eigenfunction $|\, lm\rangle$ of Eqs. (1.4) and (1.5) we note that instead of the operator $j_z(j_\pm)$ that acts on the function, the commutator $j_z(j_\pm)$ is used with an operator here.

We can easily see that linear combinations of the angular momentum components

$$j_0^{(1)} = j_z,\ j_{\pm 1}^{(1)} = \mp \frac{1}{\sqrt{2}} (j_x \pm i j_y) \tag{1.53}$$

satisfy the rules of commutation given by Eqs. (1.51) and (1.52) and, as a result, form an irreducible tensor operator of the first rank.

The $T_q^{(k)}$ operators are defined to within the accuracy of a phase factor by Eqs. (1.50) or (1.51) and (1.52). This phase factor must be chosen consistently for all operators and wave functions being used, which means that a phase system must be chosen. A standard phase system in which the relation

$$T_q^{(k)\dagger} = (-1)^{k-q} T_{-q}^{(k)}. \tag{1.54}$$

is satisfied for the Hermitian operator was used in a number of works [34, 42, 43, 49]. The pseudostandard phase system in which Eq. (1.54) is replaced by

$$T_q^{(k)\dagger} = (-1)^{q} T_{-q}^{(k)}. \tag{1.55}$$

is more commonly used [39, 36, 45–47]. This system is more convenient because the matrix elements for the spin and orbital angular momenta and some other operators are real quantities. The pseudostandard phase system will be used throughout this book.

One advantage to using operators in the form of irreducible tensors is that the Wigner–Eckart theorem is satisfied, which makes it possible to isolate that portion of a matrix element that depends on the projections of the angular momenta as the Wigner symbol,

$$\langle \gamma jm | T_q^{(k)} | \gamma' j' m' \rangle = (-1)^{j-m} \begin{pmatrix} j & k & j' \\ -m & q & m' \end{pmatrix} \langle \gamma j \| T^{(k)} \| \gamma' j' \rangle. \tag{1.56}$$

The quantity $\langle \| \quad \| \rangle$ that is independent of the projections is called the submatrix or the reduced matrix element. For $k = 0$ and keeping Eq. (1.20) in mind, we obtain

$$\langle \gamma jm | T^{(0)} | \gamma' j' m' \rangle = \delta(j, j')\, \delta(m, m') [j]^{-1/2} \langle \gamma j \| T^{(0)} \| \gamma' j' \rangle, \tag{1.57}$$

i.e., the matrix element of the scalar operator is diagonal with respect to the resultant angular momentum and is independent of its projection.

From the definition of the Hermitian that corresponds to the physical quantity

$$\langle \gamma jm | T_q^{(k)\dagger} | \gamma' j' m' \rangle = \langle \gamma' j' m' | T_q^{(k)} | \gamma jm \rangle^*, \tag{1.58}$$

using Eq. (1.55) for the left-hand side, and applying the Wigner–Eckart theorem to both sides, we obtain the rule for transposing the submatrix element of the Hermitian:

$$\langle \gamma j \| T^{(k)} \| \gamma' j' \rangle = (-1)^{j-j'} \langle \gamma' j' \| T^{(k)} \| \gamma j \rangle^*. \tag{1.59}$$

Ordinarily, the submatrix element is a real quantity and the complex conjugate symbol can be omitted.

We find for the submatrix elements of the orbital and spin angular momenta of an electron that, on the basis of Eqs. (1.4) and (1.53) and the algebraic expressions for the Wigner symbol,

$$\langle l \| l^{(1)} \| l' \rangle = \delta(l, l')\hbar[l(l+1)(2l+1)]^{1/2}, \tag{1.60}$$

$$\langle s \| s^{(1)} \| s \rangle = \hbar\sqrt{\frac{3}{2}}. \tag{1.61}$$

The irreducible tensors, as are the angular momenta, are coupled by means of the vector coupling coefficients:

$$[T^{(k_1)} \times W^{(k_2)}]^{(k)}_q = \sum_{q_1 q_2} \begin{bmatrix} k_1 & k_2 & k \\ q_1 & q_2 & q \end{bmatrix} T^{(k_1)}_{q_1} W^{(k_2)}_{q_2}. \tag{1.62}$$

The submatrix element of the tensor product of the operators is given in terms of the submatrix elements of the individual operators:

$$\langle \gamma j \| [T^{(k_1)} \times W^{(k_2)}]^{(k)} \| \gamma' j' \rangle = (-1)^{j+j'+k}[k]^{\frac{1}{2}} \times \tag{1.63}$$
$$\times \sum_{\gamma'' j''} \langle \gamma j \| T^{(k_1)} \| \gamma'' j'' \rangle \langle \gamma'' j'' \| W^{(k_2)} \| \gamma' j' \rangle \begin{Bmatrix} k_1 & k_2 & k \\ j' & j & j'' \end{Bmatrix}.$$

If the resultant rank is $k = 0$, then the tensor product of the tensors transforms into a scalar product that is defined as follows:

$$(T^{(k)} \cdot W^{(k)}) = \sum_q (-1)^q T^{(k)}_q W^{(k)}_{-q} = (-1)^k [k]^{1/2} [T^{(k)} \times W^{(k)}]^{(0)}_0. \tag{1.64}$$

Equation (1.63) is then simplified:

$$\langle \gamma j \| (T^{(k)} \cdot W^{(k)}) \| \gamma' j' \rangle = \delta(j, j')[j]^{-1/2} \times$$

$$\times \sum_{\gamma'' j''} (-1)^{j-j''} \langle \gamma j \| T^{(k)} \| \gamma'' j'' \rangle \langle \gamma'' j'' \| W^{(k)} \| \gamma' j' \rangle . \quad (1.65)$$

If the operators $T^{(k)}$ and $W^{(k)}$ act in different subspaces (orbital and spin), or on different subsystems (nonequivalent electrons, shells, or groups of shells), then the submatrix elements of their tensor and scalar products can be factorized:

$$\langle \gamma_1 j_1 \gamma_2 j_2 j \| [T^{(k_1)} \times W^{(k_2)}]^{(k)} \| \gamma_1' j_1' \gamma_2' j_2' j' \rangle =$$
$$= [k, j, j']^{1/2} \begin{Bmatrix} j_1 & j_1' & k_1 \\ j_2 & j_2' & k_2 \\ j & j' & k \end{Bmatrix} \langle \gamma_1 j_1 \| T^{(k_1)} \| \gamma_1' j_1' \rangle \langle \gamma_2 j_2 \| W^{(k_2)} \| \gamma_2' j_2' \rangle, \quad (1.66)$$

$$\langle \gamma_1 j_1 \gamma_2 j_2 j \| (T^{(k)} \cdot W^{(k)}) \| \gamma_1' j_1' \gamma_2' j_2' j' \rangle =$$
$$= \delta(j, j')(-1)^{j_1'+j_2+j} [j]^{1/2} \begin{Bmatrix} j_1 & j_2 & j \\ j_2' & j_1' & k \end{Bmatrix} \times$$
$$\times \langle \gamma_1 j_1 \| T^{(k)} \| \gamma_1' j_1' \rangle \langle \gamma_2 j_2 \| W^{(k)} \| \gamma_2' j_2' \rangle . \quad (1.67)$$

In the special case when $k_2 = 0$ and the operator acts only in the space of the $|\gamma_1 j_1 m_1\rangle$, we find from Eq. (1.66) that

$$\langle \gamma_1 j_1 \gamma_2 j_2 j \| T^{(k)} \| \gamma_1' j_1' \gamma_2' j_2' j' \rangle = \delta(\gamma_2, \gamma_2') \delta(j_2, j_2') \times$$
$$\times (-1)^{j_1+j_2+j'+k} [j, j']^{1/2} \begin{Bmatrix} j_1 & j & j_2 \\ j' & j_1' & k \end{Bmatrix} \langle \gamma_1 j_1 \| T^{(k)} \| \gamma_1' j_1' \rangle . \quad (1.68)$$

In a similar fashion, for an operator that acts only on $|\gamma_2 j_2 m_2\rangle$, we have

$$\langle \gamma_1 j_1 \gamma_2 j_2 j \| T^{(k)} \| \gamma_1' j_1' \gamma_2' j_2' j' \rangle = \delta(\gamma_1, \gamma_1') \delta(j_1, j_1') \times$$

$$\times (-1)^{j_1+j_2'+j+k} [j, j']^{1/2} \begin{Bmatrix} j_2 & j & j_1 \\ j' & j_2' & k \end{Bmatrix} \langle \gamma_2 j_2 \| T^{(k)} \| \gamma_2' j_2' \rangle . \quad (1.69)$$

Any operator in Eqs. (1.62)–(1.69) can have an internal structure that is revealed by iteration of these formulas.

The expression for the submatrix element of a tensor operator is basically determined by the ranks of the operator and only in terms of one- or two-electron matrix elements – its concrete form. For this reason, we introduce a standard unit operator whose one-electron submatrix element is unity:

$$\langle l \| u^{(k)} \| l' \rangle = \delta(l, l')\{l\ k\ l'\}. \tag{1.70}$$

An operator $U^{(k)}$ is composed from $u_{qi}{}^{(k)}$ operators that act on the coordinates of the various electrons in the l^N shell (we will omit the projections of the ranks in the subsequent formulas that appear in this section):

$$U^{(k)} = \sum_{i=1}^{N} u_i^{(k)}, \tag{1.71}$$

where i is the number of the electron's coordinate.

We can introduce a more general operator, the double tensor operator $V^{(kk')}$ that acts on not only orbital, but on the spin functions as well:

$$V^{(kk')} = \sum_{i=1}^{N} v_i^{(kk')}, \tag{1.72}$$

$$\langle ls \| v^{(kk')} \| l's \rangle = \delta(l, l')\{l\ k\ l'\}\{s\ k'\ s\}, \tag{1.73}$$

$$V^{(k0)} = \frac{1}{\sqrt{2}}\, U^{(k)}. \tag{1.74}$$

The triad condition $\{s\ k'\ s\}$ limits the spin rank k' to values of 0 and 1.

As an example of using irreducible operators, we will derive an expression for the one-shell submatrix element of the $V^{(k1)}$ operator. Since the electrons are equivalent, all terms in the operator will make the same contribution to the matrix element:

$$\begin{aligned} &\langle l^N \gamma LSM_L M_S | V^{(kk')} | l^N \gamma' L' S' M_{L'} M_{S'} \rangle = \\ &= N \langle l^N \gamma LSM_L M_S | v_N^{(kk')} | l^N \gamma' L' S' M_{L'} M_{S'} \rangle, \end{aligned} \tag{1.75}$$

where N is the number of electrons in the shell, and v_N is the one-electron operator that acts on the Nth set of coordinates. Using coefficients of fractional parentage, we separate out in the bra and ket wave functions for a single electron (to which we assign the coordinates x_N):

$$\langle l^N\gamma LSM_L M_S | V^{(kk')} | l^N\gamma' L' S' M_{L'} M_{S'}\rangle = N \sum_{\bar\gamma \bar L \bar S \bar\gamma' \bar L' \bar S'} (l^N\gamma LS \| l^{N-1}\bar\gamma \bar L \bar S l) \times$$
$$\times (l^{N-1}\bar\gamma' \bar L' \bar S' l \| l^N\gamma' L' S') \times$$
$$\times (l^{N-1}\bar\gamma \bar L \bar S l LSM_L M_S | v_N^{(kk')} | l^{N-1}\bar\gamma' \bar L' \bar S' l L' S' M_{L'} M_{S'}). \tag{1.76}$$

Using the Wigner–Eckart theorem to make the transition from matrix elements to submatrix elements, applying the equation (1.69) to $v_N^{(kk')}$ that acts only on the wave function of a decoupled electron, and giving k' a specific value, we finally obtain

$$\langle l^N\gamma LS \| V^{(k1)} \| l^N\gamma' L' S'\rangle = N [L, L', S, S']^{1/2} \sum_{\bar\gamma \bar L \bar S} (-1)^{\bar L + \bar S + L + S} \times$$
$$\times (-1)^{l-1/2+k} (l^N\gamma LS \| l^{N-1}\bar\gamma \bar L \bar S l)(l^{N-1}\bar\gamma \bar L \bar S l \| l^N\gamma' L' S') \times \tag{1.77}$$
$$\times \begin{Bmatrix} l & l & k \\ L & L' & \bar L \end{Bmatrix} \begin{Bmatrix} 1/2 & 1/2 & 1 \\ S & S' & \bar S \end{Bmatrix},$$

$$\langle l^N\gamma LS \| U^{(k)} \| l^N\gamma' L' S'\rangle = \sqrt{2}\, \langle l^N\gamma LS \| V^{(k0)} \| l^N\gamma' L' S'\rangle =$$
$$= \delta(S, S') N [L, L', S]^{1/2} \sum_{\bar\gamma \bar L \bar S} (-1)^{\bar L + l + L + k} (l^N\gamma LS \| l^{N-1}\bar\gamma \bar L \bar S l) \times$$
$$\times (l^{N-1}\bar\gamma \bar L \bar S l \| l^N\gamma' L' S') \begin{Bmatrix} l & l & k \\ L & L' & \bar L \end{Bmatrix}. \tag{1.78}$$

In special situations, Eqs. (1.77) and (1.78) can be simplified:

$$\langle l^N\gamma LS \| U^{(0)} \| l^N\gamma' L' S'\rangle = \delta(\gamma LS, \gamma' L' S') N [L, S]^{1/2} [l]^{-1/2}, \tag{1.79}$$

$$\langle l^N\gamma LS \| V^{(01)} \| l^N\gamma' L' S'\rangle =$$
$$= \delta(\gamma LS, \gamma' L' S') \left[\frac{2(2L+1)(2S+1)S(S+1)}{3(2l+1)}\right]^{1/2}, \tag{1.80}$$

$$\langle l^N \gamma LS \| U^{(1)} \| l^N \gamma' L' S' \rangle =$$
$$= \delta(\gamma LS, \gamma' L' S') \left[\frac{L(L+1)(2L+1)(2S+1)}{l(l+1)(2l+1)} \right]^{1/2}. \quad (1.81)$$

The equation

$$\langle l^{4l+2-N} \gamma \upsilon LS \| V^{(kk')} \| l^{4l+2-N} \gamma' \upsilon' L' S' \rangle = (-1)^{k+k'+1+\frac{\upsilon-\upsilon'}{2}} \times$$
$$\times \langle l^N \gamma \upsilon LS \| V^{(kk')} \| l^N \gamma' \upsilon' L' S' \rangle, \quad k+k' \neq 0. \quad (1.82)$$

is obtained, which is similar to the relation between the coefficients of fractional parentage for partially and almost-filled shells given by Eq. (1.45) and consistent with it. When $k + k' = 0$, Eq. (1.82) loses its validity and we must then use Eqs. (1.79) and (1.74) for all N.

Tables were presented in [51] for the submatrix elements of the $U^{(k)}$ and $V^{(11)}$ operators for the p^N-, d^N-, and f^N-shells ($N \leq 2l + 1$) and, in [52], $U^{(k)}$ and $V^{(k1)}$ for the p^N, d^N ($N \leq 2l + 1$) and f^2, f^3, and f^4 shells (their definitions differ somewhat from those used here).

The special electron creation and annihilation operators play an important role in the theory of an atom [40, 49]. The electron creation operator $a_\nu^\dagger$, which acts on the wave function for the uncoupled angular momenta of the $N - 1$ electron in which the state ν ($\nu \equiv nlm\mu$) is absent, transforms it into a wave function for the N electrons that have an occupied state ν. The electron annihilation operator a_ν, on the other hand, excludes the ν state from the wave function. Thus, we can derive the eigenfunction for an N-electron atom by having the N operators for the creation of an electron act on the vacuum state $|0\rangle$:

$$a_\nu^\dagger a_\varkappa^\dagger \ldots a_\eta^\dagger |0\rangle. \quad (1.83)$$

The $a^\dagger$ and a operators are subject to the following rules of anticommutation:

$$a_\nu^\dagger a_\eta^\dagger + a_\eta^\dagger a_\nu^\dagger = 0, \quad a_\nu a_\eta + a_\eta a_\nu = 0,$$
$$a_\nu a_\eta^\dagger + a_\eta^\dagger a_\nu = \delta(\nu, \eta). \quad (1.84)$$

The operators for the atomic quantities are expressed in terms of the $a^\dagger$ and a [40, 49]. For example, the most widely used one- and two-electron operators

$$F=\sum_i f_i, \quad G=\sum_{i<j} g_{ij} \tag{1.85}$$

are written

$$F=\sum_{\nu\eta} a_\nu^\dagger \langle \nu | f | \eta \rangle a_\eta, \tag{1.86}$$

$$G=\frac{1}{2}\sum_{\nu\xi\zeta\eta} a_\nu^\dagger a_\xi^\dagger \langle \nu\xi | g_{12} | \zeta\eta \rangle a_\eta a_\zeta. \tag{1.87}$$

The two-electron element in the right-hand side of Eq. (1.87) is defined relative to the nonantisymmetric wave functions, since the Pauli principle is automatically satisfied by the anticommutation rules of Eq. (1.84). The summations in Eqs. (1.86) and (1.87) are done over all possible one-electron states.

The use of $a^\dagger$ and a operators to construct wave functions and expressions for the physical operators corresponds to making the transition to second quantization representation [40, 49].

An examination of the commutation of $a^\dagger$ and a operators with the orbital and spin angular momenta shows [40] that the sets $a_\nu{}^\dagger$ and $\tilde{a}_\nu = (-1)^{l+s-m-\mu} a_{\bar{\nu}}$ ($\nu = nlm\mu$, $\bar{\nu} \equiv nl-m-\mu$) with given set quantum numbers ls and all possible projections of them are irreducible tensor operators of rank l in orbital space and rank s in spin space. This allows us to combine $a^\dagger$ and $\tilde{a}$ operators of different rank and to act on the coupled-angular momenta functions with them. In particular, the coefficients of fractional parentage turns out to be proportional to the submatrix element of the $a^\dagger$ or $\tilde{a}$ operators [40]:

$$\begin{aligned}(l^N\gamma LS \| l^{N-1}\bar{\gamma}\bar{L}\bar{S}l) &= (-1)^N N^{-1/2}[L, S]^{-1/2}\langle l^N\gamma LS \| a^\dagger \| l^{N-1}\bar{\gamma}\bar{L}\bar{S}\rangle = \\ &= (-1)^{N+L+S+l+\frac{1}{2}+\bar{L}-\bar{S}} N^{-1/2}[L, S]^{-1/2}\langle l^{N-1}\bar{\gamma}\bar{L}\bar{S} \| \tilde{a} \| l^N\gamma LS\rangle.\end{aligned} \tag{1.88}$$

According to Eqs. (1.86), (1.56), and (1.73) the unit tensor $v^{(kk)}$ is, to within the accuracy of a simple factor, equal to the vector product of the creation and annihilation operators of an electron [40]

$$v^{(kk')} \sim [a^{(ls)\dagger} \times \tilde{a}^{(ls)}]^{(kk')}. \tag{1.89}$$

From $a^\dagger$ and $\tilde{a}$ we can generate a so-called quasispin operator [40, 49] whose components

$$Q_+ = \frac{1}{2}\,[l,\,s]^{1/2}\,[a^{(ls)\dagger} \times a^{(ls)\dagger}]^{(00)}, \quad Q_- = -\frac{1}{2}\,[l,\,s]^{1/2}\,[\tilde{a}^{(ls)} \times \tilde{a}^{(ls)}]^{(00)},$$
$$Q_z = -\frac{1}{4}\,[l,\,s]^{1/2}\{[a^{(ls)\dagger} \times \tilde{a}^{(ls)}]^{(00)} + [\tilde{a}^{(ls)} \times a^{(ls)\dagger}]^{(00)}\} \tag{1.90}$$

commute in the same way as do the components of the spin angular momentum. The eigenvalues of the quasispin and its projection are

$$Q = \frac{1}{2}\,(2l+1-v), \quad M_Q = -\frac{1}{2}\,(2l+1-N), \tag{1.91}$$

i.e., they are uniquely associated with the quantum number v and the number of electrons N.

An examination of the commutation of $a^\dagger$ and $\tilde{a}$ operators with the quasispin shows that these operators behave like the components of a tensor operator with rank $q = 1/2$:

$$a^{(ls)\dagger}_{m\mu} \equiv a^{(qls)}_{1/2m\mu}, \quad \tilde{a}^{(ls)}_{m\mu} = (-1)^{l+s-m-\mu}\,a^{(ls)}_{-m-\mu} \equiv a^{(qls)}_{-1/2m\mu} \tag{1.92}$$

and from these the triple unit tensors

$$v^{(k_1 k_2 k_3)} \sim [a^{(qls)} \times a^{(qls)}]^{(k_1 k_2 k_3)}, \tag{1.93}$$

are generated, where k_1 is the rank of the operator in the auxiliary quasispin space. Since both physical operators and wave functions can be constructed via the $\tilde{a}^\dagger$ and $\tilde{a}$ operators, irreducible tensors can also be used in the quasispin space. In particular, using the Wigner—Eckart theorem in this space allows us to write the dependence of a matrix element or coefficient of fractional parentage on the number of electrons (the projection of the quasispin) as a Wigner symbol.

A second quantization representation is convenient for deriving relationships between the matrix elements, carrying out summations over them, and examining correlation effects [40, 49].

1.4. Interactions in a Nonrelativistic Atom

In nonrelativistic theory, the Hamiltonian for an atom with many electrons is

$$H = H^k + H^p + H^e, \tag{1.94}$$

$$H^k = -\frac{\hbar^2}{2m} \sum_i \nabla_i^2, \quad H^p = -\sum_i \frac{Ze^2}{r_i}, \tag{1.95}$$

$$H^e = \sum_{i<j} h_{ij}^e = \sum_{i<j} \frac{e^2}{r_{ij}} \quad (r_{ij} = |\mathbf{r}_i - \mathbf{r}_j|). \tag{1.96}$$

Here, H^k and H^p are, respectively, the kinetic and potential energy operators for the electrons in the field of the nuclear charge; H^e is the electrostatic interaction operator between the electrons; m is the electron mass, and the intervals of summation cover the coordinates of all electrons or pairs of electrons.

The H^k and H^p operators are scalars; therefore, their matrix elements are diagonal and independent of the many-electron quantum numbers. The eigenvalue of H^k is positive, that of H^p is negative and, according to the Clausius virial theorem, the absolute value of E^p is twice that of E^k. The total matrix element is equal to

$$\langle K\gamma JM | (H^k + H^p) | K\gamma' JM \rangle = \delta(\gamma, \gamma') \sum_{nl} N_{nl} I(nl), \tag{1.97}$$

where

$$I(nl) = -\frac{1}{2} \int_0^\infty P_{nl}(r) \left[\frac{d^2}{dr^2} + \frac{2Z}{r} - \frac{l(l+1)}{r^2} \right] P_{nl}(r)\, dr. \tag{1.98*}$$

In Eq. (1.97), N_{nl} is the number of electrons in the nl shell, K is the configuration, and summation is carried out over all shells in this configuration.

The integral $I(nl)$ for hydrogenlike radial functions is proportional to Z^2 [54]; with due regard for shielding of the shell by other electrons, Z must be replaced by the effective charge Z^*. Thus, the kinetic and potential energies of the electrons increase quadratically as the effective nuclear charge increases.

---

*An asterisk following the formula number indicates that this formula is given in atomic units.

The operator of electrostatic interaction between the *i*th and *j*th electrons h_{ij}^e when the expansion r_{ij}^{-1} over spherical harmonics is used is presented in the form of irreducible tensors:

$$h_{ij}^e = e^2 \sum_k \frac{r_<^k}{r_>^{k+1}} (C_i^{(k)} \cdot C_j^{(k)}), \tag{1.99}$$

where $r_<$ and $r_>$ are, respectively, the smaller and larger absolute values of the vectors $\mathbf{r}_i$ and $\mathbf{r}_j$;

$$C_q^{(k)}(\theta, \varphi) = \left[\frac{4\pi}{2k+1}\right]^{1/2} Y_{kq}(\theta, \varphi). \tag{1.100}$$

The contribution made by electrostatic repulsion between electrons to the atom's energy is positive, even though individual terms in the matrix element may be negative. Using Eq. (1.57) on H^e and writing the dependence of the matrix element on S and J via Eq. (1.68) (H^e does not act on the spin variables), we obtain

$$\begin{aligned} &\langle \gamma LSJM | H^e | \gamma' L' S' J' M' \rangle = \\ &= \delta(J, J')\delta(M, M')\delta(L, L')\delta(S, S')[L, S]^{-1/2} \langle \gamma LS \| H^e \| \gamma' LS \rangle. \end{aligned} \tag{1.101}$$

The matrix elements of the electrostatic interaction operator are independent of the quantum numbers JM (or M_L, M_S) and are diagonal with respect to the resultants L and S.

The one-electron submatrix element for the $C^{(k)}$ is given in terms of the Wigner symbol with zero projections:

$$\langle l \| C^{(k)} \| l' \rangle = (-1)^l [l, l']^{1/2} \begin{pmatrix} l & k & l' \\ 0 & 0 & 0 \end{pmatrix} = (-1)^{l+g} [l, l']^{1/2}$$

$$\frac{g!}{(g-l)!\,(g-k)!\,(g-l')!} \left[\frac{(2g-2l)!\,(2g-2k)!\,(2g-2l')!}{(2g+1)!}\right]^{1/2}. \tag{1.102}$$

The submatrix element will not disappear if $2g = l + k + l'$ is an odd number. The rule for transposing this quantity is as follows:

$$\langle l \| C^{(k)} \| l' \rangle = (-1)^{l+l'} \langle l' \| C^{(k)} \| l \rangle. \tag{1.103}$$

Values of the submatrix elements for the $C^{(k)}$ operator for $l, l \leq 4$ are shown in Table 1. In the simplest cases they are

$$\langle l \| C^{(k)} \| 0 \rangle = \delta(l, k), \quad \langle l \| C^{(0)} \| l' \rangle = \delta(l, l')[l]^{1/2}, \tag{1.104}$$

$$\langle l \| C^{(1)} \| l' \rangle = (-1)^{l+g} \sqrt{l_>}\, \{l \; 1 \; l'\}, \quad l_> = \max(l, l'). \tag{1.105}$$

The two-electron matrix element of the h^e operator for the wave functions given by Eq. (1.33) consists of two F-type terms and two G-type terms (in one function the quantum numbers of the electrons are interchanged). Because of the Hermite nature of the operator, both F-type (and G-type) terms are equal among themselves and are given in terms of the one-electron submatrix elements according to Eq. (1.67):

$$\langle n_1 l_1\, n_2 l_2 LS | h^e | n_1 l_1\, n_2 l_2 LS \rangle = 2N^2_{n_1 l_1, n_2 l_2} \sum_k \Bigg\{ (-1)^{l_1+l_2+L} \times$$

$$\times \langle l_1 \| C^{(k)} \| l_1 \rangle \langle l_2 \| C^{(k)} \| l_2 \rangle \begin{Bmatrix} l_1 & l_1 & k \\ l_2 & l_2 & L \end{Bmatrix} F^k(n_1 l_1, n_2 l_2) +$$

$$+ (-1)^S \langle l_1 \| C^{(k)} \| l_2 \rangle^2 \begin{Bmatrix} l_1 & l_2 & k \\ l_1 & l_2 & L \end{Bmatrix} G^k(n_1 l_1, n_2 l_2) \Bigg\}. \tag{1.106}$$

Here, $N_{n_1 l_1, n_2 l_2}$ is a normalizing factor,

$$N_{n_1 l_1, n_2 l_2} = \begin{cases} 1/2, & \text{if} \quad n_1 l_1 = n_2 l_2, \\ 1/\sqrt{2}, & \text{if} \quad n_1 l_1 \neq n_2 l_2; \end{cases} \tag{1.107}$$

and F^k and G^k are, respectively, the F-type and G-type radial integrals:

$$F^k(n_1 l_1, n_2 l_2) = \int_0^\infty P^2_{n_1 l_1}(r_1)\, r_1^{-1}\, Y^k_{n_2 l_2, n_2 l_2}(r_1)\, dr_1, \tag{1.108}$$

Table 1.1. One-Electron Submatrix Elements for the Spherical Function Operator $C(k)$

l	l'	k	$\langle l \| C^{(k)} \| l' \rangle$	l	l'	k	$\langle l \| C^{(k)} \| l' \rangle$
0	k	k	$(-1)^k$	3	5	8	$-2\cdot7\sqrt{2}/\sqrt{13\cdot17}$
1	1	2	$-\sqrt{2\cdot3}/\sqrt{5}$	3	6	3	$-2\cdot5/\sqrt{3\cdot11}$
1	2	1	$-\sqrt{2}$	3	6	5	$7/\sqrt{3\cdot11}$
1	2	3	$3\sqrt{7}$	3	6	7	$-2\cdot7\sqrt{2\cdot3}/\sqrt{5\cdot11\cdot17}$
1	3	2	$3\sqrt{5}$	3	6	9	$2\cdot7\sqrt{3}/\sqrt{17\cdot19}$
1	3	4	$-2\sqrt{3}$	4	4	2	$-2\cdot3\sqrt{5}/\sqrt{7\cdot11}$
1	4	3	$-2\sqrt{3}/\sqrt{7}$	4	4	4	$3\cdot9\sqrt{2}/\sqrt{7\cdot11\cdot13}$
1	4	5	$\sqrt{3\cdot5}/\sqrt{11}$	4	4	6	$-2\cdot3\sqrt{5}/\sqrt{11\cdot13}$
1	5	4	$\sqrt{5}/\sqrt{3}$	4	4	8	$3\cdot7\sqrt{2\cdot5}/\sqrt{11\cdot13\cdot17}$
1	5	6	$-3\sqrt{2}/\sqrt{13}$	4	5	1	$-\sqrt{5}$
1	6	5	$-3\sqrt{2}/\sqrt{11}$	4	5	3	$2\cdot3\sqrt{5}/\sqrt{7\cdot13}$
1	6	7	$\sqrt{7}/\sqrt{5}$	4	5	5	$-3\sqrt{2}/\sqrt{13}$
2	2	2	$-\sqrt{2\cdot5}/\sqrt{7}$	4	5	7	$2\sqrt{2\cdot5\cdot7}/\sqrt{13\cdot17}$
2	2	4	$\sqrt{2\cdot5}/\sqrt{7}$	4	5	9	$-9.7\sqrt{2}/\sqrt{13\cdot17\cdot19}$
2	3	1	$-\sqrt{3}$	4	6	2	$3\sqrt{5}/\sqrt{11}$
2	3	3	$2/\sqrt{3}$	4	6	4	$-2\sqrt{5}/\sqrt{11}$
2	3	5	$-5\sqrt{2}/\sqrt{3\cdot11}$	4	6	6	$2\cdot3\sqrt{7}/\sqrt{11\cdot17}$
2	4	2	$3\sqrt{2}/\sqrt{7}$	4	6	8	$-2\cdot9\sqrt{2\cdot7}/\sqrt{11\cdot17\cdot19}$
2	4	4	$-2\cdot5/\sqrt{7\cdot11}$	4	6	10	$3\sqrt{2\cdot5\cdot7}/\sqrt{17\cdot19}$
2	4	6	$3\cdot5/\sqrt{11\cdot13}$	5	5	2	$-\sqrt{2\cdot5\cdot11}/\sqrt{3\cdot13}$
2	5	3	$-5\sqrt{2}/\sqrt{3\cdot7}$	5	5	4	$\sqrt{2\cdot11}/\sqrt{13}$
2	5	5	$5\sqrt{2}/\sqrt{3\cdot13}$	5	5	6	$-4\sqrt{5\cdot11}/\sqrt{3\cdot13\cdot17}$
2	5	7	$-\sqrt{3\cdot7}/\sqrt{13}$	5	5	8	$7\sqrt{2\cdot5\cdot11}/\sqrt{13\cdot17\cdot19}$
2	6	4	$5/\sqrt{11}$	5	5	10	$-2\cdot3\sqrt{3\cdot7\cdot11}/\sqrt{13\cdot17\cdot19}$
2	6	6	$-\sqrt{2\cdot7}/\sqrt{11}$	5	6	1	$-\sqrt{2\cdot3}$
2	6	8	$2\sqrt{7}/\sqrt{17}$	5	6	3	$\sqrt{7}/\sqrt{3}$
3	3	2	$-2\sqrt{7}/\sqrt{3\cdot5}$	5	6	5	$-4\sqrt{5}/\sqrt{3\cdot17}$
3	3	4	$\sqrt{2\cdot7}/\sqrt{11}$	5	6	7	$2\sqrt{3\cdot5\cdot7}\cdot\sqrt{17\cdot19}$
3	3	6	$-2\cdot5\sqrt{7}/\sqrt{3\cdot11\cdot13}$	5	6	9	$-2\sqrt{3\cdot5\cdot7}/\sqrt{17\cdot19}$

Table 1.1 (continued)

l	l'	k	$\langle l \| C^{(k)} \| l' \rangle$	l	l'	k	$\langle l \| C^{(k)} \| l' \rangle$
3	4	1	-2	5	6	11	$3\cdot 11\sqrt{2\cdot 7}/\sqrt{17\cdot 19\cdot 23}$
3	4	3	$3\sqrt{2}/\sqrt{11}$	6	6	2	$-\sqrt{2\cdot 7\cdot 13}/\sqrt{5\cdot 11}$
3	4	5	$-2\cdot 3\sqrt{5}/\sqrt{11\cdot 13}$	6	6	4	$2\sqrt{7\cdot 13}/\sqrt{11\cdot 17}$
3	4	7	$7\sqrt{5}/\sqrt{11\cdot 13}$	6	6	6	$-4\cdot 5\sqrt{13}/\sqrt{11\cdot 17\cdot 19}$
3	5	2	$\sqrt{2\cdot 5}/\sqrt{3}$	6	6	8	$5\sqrt{2\cdot 7\cdot 13}/\sqrt{11\cdot 17\cdot 19}$
3	5	4	$-2\sqrt{5}/\sqrt{13}$	6	6	10	$-2\cdot 3\sqrt{3\cdot 7\cdot 13}/\sqrt{17\cdot 19\cdot 23}$
3	5	6	$7\sqrt{3\cdot 13}$	6	6	12	$2\cdot 3\cdot 11\sqrt{7\cdot 13}/5\sqrt{17\cdot 19\cdot 23}$

$$G^k(n_1 l_1, n_2 l_2) = \int_0^\infty P_{n_1 l_1}(r_1) P_{n_2 l_2}(r_1) r_1^{-1} Y^k_{n_1 l_1, n_2 l_2}(r_1)\, dr_1, \tag{1.109*}$$

where

$$Y^k_{n_1 l_1, n_2 l_2}(r_1) = r_1 \int_0^\infty \frac{r_<^k}{r_>^{k+1}} P_{n_1 l_1}(r_2) P_{n_2 l_2}(r_2)\, dr_2. \tag{1.110*}$$

The integrals F^k and G^k are symmetric with respect to the replacements of the quantum numbers $n_1 l_1$ and $n_2 l_2$:

$$F^k(n_1 l_1, n_2 l_2) = F^k(n_2 l_2, n_1 l_1), \quad G^k(n_1 l_1, n_2 l_2) = G^k(n_2 l_2, n_1 l_1). \tag{1.111}$$

They are positive and satisfy the equation

$$F^k > F^{k+1} > 0, \quad G^k/(2k+1) \geqslant G^{k+1}/(2k+3) > 0. \tag{1.112}$$

Ordinarily, the inequality $G^k > G^{k+1}$ is valid for the integral G^k. Values of G^k and the ratio F^k/F^{k+2} fall off rapidly as the degree of overlap between the wave functions $P_{n_1 l_1}(r)$ and $P_{n_2 l_2}(r)$ decreases [47].

If $n_1 l_1 = n_2 l_2$, then $F^k = G^k$ and Eq. (1.106) becomes

$$\langle nl^2 LS | h^e | nl^2 LS \rangle = (-1)^L \frac{1}{2} [1 + (-1)^{L+S}] \sum_k \langle l \| C^{(k)} \| l \rangle^2 \times$$
$$\times \begin{Bmatrix} l & l & L \\ l & l & k \end{Bmatrix} F^k(nl, nl). \qquad (1.113)$$

This matrix element can be put into the form

$$\langle nl^2 LS | h^e | nl^2 LS \rangle =$$
$$= \sum_k \langle l^2 LS | (u^{(k)} \cdot u^{(k)}) | l^2 LS \rangle \langle l \| C^{(k)} \| l \rangle^2 F^k(nl, nl). \qquad (1.114)$$

Writing this equation in operator form in terms of the unit operators $u^{(k)}$, Eq. (1.70),

$$h^e_{ij} = \sum_k (u_i^{(k)} \cdot u_j^{(k)}) \langle l \| C^{(k)} \| l \rangle^2 F^k(nl, nl), \qquad (1.115)$$

making the substitution in Eq. (1.96), and separating the summations over i and j, gives us the following expression for the electrostatic interaction operator that acts within the shell of equivalent electrons in terms of operators composed from unit tensors:

$$H^e = \frac{1}{2} \sum_k \left[(U^{(k)} \cdot U^{(k)}) - \sum_i (u_i^{(k)} \cdot u_i^{(k)}) \right] \langle l \| C^{(k)} \| l \rangle^2 F^k(nl, nl). \qquad (1.116)$$

The one-shell matrix element of H^e is found via Eqs. (1.57), (1.65), and (1.75). In this element the spin-angular coefficient at the integral $F^k(nl,nl)$ is

$$f_k(l^N \gamma LS, \gamma' LS) = \frac{1}{2} \langle l \| C^{(k)} \| l \rangle^2 \Bigg\{ [L, S]^{-1} \sum_{\gamma'' L''} \times$$
$$\times \langle l^N \gamma LS \| U^{(k)} \| l^N \gamma'' L'' S \rangle \langle l^N \gamma' LS \| U^{(k)} \| l^N \gamma'' L'' S \rangle - \delta(\gamma, \gamma') \frac{N}{2l+1} \Bigg\}. \qquad (1.117)$$

In special cases,

$$f_0(l^N \gamma LS, \gamma' LS) = \delta(\gamma, \gamma') \frac{N(N-1)}{2}, \qquad (1.118)$$

$$f_k(l^{4l+2}\,{}^1S) = -\langle l \| C^{(k)} \| l \rangle^2 \quad (k > 0), \tag{1.119}$$

$$f_k(l^{4l+1}\,{}^2l) = -\frac{2l}{2l+1}\langle l \| C^{(k)} \| l \rangle^2 \quad (k > 0). \tag{1.120}$$

If there are two open shells in the configuration, electrostatic interaction between the electrons can be divided into an interaction between the electrons of both shells and an interaction within the shells. The latter is given by Eq. (1.117). A consequence of the fact that H^e is a scalar in the orbital and spin spaces we can omit the quantum numbers of the shells, including open shells, on which the operator does not act, in its matrix elements according to Eqs. (1.57) and (1.68) at $k = 0$. A matrix element that corresponds to an interaction between shells is

$$\langle l_1^{N_1} l_2^{N_2} | H^e | l_1^{N_1} l_2^{N_2} \rangle = \sum_k [f_k F^k(l_1, l_2) + g_k G^k(l_1, l_2)], \tag{1.121}$$

where f_k and g_k are coefficients at the F-type and G-type integrals.

The operator h_{ij}^e that describes the interaction between nonequivalent electrons is also given in terms of unit tensors, but transforming the G term in Eq. (1.106) requires that another summation be introduced:

$$h_{ij}^e = \sum_k \Bigg\{ (u_i^{(k)} \cdot u_j^{(k)}) \langle l_1 \| C^{(k)} \| l_1 \rangle \langle l_2 \| C^{(k)} \| l_2 \rangle F^k(n_1 l_1, n_2 l_2) -$$

$$- \sum_{xx'} (-1)^x [x, x'] \begin{Bmatrix} l_1 & l_2 & k \\ l_2 & l_1 & x \end{Bmatrix} (v_i^{(xx')} \cdot v_j^{(xx')}) \langle l_1 \| C^{(k)} \| l_2 \rangle^2 G^k(n_1 l_1, n_2 l_2) \Bigg\}. \tag{1.122}$$

The corresponding operator H^e is obtained by making the substitutions $u_i^{(k)} \to U^{(k)}$ and $v_i^{(xx')} \to V^{(xx')}$, whereby the condition that $i < j$ is automatically satisfied since the i and j cover the coordinates of the electrons in the different shells. Making use of Eq. (1.67), we find expressions for the f_k and g_k coefficients:

$$f_k(l_1^{N_1} l_2^{N_2} \gamma_1 L_1 S_1 \gamma_2 L_2 S_2 LS, \gamma_1' L_1' S_1' \gamma_2' L_2' S_2' LS) = \delta(S_1, S_1')\delta(S_2, S_2')$$

$$(-1)^{L_1' + L_2 + L} [S_1, S_2]^{-1/2} \langle l_1 \| C^{(k)} \| l_1 \rangle \langle l_2 \| C^{(k)} \| l_2 \rangle \times \begin{Bmatrix} L_1 & L_2 & L \\ L_2' & L_1' & k \end{Bmatrix}$$

$$\langle l_1^{N_1} \gamma_1 L_1 S_1 \| U^{(k)} \| l_1^{N_1} \gamma_1' L_1' S_1' \rangle \times \langle l_2^{N_2} \gamma_2 L_2 S_2 \| U^{(k)} \| l_2^{N_2} \gamma_2' L_2' S_2' \rangle. \tag{1.123}$$

$$g_k(l_1^{N_1} l_2^{N_2} \gamma_1 L_1 S_1\, \gamma_2 L_2 S_2 LS,\ \gamma_1' L_1' S_1'\, \gamma_2' L_2' S_2' LS) =$$

$$= (-1)^{L_1' + L_2 + L + S_1' + S_2 + S + 1} \langle l_1 \| C^{(k)} \| l_2 \rangle^2 \sum_{xx'} (-1)^x [x,\ x'] \times$$

$$\times \begin{Bmatrix} l_1 & l_2 & k \\ l_2 & l_1 & x \end{Bmatrix} \begin{Bmatrix} L_1 & L_2 & L \\ L_2' & L_1' & x \end{Bmatrix} \begin{Bmatrix} S_1 & S_2 & S \\ S_2' & S_1' & x' \end{Bmatrix} \times$$

$$\times \langle l_1^{N_1} \gamma_1 L_1 S_1 \| V^{(xx')} \| l_1^{N_1} \gamma_1' L_1' S_1' \rangle \langle l_2^{N_2} \gamma_2 L_2 S_2 \| V^{(xx')} \| l_2^{N_2} \gamma_2' L_2' S_2' \rangle. \quad (1.124)$$

The coefficients are diagonal with respect to the many-electron quantum numbers and are independent of them in the following situations:

$$f_0(l_1^{N_1} l_2^{N_2}) = N_1 N_2, \quad (1.125)$$

$$f_k(l_1^{4l_1+2} l_2^{N_2}) = 0 \quad (k > 0), \quad (1.126)$$

$$g_k(l_1^{4l_1+2} l_2^{N_2}) = -\frac{N_2}{2l_2+1} \langle l_1 \| C^{(k)} \| l_2 \rangle^2. \quad (1.127)$$

Between the coefficients for partially and almost filled shells, the relations ($k \neq 0$):

$$f_k(l^{4l+2-N} \gamma v LS,\ \gamma' v' LS) = (-1)^{(v-v')/2} f_k(l^N \gamma v LS,\ \gamma' v' LS)$$

$$- \delta(\gamma v,\ \gamma' v') \frac{2l+1-N}{2l+1} \langle l \| l^{(k)} \| C \rangle^2 \quad (1.128)$$

$$f_k\left(l_1^{4l_1+2-N_1} l_2^{N_2}\right) = (-1)^{\frac{v_1 - v_1'}{2}+1} f_k\left(l_1^{N_1} l_2^{N_2}\right), \quad (1.129)$$

$$f_k\left(l_1^{4l_1+2-N_1} l_2^{4l_2+2-N_2}\right) = (-1)^{\frac{v_1 - v_1' + v_2 - v_2'}{2}} f_k\left(l_1^{N_1} l_2^{N_2}\right), \quad (1.130)$$

$$g_k\left(l_1^{4l_1+2-N_1} l_2^{4l_2+2-N_2}\right) = (-1)^{\frac{\upsilon_1-\upsilon_1'+\upsilon_2-\upsilon_2'}{2}} g_k\left(l_1^{N_1} l_2^{N_2}\right)$$
$$+ \delta(\gamma_1\upsilon_1L_1S_1, \gamma_1'\upsilon_1'L_1'S_1')\,\delta(\gamma_2\upsilon_2L_2S_2, \gamma_2'\upsilon_2'L_2'S_2')\left[-2+\frac{N_1}{2l_1+1}+\frac{N_2}{2l_2+1}\right] \quad (1.31)$$
$$\times \langle l_1 \| C^{(k)} \| l_2 \rangle^2$$

are satisfied, according to Eq. (1.82). For the sake of brevity the quantum numbers in the f_k and g_k coefficients have been omitted in Eqs. (1.128)-(1.131).

If the configuration contains more than two open shells, we can omit the designations of open shells upon which the operator does not act from the matrix element of the electrostatic interaction operator (they must be diagonal in quantum numbers) when we examine an interaction within a shell or between shells whose angular momenta are directly coupled to one another. If the angular momenta are otherwise coupled, the dependence of the "passive" shells on the quantum numbers is determined via the transformation matrices and Eqs. (1.68) and (1.69). General expressions for configurations having any number of open shells can be found in [46].

The F^k and G^k integrals for one-electron functions have algebraic expressions [54] – they are directly proportional to Z. Consequently, the electrostatic interaction increases linearly when the effective nuclear charge increases.

Electrostatic interaction between electrons splits the eigenvalue of the $H^k + H^p$ operators into the energy terms, but does not reduce degeneration over the total angular momentum J. The energies of the levels are determined by the spin-orbit interaction. This interaction follows from Dirac's relativistic theory, although if we postulate the existence of an electron spin and an associated magnetic moment, an operator of the spin—orbit interaction is obtained in nonrelativistic theory when the interaction between the spin magnetic moment and the magnetic field, created by the orbital motion of an electron in the Coulomb field of the nucleus and other electrons $V(r)$, is examined:

$$H^{so} = \frac{1}{2m^2c^2} \sum_{i=1}^{N} \frac{1}{r_i} \frac{dV(r_i)}{dr_i} (l_i^{(1)} \cdot s_i^{(1)}), \quad (1.132)$$

where $l_i^{(1)}$ and $s_i^{(1)}$ are the operators of the orbital and spin angular momenta of the ith electron.

If we consider only the field of the nucleus, then

$$V(r) = -\frac{Ze^2}{r}, \quad \frac{1}{r}\frac{dV(r)}{dr} = \frac{Ze^2}{r^3}. \quad (1.133)$$

The operator H^{so} has ranks of unity for the orbital and spin angular momenta and zero for the total angular momentum. According to Eqs. (1.57) and (1.67), its matrix element is given by

$$\langle \gamma LSJM | H^{so} | \gamma' L' S' J' M' \rangle = \delta(J, J')\delta(M, M') \times$$
$$\times (-1)^{L'+S+J} \begin{Bmatrix} L & S & J \\ S' & L' & 1 \end{Bmatrix} \langle \gamma LS \| H^{so} \| \gamma' L' S' \rangle. \quad (1.134)$$

For a single open shell the submatrix element is directly proportional to the submatrix element of the $V^{(11)}$ operator:

$$\langle nl^N \gamma LS \| H^{so} \| nl^N \gamma' L' S' \rangle =$$
$$= \left[\frac{3}{2} l(l+1)(2l+1) \right]^{1/2} \langle l^N \gamma LS \| V^{(11)} \| l^N \gamma' L' S' \rangle \zeta_{nl}, \quad (1.135)$$

where ζ_{nl} is the spin–orbit interaction

$$\zeta_{nl} = \frac{\hbar^2}{2m^2 c^2} \int_0^\infty \frac{1}{r} \frac{dV(r)}{dr} P_{nl}^2(r)\, dr. \quad (1.136)$$

The potential that is produced by other electrons is frequently included in $V(r)$; however, it has a complex, nonlocal form. A more accurate and consistent method is to consider only the field of the nucleus in Eq. (1.136) and to take into account the spin—orbit interaction with other electrons as a matrix element for the "spin–other orbit" (see Sec. 1.6). The basic term in its matrix element is the very same spin-angular dependence found in Eq. (1.135) which, when taken into account, will lead only to a change in ζ_{nl}.

According to Eqs. (1.82) and (1.135), the matrix elements of the H^{so} operator for almost- and partially filled shells differ only in algebraic sign (for equal radial integrals):

$$\langle l^{4l+2-N} \gamma J | H^{so} | l^{4l+2-N} \gamma' J \rangle = (-1)^{\frac{v-v'}{2}+1} \langle l^N \gamma J | H^{so} | l^N \gamma' J \rangle, \quad (1.137)$$

and the diagonal matrix element disappears when $N = 2l + 1$.

Using the expression given by Eq. (1.25) for the 6j-symbol for the case of a single electron, we obtain

$$\langle nlsjm_j | h^{so} | nlsj'm'_j \rangle = \\ = \delta(j, j')\,\delta(m_j, m'_j)\,[j(j+1) - l(l+1) - s(s+1)]\,\zeta_{nl}/2. \tag{1.138}$$

According to Eqs. (1.137) and (1.138), the electron and vacancy levels are arranged in reverse order [55]:

$$E^{so}(nlj) = \begin{cases} \frac{l}{2}\zeta_{nl}, & j = l + \frac{1}{2}, \\ -\frac{l+1}{2}\zeta_{nl}, & j = l - \frac{1}{2}; \end{cases} \tag{1.139}$$

$$E^{so}(nl^{4l+1}j) \equiv E^{so}(nlj^{-1}) = \begin{cases} -\frac{l}{2}\zeta_{nl}, & j = l + \frac{1}{2}, \\ \frac{l+1}{2}\zeta_{nl}, & j = l - \frac{1}{2}, \end{cases} \tag{1.140}$$

but the distance between the levels with $j = l \pm 1/2$ have the same value

$$|\Delta E| = \frac{2l+1}{2}\zeta_{nl}. \tag{1.141}$$

The integral $\langle nl \mid r^{-3} \mid nl \rangle$ for hydrogenlike wave functions is proportional to Z^3 [54]; therefore, the spin—orbit interaction increases as the effective nuclear charge increases with the fourth degree.

When there are two or more open shells, the matrix element of the H^{so} operator in a JJ-coupling is equal to the sum of the one-shell elements (the operator is a scalar in the total angular momentum space):

$$\langle l_1^{N_1}\gamma_1 J_1 l_2^{N_2}\gamma_2 J_2 J | H^{so} | l_1^{N_1}\gamma'_1 J'_1 l_2^{N_2}\gamma'_2 J'_2 J \rangle = \\ = \delta(\gamma_2 J_2, \gamma'_2 J'_2)\,\langle l_1^{N_1}\gamma_1 J_1 | H^{so} | l_1^{N_1}\gamma'_1 J'_1 \rangle + \\ + \delta(\gamma_1 J_1, \gamma'_1 J'_1)\,\langle l_2^{N_2}\gamma_2 J_2 | H^{so} | l_2^{N_2}\gamma'_2 J'_2 \rangle, \tag{1.142}$$

and in an LS-coupling, because the orbital and spin ranks are not zero, the matrix element contains additional $6j$-symbols [46, 52].

If spin–orbit interaction is taken into consideration, the terms split up into levels. The degeneration of the levels in M is removed only by an external field.

A fundamental integral characteristic of the energy spectrum is the average energy of the configuration

$$\bar{E}(K)=g(K)^{-1}\sum_{\gamma JM}\langle K\gamma J|H|K\gamma J\rangle=g(K)^{-1}\sum_{\gamma J}[J]\langle K\gamma J\|H\|K\gamma J\rangle, \tag{1.143}$$

where $g(K)$ is the number of states in a given configuration K or its statistical weight. This statistical weight is equal to the product of the statistical weights of all the open shells.

Spin–orbit interaction makes no contribution to the average energy [when the matrix element is summed over J according to Eq. (A2.4) a $\delta(1,0)$ appears]. An expression for the average electrostatic energy is obtained by multiplying the average energy of an electron pair by the number of such pairs [35, 49]:

$$\begin{aligned}\bar{E}(K)=\sum_i\Bigg\{&N_i I(n_i l_i)+\frac{N_i(N_i-1)}{2}F^0(n_i l_i, n_i l_i)-\\&-\frac{N_i(N_i-1)}{(4l_i+2)(4l_i+1)}\sum_{k>0}\langle l_i\|C^{(k)}\|l_i\rangle^2 F^k(n_i l_i, n_i l_i)+N_i\sum_{j(>i)}N_j F^0(n_i l_i, n_j l_j)-\\&-\frac{N_i}{4l_i+2}\sum_{j(>i)}\sum_k\frac{N_j}{2l_j+1}\langle l_i\|C^{(k)}\|l_j\rangle^2 G^k(n_i l_i, n_j l_j)\Bigg\}.\end{aligned} \tag{1.144}$$

The summation is carried out over all shells and all pairs of shells.

1.5. The Hartree–Fock Method

Let Ψ be an approximate wave function of an atom. We will expand this function in a basis of exact wave functions Φ for the atom's Hamiltonian H:

$$\Psi=\sum_i c_i\Phi_i,\quad H\Phi_i=\mathscr{E}_i\Phi_i. \tag{1.145}$$

Then,

$$E=\int\Psi^*H\Psi d\tau=\sum_i|c_i|^2\mathscr{E}_i. \tag{1.146}$$

For the ground state of an atom $\varepsilon_0 \leq \varepsilon_i$ and, by replacing all ε_i with ε_0 and using $\Sigma_i \mid c_i \mid ^2 = 1$, we obtain the inequality

$$\mathscr{E}_0 \leqslant \int \Psi^* H \Psi d\tau. \tag{1.147}$$

Equality is attained only when $\Psi = \Phi_0$.

In order that Eq. (1.147) be satisfied for the kth excited state, the auxiliary conditions [56]

$$c_i = \int \Psi \Phi_i^* \, d\tau = 0 \tag{1.148}$$

for $i < k$ must be satisfied.

Thus, the energy of an atom as a functional of Ψ has an absolute minimum and the wave functions can be found, starting from the variational principle which states that in the neighborhood of an extremum for the functional the variation $\delta\Psi$ in a wave function (a small change $\Psi \rightarrow \Psi + \delta\Psi$) changes the functional E which variation is zero in the first order in $\delta\Psi$ [50, 56]:

$$\delta E = 0. \tag{1.149}$$

Using the expression $E^k + E^p + E^e$, varying the radial wave functions $P_{nl}(r)$, and considering their variations to be independent in the one-configuration, nonrelativistic theory of a spherically symmetric field, we obtain the Hartree–Fock equations.

In order that the radial functions be orthonormalized

$$\int_0^\infty P_{nl}(r) P_{n'l}(r) \, dr = \delta(n, n'), \tag{1.150}$$

the diagonal ($\lambda_{nl,nl}$) and the nondiagonal ($\lambda_{nl,n'l}$) factors of a Lagrangian multiplier must be added to the functional and, instead of Eq. (1.149), the conditions [37, 45]

$$\delta\left(E + \sum_{nln'} \lambda_{nl,n'l} \int_0^\infty P_{nl}(r) P_{n'l}(r) \, dr\right) = 0. \tag{1.151}$$

must be satisfied.

The Hartree–Fock equations are derived, for example, in [37, 45, 39]. They are a system of integrodifferential equations, the number of which is equal to the number of shells in the configuration. An equation for one $P_{nl}(r)$ function is

$$\left(\frac{d^2}{dr^2} - 2V_{nl}(r) - \frac{l(l+1)}{r^2} - \varepsilon_{nl,nl}\right) \tag{1.152*}$$

$$P_{nl}(r) = 2X_{nl}(r) + \sum_{n'(\neq n)} \varepsilon_{nl,n'l} P_{n'l}(r), \quad (145)^*$$

where

$$V_{nl}(r) = \frac{1}{r}\left\{ -Z + \sum_k \frac{2f_k(l^N)}{N} Y^k_{nl,nl}(r) + \right.$$

$$\left. + \sum_k \sum_{n'l'(\neq nl)} \frac{f_k(l^N l'^{N'})}{N} Y^k_{n'l',n'l'}(r) \right\}; \tag{1.153*}$$

$$X_{nl}(r) = \frac{1}{r} \sum_k \sum_{n'l'(\neq nl)} \frac{g_k(l^N l'^{N'})}{N} Y^k_{nl,n'l'}(r) P_{n'l'}(r); \tag{1.154*}$$

$$\varepsilon_{nl,n'l} = -\frac{2\lambda_{nl,n'l}}{N}. \tag{1.155}$$

The integral function $Y_{nl,n'l'}{}^k(r)$ is defined according to Eq. (1.110). The f_k and g_k are spin-angular coefficients in the matrix element of the electrostatic interaction operator for a specific term (we then have Hartree—Fock equations for the term, HF-t) or the coefficients in the expression for $\bar{E}(K)$, Eq. (1.144) (which are then the equations for the average energy, HF-av.).

The term $V_{nl}(r)$ is a potential function (in the atomic system of units it is equal to the potential defined for a unit negative charge). If the nonspherical terms with $k > 0$ are omitted from Eq. (1.153), then $V_{nl}(r)$ corresponds to the classical electrostatic energy of an nl-electron in the field of the nucleus and other electrons, the charge of each of which has been "spread" over the volume of an atom having a radial density of $-eP_{nl}{}^2(r)$.

The $-l(l + 1)/r^2$ term is the centrifugal energy multiplied by -2. It increases with increasing orbital moment and prevents an electron with $l > 0$ from penetrating to the nucleus. The Hartree—Fock parameter $\varepsilon_{nl,nl}$ is associated with the one-electron energy. If the equation for HF-av. is multiplied from the left by $P_{nl}(r)$ and integrated over dr, then $\varepsilon_{nl,nl}$ is given in terms of the integrals F^k, G^k, and I and is written in the form

$$\frac{1}{2}\varepsilon_{nl,\,nl}=\bar{E}'(Knl^{-1})-\bar{E}(K)\approx I_{nl}. \tag{1.156}$$

Here, $\bar{E}(K)$ is the average energy of the atom, and $\bar{E}'(Knl^{-1})$ is the average energy of an ion with a vacancy nl^{-1}, calculated with the atom's frozen wave functions. Thus, the quantity $-{}^1/{}_2\varepsilon_{nl,nl}$ has the sense of the one-electron energy ε_{nl} and its absolute value is approximately equal to the binding energy I_{nl} of an nl electron. Equation (1.156) is called Koopmans' theorem.

The term $X_{nl}(r)$ is an exchange function that proceeds from the exchange portion of the matrix element of the H^e operator. The presence of $X_{nl}(r)$, and a term containing $\varepsilon_{nl,n'l'}$ makes Eq. (1.152) nonhomogeneous. This equation can be replaced by a simple, approximate equation if $X_{nl}(r)$ can be approximated by a local potential function, e.g., that obtained by Slater from the statistical model of an electron gas:

$$V_{nl}^{ex}(r)=-\frac{3}{2}\left(\frac{3\rho(r)}{\pi}\right)^{1/3}, \tag{1.157*}$$

where $\rho(r)$ is the spherically averaged probability density of finding electrons on the atom [$-e\rho(r)$ is the distribution density of the electron charge]:

$$\rho(r)=\frac{1}{4\pi r^2}\sum_{nl}N_{nl}P_{nl}^2(r),\quad \int\rho(r)\,d\mathbf{r}=\sum_{nl}N_{nl}=N. \tag{1.158}$$

The exchange function given by Eq. (1.157) has an inexact asymptote as $r\to\infty$; therefore, from some $r = r_1$ it merges with a proper asymptotic function. In addition, the term $V_{nl}^{ex}(r)$ includes the self-action of the electron [$\rho(r)$ also contains the density of the electron being examined] and is less accurate in an atom's interior where the charge of the electron shells has the highest density and the electron gas model is less appropriate. Various modifications of Slater's potential function are used, which exclude self-action and refine the function through multiplication by a semiempirical parameter β, chosen in the interval $2/3 \leq \beta \leq 1$, or a polynomial in r [57, 46].

Hartree–Fock radial wave functions correspond to an approximate, one-electron Hamiltonian:

$$H_0=\sum_{i=1}^{N}\left(-\frac{\hbar^2}{2m}\nabla_i^2-\frac{Ze^2}{r_i}+u(r_i)\right), \tag{1.159}$$

where $u(r)$ is the Hartree–Fock one-electron potential energy that consists of an F-type and G-type part. The sum $\Sigma_i u(r_i)$ includes the fundamental, spherically symmetric part of an electrostatic interaction between electrons. Its other part

$$V = \sum_{i<j} \frac{e^2}{r_{ij}} - \sum_i u(r_i) \tag{1.160}$$

is taken into consideration in first-order perturbation theory when energy levels are calculated.

Hartree–Fock equations are solved by the method of self-consistent field: using initial wave functions, the functions $V_{nl}(r)$ and $X_{nl}(r)$ for an nl-electron are calculated by numerical integration, and its radial wave function that satisfies the condition given by Eq. (1.150) and the boundary conditions

$$P_{nl}(0) = P_{nl}(\infty) = 0 \tag{1.161}$$

and having $n - l - 1$ nodes is found by numerical integration [45].

If the direct and exchange parts of the potential in the Hartree–Fock equations are approximated by the local potential which is the same for all electrons, then, in the general case, for a configuration with open shells the functions $P_{nl}(r)$ that have the same l are automatically obtained orthogonal without having to introduce nondiagonal Lagrangian multipliers because they appear in the same orthonormalized system of radial functions. Although a simplification of this kind reduces the accuracy of the wave functions, it does make them more useful as basis functions.

For a configuration with one open shell, or two comparatively weakly electrostatically interacting open shells, the dependence of the wave functions on the term is usually weak; therefore, HF-av. equations are solved that will yield one set of radial wave functions for a given configuration. These functions make subsequent calculation of the physical quantities much easier. For some configurations with open shells, especially for configurations that contain a collapsing electron (Sec. 3.3), the solutions of the HF-t equations must be examined. If there are large nondiagonal matrix elements for the H^e operator in this configuration, the functional must be defined for the intermediate coupling functions that diagonalize the energy matrix, and the more general equations are solved self-consistently with the diagonalization of the matrix [58].

When Hartree–Fock functions are obtained, the functional is varied without putting $P_{nl}(r)$ into a specific form; therefore, the solutions of these equations are the best one-electron functions. We can give expressions for $P_{nl}(r)$ that have free parameters and choose them so that they minimize the energy [59]. Analytical functions such as these with several parameters are only slightly more accurate as numerical functions, although the difficulty in calculating analytical functions increases rapidly as the number of nodes in them increases.

As we mentioned at the beginning of this section, the energy of an excited state has an absolute minimum only when the auxiliary condition that the wave function be orthogonal to the exact wave functions of the lower states is satisfied. This condition is usually satisfied automatically because the angular or spin wave functions are orthogonal (if at least one electron in both states differs by the quantum number l or the states have different resultant angular momenta L, S, or J). However, the terms for two open shells are characterized not only by resultant angular momenta, but by the angular momenta of the individual shells, relative to which the terms cannot be divided into those contained and not contained in lower-lying configurations of the same type. To carry out this kind of separation, a basis of wave functions was suggested in [60, 61] in which the terms of the $l^{N_1}l^{N_2}$ configuration are described by their genealogy in the series of configurations $n_1l^{N_1}$, $n_2l^{N_2}$, $n_1l^{N_1+1}$, $n_2l^{N_2-1}$, . . . or by the isospin quantum number [61]. In this basis we can average the energy over the terms that are not contained in the lower-lying configurations of this series and thus obtain a functional that will ensure the aforementioned orthogonality in the wave functions [62, 60]. In the given functional the coefficients at the integrals for the electrostatic interaction are [60]

$$\tilde{f}_k(l^{N_1}l^{N_2}) = \frac{N_2(4l+2-N_1)}{(2l+1)(4l+1)(4l+3)}\langle l \| C^{(k)} \| l\rangle^2, \quad k>0, \tag{1.162}$$

$$\tilde{g}_k(l^{N_1}l^{N_2}) = -\frac{N_2[(4l+2)N_1-1]}{(2l+1)(4l+1)(4l+3)}\langle l \| C^{(k)} \| l\rangle^2, \quad k>0, \tag{1.163}$$

$$\tilde{g}_0(l^{N_1}l^{N_2}) = -N_2. \tag{1.164}$$

The expressions for the coefficients $\tilde{f}_0(l^{N_1}l^{N_2})$ and $\tilde{f}_k(l^N)$ coincide with the corresponding formulas for the average energy of the configuration, Eq. (1.144).

The use of the coefficients given by Eqs. (1.162)–(1.164) in the Hartree–Fock equations also makes it possible to take some correlation effects into account [62, 60].

Another method for ensuring the orthogonality of a wave function to functions of lower states having the same symmetry is to use identical "frozen" wave functions for the core; however, this produces significant errors, especially for configurations with several open shells.

Several attempts were made to account for the condition given by Eq. (1.148) by way of incorporating the respective Lagrangian multipliers into the Hartree–Fock equations, but this substantially complicates the calculations and leads to an increase in the error [63]. For the case of highly excited states with internal vacancies, which have a large number of lower states with the same symmetry, this method is unusable in practice. The orthogonality condition can also be satisfied when a special function that is nonlinear in H is used [64], but this involves significant labor in the calculations.

In an atom with many electrons, the electron density of the core when a transition is made from the ground to an excited configuration is changed only slightly; therefore, $P_{nl}(r)$ functions having the same l and $n \neq n'$ obtained for various configurations are approximately orthogonal [65]. This means the coefficients c_i in Eq. (1.146) are small for all $i < k$ (k is the state being examined); also, when an inequality similar to Eq. (1.147) is obtained, the $\mathcal{E}_i$ are replaced not by the average value, but with an eigenvalue, starting from where the series begins. This explains why the wave functions for excited states obtained without regard for the conditions given by Eq. (1.148) and used in many works yield results that are physically reliable.

The variation principle can even be used to find the wave function of a free electron having kinetic energy ε, orbital angular momentum l, and moving in the field of an atom. The Hartree–Fock equation for $P_{\varepsilon l}(r)$ is similar to (1.152) (only the negative one-electron energy $-{}^1/_2\varepsilon_{nl,nl}$ changes by the positive energy ε)

$$\left(\frac{d^2}{dr^2} - 2V_{\varepsilon l}(r) - \frac{l(l+1)}{r^2} + 2\varepsilon\right) P_{\varepsilon l}(r) = 2X_{\varepsilon l}(r) + \sum_n \lambda_{\varepsilon l,\,nl} P_{nl}(r). \quad (1.165)^*$$

The expressions for $V_{\varepsilon l}(r)$ and $X_{\varepsilon l}(r)$ correspond to the replacement $n \to \varepsilon$ and the substitution $N = 1$ in Eqs. (1.153) and (1.154). They are calculated with the "frozen" one-electron wave functions of an atom, since the field of a free electron that has been averaged with respect to motion vanishes [66]. The function $P_{\varepsilon l}(r)$ must satisfy an asymptotic condition as $r \to \infty$:

$$P_{\varepsilon l}(r) \xrightarrow[r\to\infty]{} \left(\frac{2}{\pi^2 \varepsilon}\right)^{1/4} \sin\left[\sqrt{2\varepsilon}\, r - \frac{1}{2}\, l\pi + \eta_l^C(\varepsilon) + \delta_l(\varepsilon)\right], \quad (1.166)^*$$

Here δ_l is the scattering phase or the phase shift and η_l^C is the additional Coulomb shift that appears because the potential of an ion decays slowly according to r^{-1}; it is absent for the field of a neutral atom. The function $P_{\varepsilon l}(r)$ has been normalized to a Dirac δ-function:

$$\int_0^\infty P_{\varepsilon l}(r) P_{\varepsilon' l}(r)\, dr = \delta(\varepsilon - \varepsilon'). \quad (1.167)$$

The functions $P_{ql}(r)$ normalized to $\delta(q - q\ )$ are often used (q is the absolute value of the wave vector and is equal to $\hbar^{-1}\sqrt{2m\varepsilon}$), or $P_{\varepsilon l}^{u}(r)$, normalized to unit electron flux density. They are related to $P_{\varepsilon l}(r)$ by the relations

$$P_{ql}(r) = \hbar\sqrt{\frac{q}{m}}\, P_{\varepsilon l}(r), \tag{1.168}$$

$$P_{\varepsilon l}^{u}(r) = (2\pi\hbar)^{1/2} P_{\varepsilon l}(r). \tag{1.169}$$

The dependence of the radial functions for free and bound electrons on the nuclear charge corresponds approximately to the law for transforming hydrogenlike functions [37, 39]

$$P_{nl}^{Z}(r) = Z^{1/2} P_{nl}^{\mathrm{H}}(Zr), \quad P_{\varepsilon l}^{Z}(r) = Z^{-1/2} P_{\varepsilon l}^{\mathrm{H}}(Zr). \tag{1.170}$$

The superscript H denotes a hydrogen function and Z denotes the function for an atom having a nuclear charge of Z.

1.6. Relativistic Theory

A one-electron atom is described in the relativistic theory by the stationary Dirac equation:

$$h^{D}\Phi = \varepsilon\Phi, \tag{1.171}$$

$$h^{D} = c\,\boldsymbol{\alpha}\cdot\mathbf{p} + \beta mc^{2} - \frac{Ze^{2}}{r}, \tag{1.172}$$

where α and β are fourth-order Dirac matrices that are given in terms of the Pauli matrix σ and a second-order unit matrix I:

$$\boldsymbol{\alpha} = \begin{pmatrix} 0 & \sigma \\ \sigma & 0 \end{pmatrix}, \quad \beta = \begin{pmatrix} I & 0 \\ 0 & -I \end{pmatrix}; \tag{1.173}$$

Φ is the one-electron, four-component wave function and m is the rest mass of the electron. The quantum numbers l, m_l, and μ do not describe the characteristics of single-electron state well because the Hamiltonian of the Dirac equation h^D does not commute with $\mathbf{l}^2$, l_z, and s_z but only with j^2, j_z, $\mathbf{s}^2$, and the operator

$$\mathcal{K} = \hbar^{-2}(\mathbf{l}^2 + \mathbf{s}^2 - \mathbf{j}^2) - 1 = -(1 + 2\hbar^{-2}\mathbf{s}\cdot\mathbf{l}). \tag{1.174}$$

Its eigenvalues are equal to

$$\varkappa=\begin{cases} l, & \text{if} \quad j=l-\frac{1}{2}, \\ -l-1, & \text{if} \quad j=l+\frac{1}{2}. \end{cases} \tag{1.175}$$

The quantum numbers $n\kappa m_j$ (m_j is the projection of j) are sufficient to uniquely designate the one-electron wave functions; however, by analogy with the nonrelativistic theory and to point out the parity of the wave function (it is defined by the quantum number l) the equivalent set $nljm_j$ is frequently used.

In a spherically symmetric field the one-electron wave function has the form

$$\Phi_{nljm_j}(\mathbf{r},\ \sigma)=\frac{1}{r}\begin{pmatrix} P_{nlj}(r)\,\Omega_{jlm_j}(\theta,\ \varphi,\ \sigma) \\ iQ_{n\bar{l}j}(r)\,\Omega_{j\bar{l}m_j}(\theta,\ \varphi,\ \sigma) \end{pmatrix}. \tag{1.176}$$

Here, $\bar{l} = 2j - l$, and P and Q are, respectively, the large and small components of the radial function (both are taken to be real). When $v/c \to 0$, P_{nlj} becomes the nonrelativistic function P_{nl}, and Q_{nlj} goes to zero; Ω is the generalized spherical function

$$\Omega_{jlm_j}(\theta,\ \varphi,\ \sigma)=\sum_{\mu}\begin{bmatrix} l & \frac{1}{2} & j \\ m_j-\mu & \mu & m_j \end{bmatrix} Y_{lm_j-\mu}(\theta,\ \varphi)\,\chi_{\mu}(\sigma), \tag{1.177}$$

where Y_{lm} is the spherical, and χ_{μ} is the spin two-component function.

In the relativistic theory, an electron shell is divided into two subshells with $j = l - 1/2$ and $l + 1/2$. A closed subshell contains $2j + 1$ electrons. This division is unique only for a closed shell and for a shell with one vacancy having given j. In the general case, several relativistic subconfigurations with different electron distributions in the subshells correspond to a single relativistic configuration. In the case of open intermediate and outer shells, in which Coulomb repulsion prevails over the spin–orbit interaction, these subconfigurations are strongly mixed among themselves. Thus, the relativistic one-configuration theory takes correlation effects into account less well than does the nonrelativistic one-configuration approximation.

The electrons in each subshell are described by two different radial wave functions, which means that the number of radial functions in the relativistic theory is quadrupled. Another trouble with this theory is that the only possible coupling is the jj-coupling, whereas in atoms, especially within the shells, another type of coupling is often realized.

On the other hand, the description of a subshell with fewer electrons is simpler than for an entire shell: the quantum number J is sufficient for the j^N states up to $(7/2)^4$, and, in most cases, the auxiliary number υ [67].

The exact relativistic Hamiltonian for a two-electron atom is unknown; it is only approximated by a series from perturbation theory. In a first-order approximation, when the exchange of two electrons with a virtual photon is considered, a two-electron relativistic Hamiltonian in a Coulomb gauge has the form [67, 68]

$$h_{ij}^{rel} = \frac{e^2}{r_{ij}} \left[1 - \boldsymbol{\alpha}_i \cdot \boldsymbol{\alpha}_j \cos(kr_{ij}) + (\boldsymbol{\alpha}_i \cdot \nabla_i)(\boldsymbol{\alpha}_j \cdot \nabla_j) \frac{\cos(kr_{ij}) - 1}{k^2} \right], \tag{1.178}$$

where $k = \omega/c$ is the wave number of a virtual photon; α_i is the Dirac matrix, Eq. (1.173), and $r_{ij} = | \mathbf{r}_i - \mathbf{r}_j |$.

Breit's Hamiltonian [54] is often used in the relativistic description of an atom. This Hamiltonian is derived from the h_{ij}^{rel} operator when expanding the cosine in powers of kr_{ij} and maintaining the first terms in Eq. (1.178) that are independent of k (also, the instantaneous Coulomb interaction operator is excluded from the operator):

$$h_{ij}^{B} = -\frac{e^2}{2r_{ij}} \left[\boldsymbol{\alpha}_i \cdot \boldsymbol{\alpha}_j + \frac{(\boldsymbol{\alpha}_i \cdot \mathbf{r}_{ij})(\boldsymbol{\alpha}_j \cdot \mathbf{r}_{ij})}{r_{ij}^2} \right] = \tag{1.179a}$$

$$= \frac{e^2}{r_{ij}} \left[\boldsymbol{\alpha}_i \cdot \boldsymbol{\alpha}_j + \frac{1}{2} (\boldsymbol{\alpha}_i \cdot \nabla_i)(\boldsymbol{\alpha}_j \cdot \nabla_j) r_{ij}^2 \right]. \tag{1.179b}$$

The first term in Eq. (1.179b) is the operator for instantaneous magnetic interaction between two Dirac currents, and the second term, roughly an order of magnitude smaller, approximately describes the retardation of the Coulomb interaction between the electrons.

Breit's operator considers two-electron relativistic effects to within an order of $(Z\alpha)^2$; however, in practice, it does not yield sufficiently accurate results for heavy atoms. For the binding energies of electrons closest to the nucleus in heavy atoms, a substantial contribution is made by corrections that have a higher order in α – the self-energy and the vacuum polarization (Sec. 1.7).

The Hamiltonian for a system of many electrons is the sum – over all electrons and electron pairs – of the Hamiltonian for the Dirac equation h_i^D (1.172), the operator for an instantaneous Coulomb interaction h_{ij}^e (1.96), and h_{ij}^B (1.179):

$$H^{rel} = H^D + H^e + H^B = \sum_i h_i^D + \sum_{i<j} h_{ij}^e + \sum_{i<j} h_{ij}^B . \tag{1.180}$$

the same way as for the nonrelativistic case, although the expressions are somewhat more complicated because of the use of four-component wave functions, which is thoroughly discussed in [67, 69, 48]. Here, we will present only the final formulas.

The matrix element for the H^D operator is independent of the many-electron quantum numbers, being the sum of the one-electron matrix elements*

$$I(nlj)=(nljm_j\,|\,h^D\,|\,nljm_j)=\int\limits_0^\infty dr\left\{c^2(P^2-Q^2)+cQ\left(\frac{dP}{dr}+\frac{\varkappa}{r}P\right)-\right.$$
$$\left.-cP\left(\frac{dQ}{dr}-\frac{\varkappa}{r}Q\right)-\frac{Z}{r}(P^2+Q^2)\right\}. \qquad (1.181)^*$$

If the energy corresponding to the rest mass of an electron is used as the zero value of energy, then the term $c^2(P^2 - Q^2)$ must be replaced by $-2c^2Q^2$.

For the case of a single subshell, the matrix elements for the H^e and H^B operators are

$$(nlj^N\gamma J\,|\,H^e\,|\,nlj^N\gamma' J)=\sum_{k}^{\text{четн.}} f_k(j^N\gamma\gamma' J)\,F^k(\lambda,\lambda), \qquad (1.182)$$

$$f_k(j^N\gamma\gamma' J)=\frac{1}{2}[j]^2\begin{pmatrix} j & k & j\\ -\frac{1}{2} & 0 & \frac{1}{2}\end{pmatrix}^2\left\{[J]^{-1}\sum_{\gamma'' J''}(j^N\gamma J\,\|\,U^{(k)}\,\|\,j^N\gamma'' J'')\times\right.$$
$$\left.\times(j^N\gamma' J\,\|\,U^{(k)}\,\|\,j^N\gamma'' J'')-\delta(\gamma,\gamma')\frac{N}{2j+1}\right\}, \qquad (1.183)$$

$$(nlj^N\gamma J\,|\,H^B\,|\,nlj^N\gamma' J)=-4[j]^2\sum_k\frac{f_k(j^N\gamma\gamma' J)}{k(k+1)}R^k(\lambda\lambda,\bar{\lambda}\bar{\lambda}), \qquad (1.184)$$

---

*The matrix and submatrix elements for relativistic wave functions will be designated (| |) and (‖ ‖).

where $\lambda \equiv nlj$. The bar over the λ means that the function $Q_{n\bar{l}_j}(r)$ is used in the radial integral; no bar means that the function $P_{nlj}(r)$ is used. The relation

$$(-1)^{l_1}\begin{Bmatrix} l_1 & j_1 & 1/2 \\ j_2 & l_2 & t \end{Bmatrix}\langle l_1\|C^{(t)}\|l_2\rangle = -\begin{pmatrix} j_1 & t & j_2 \\ -1/2 & 0 & 1/2 \end{pmatrix}(l_1\, t\, l_2), \tag{1.185}$$

which makes it possible for us to eliminate the dependence on orbital quantum numbers in relativistic (and in jj-coupling for some nonrelativistic) matrix elements, was used to derive Eq. (1.183); the symbol $(l_1\ t\ l_2)$ means that the respective parameters form a triad whose sum $l_1 + t + l_2$ is an even number.

The matrix elements for the operators H^e and H^B are given in terms of the same radial integrals R^k:

$$F^k(\lambda_1, \lambda_2) = R^k(\lambda_1\lambda_2, \lambda_1\lambda_2) + R^k(\lambda_1\bar{\lambda}_2, \lambda_1\bar{\lambda}_2) + \\ + R^k(\bar{\lambda}_1\lambda_2, \bar{\lambda}_1\lambda_2) + R^k(\bar{\lambda}_1\bar{\lambda}_2, \bar{\lambda}_1\bar{\lambda}_2), \tag{1.186}$$

$$R^k(\lambda_1\lambda_2, \lambda_3\lambda_4) = (l_1 l_3 k)(l_2 l_4 k) \times \\ \times \int_0^\infty\!\!\int f_{\lambda_1}(r_1) f_{\lambda_2}(r_2) \frac{r_<^k}{r_>^{k+1}} f_{\lambda_3}(r_1) f_{\lambda_4}(r_2)\, dr_1\, dr_2. \tag{1.187*}$$

Here $f_\lambda \equiv P_{nlj}(r)$ and $f_{\bar{\lambda}} \equiv Q_{n\bar{l}j}(r)$.

The $U^{(k)}$ operator is given in terms of unit tensors $u^{(k)}$ according to Eq. (1.71). The submatrix element $u^{(k)}$ in the relativistic theory is defined as

$$(j\,\|\,u^{(k)}\,\|\,j') = \delta(j, j')\{jkj'\}. \tag{1.188}$$

The submatrix element for the $U^{(k)}$ operator is

$$(j^N\gamma J\,\|\,U^{(k)}\,\|\,j^N\gamma' J') = N[J, J']^{1/2} \sum_{\bar{\gamma}\bar{J}} (-1)^{\bar{J}+j+J+k} \times \\ \times (j^N\gamma J\,\|\,j^{N-1}\bar{\gamma}\bar{J}j)(j^{N-1}\bar{\gamma}\bar{J}j\,\|\,j^N\gamma' J') \begin{Bmatrix} j & J & \bar{J} \\ J' & j & k \end{Bmatrix}. \tag{1.189}$$

Antisymmetric relativistic wave functions for the coupled angular momenta are constructed and expressions for the matrix elements of the operator $H^{\rm rel}$ are obtained in

It must be noted that the nonzero contribution to the matrix element given by Eq. (1.184) is made by only the magnetic interaction operator [the first term in Eq. (1.179b)] [67, 48].

When $N = 2j + 1$ or $N = 2j$, Eq. (1.183) reduces to

$$f_k(j^{2j+1}) = -\frac{1}{2}[j]^2 \begin{pmatrix} j & k & j \\ -\frac{1}{2} & 0 & \frac{1}{2} \end{pmatrix}^2, \quad k>0, \tag{1.190}$$

$$f_k(j^{2j}) = -\frac{1}{2}(2j+1)(2j-1) \begin{pmatrix} j & k & j \\ -\frac{1}{2} & 0 & \frac{1}{2} \end{pmatrix}^2, \quad k>0. \tag{1.191}$$

The coefficient f_0 is diagonal and depends only on N:

$$f_0(j^N) = \frac{N(N-1)}{2}. \tag{1.192}$$

The matrix element for the operator of electrostatic interaction between two subshells in the relativistic theory is also written as Eq. (1.121), except that the shells are replaced by subshells and nl is replaced by $nlj \equiv \lambda$. The coefficients on the integrals for the $F^k(\lambda_1, \lambda_2)$ and $G^k(\lambda_1, \lambda_2)$ are

$$f_k(j_1^{N_1} j_2^{N_2}) = (-1)^{J_1'+J_2+J+J_1+J_2+1}[j_1, j_2] \begin{pmatrix} j_1 & k & j_1 \\ -\frac{1}{2} & 0 & \frac{1}{2} \end{pmatrix} \begin{pmatrix} j_2 & k & j_2 \\ -\frac{1}{2} & 0 & \frac{1}{2} \end{pmatrix} \times$$
$$\times \begin{Bmatrix} J_2' & J_2 & k \\ J_1 & J_1' & J \end{Bmatrix} (j_1^{N_1}\gamma_1 J_1 \| U^{(k)} \| j_1^{N_1}\gamma_1' J_1')(j_2^{N_2}\gamma_2 J_2 \| U^{(k)} \| j_2^{N_2}\gamma_2' J_2'), \tag{1.193}$$

$$g_k(j_1^{N_1} j_2^{N_2}) = (-1)^{k+1+j_1+j_2+J_1'+J_2+J}[j_1, j_2] \begin{pmatrix} j_1 & k & j_2 \\ -\frac{1}{2} & 0 & \frac{1}{2} \end{pmatrix}^2 \times$$
$$\times \sum_x (-1)^x [x] \begin{Bmatrix} j_1 & j_2 & k \\ j_2 & j_1 & x \end{Bmatrix} \begin{Bmatrix} J_2' & J_2 & x \\ J_1 & J_1' & J \end{Bmatrix} (j_1^{N_1}\gamma_1 J_1 \| U^{(x)} \| j_1^{N_1}\gamma_1' J_1') \times$$
$$\times (j_2^{N_2}\gamma_2 J_2 \| U^{(x)} \| j_2^{N_2}\gamma_2' J_2'). \tag{1.194}$$

The relativistic integral G^k, just as the F^k of Eq. (1.186), is given in terms of the integrals R^k:

$$G^k(\lambda_1, \lambda_2) = [1 + \Pi(\lambda_i, \bar{\lambda}_i)][R^k(\lambda_1\lambda_2, \lambda_2\lambda_1) + R^k(\lambda_1\bar{\lambda}_2, \lambda_2\bar{\lambda}_1)]. \quad (1.195)$$

The operator $\Pi(\lambda_i, \bar{\lambda}_i)$ produces the replacement $\lambda_i \leftrightarrow \bar{\lambda}_i$.

The same coefficients f_k (only with odd k) and g_k appear in the expression for the matrix element of the H^B operator:

$$\begin{aligned}
&(n_1 l_1 j_1^{N_1}\gamma_1 J_1 n_2 l_2 j_2^{N_2}\gamma_2 J_2 J \,|\, H^B \,|\, n_1 l_1 j_1^{N_1}\gamma_1' J_1' n_2 l_2 j_2^{N_2}\gamma_2' J_2' J) = \\
&= (-1)^{-J_1 - J_2 + l_1 + l_2} \sum_k \frac{4}{k(k+1)} [j_1, j_2] f_k(j_1^{N_1} j_2^{N_2}) R^k(\lambda_1\lambda_2, \bar{\lambda}_1\bar{\lambda}_2) - \\
&- \sum_k g_k(j_1^{N_1} j_2^{N_2})[1 + \Pi(\lambda_i, \bar{\lambda}_i)] \Bigg\{ \frac{k(k+1)}{2} \Bigg[\frac{1}{2k+3} R^{k+1}(\lambda_1\lambda_2, \bar{\lambda}_2\lambda_1) - \\
&- \frac{1}{2k+3} R^{k+1}(\bar{\lambda}_1\lambda_2, \lambda_2\bar{\lambda}_1) - \frac{1}{2k-1} R^{k-1}(\lambda_1\lambda_2, \bar{\lambda}_2\bar{\lambda}_1) + \\
&+ \frac{1}{2k-1} R^{k-1}(\bar{\lambda}_1\lambda_2, \lambda_2\bar{\lambda}_1) \Bigg] + (-1)^{l_2 + J_2 + 1/2} \frac{1}{4} [(-1)^{J_1 + J_2 + k}[j_1] + [j_2]] \times \\
&\times \Bigg[\frac{3}{2k+3} R^{k+1}(\bar{\lambda}_1\lambda_2, \lambda_2\bar{\lambda}_1) - \frac{3}{2k-1} R^{k-1}(\bar{\lambda}_1\lambda_2, \lambda_2\bar{\lambda}_1) + R_1^{k+1}(\lambda_1\lambda_2, \bar{\lambda}_2\bar{\lambda}_1) - \\
&- R_2^{k+1}(\lambda_1\lambda_2, \bar{\lambda}_2\bar{\lambda}_1) - R_1^{k-1}(\lambda_1\lambda_2, \bar{\lambda}_2\bar{\lambda}_1) + R_2^{k-1}(\lambda_1\lambda_2, \bar{\lambda}_2\bar{\lambda}_1) \Bigg] + \\
&+ \frac{1}{8k(k+1)} [(-1)^{J_1 + J_2 + k}[j_1] + [j_2]]^2 \Bigg[\frac{k(k+3)}{2k+3} \big(R^{k+1}(\lambda_1\lambda_2, \bar{\lambda}_2\bar{\lambda}_1) + \\
&+ R^{k+1}(\bar{\lambda}_1\lambda_2, \lambda_2\bar{\lambda}_1) \big) + 2 \big(R^k(\lambda_1\lambda_2, \bar{\lambda}_2\bar{\lambda}_1) + R^k(\bar{\lambda}_1\lambda_2, \lambda_2\bar{\lambda}_1) \big) - \\
&- \frac{(k-2)(k+1)}{2k-1} \big(R^{k-1}(\lambda_1\lambda_2, \bar{\lambda}_2\bar{\lambda}_1) + R^{k-1}(\bar{\lambda}_1\lambda_2, \lambda_2\bar{\lambda}_1) \big) \Bigg] \Bigg\}.
\end{aligned} \quad (1.196)$$

The integrals $R_1^{\,k}$ and $R_2^{\,k}$ differ from R^k by the fact that the integration over r_1 and r_2 in them is performed not for the whole region, but is limited by the condition that $r_1 < r_2$ for $R_1^{\,k}$ and $r_1 > r_2$ for $R_2^{\,k}$.

If a subshell is closed the f_k coefficients with $k > 0$ are

$$f_k(j_1^{2j_1+1}\, j_2^{N_2}) = \delta(k, 0)\,\delta(\gamma_2 J_2, \gamma_2' J_2')[j_1] N_2 \quad (1.197)$$

and the direct part of the matrix element given by Eq. (1.196) will go to zero. Exchange Coulomb and Breit interactions will then contribute only to the total energy:

$$g_k(j_1^{2j_1+1} j_2^{N_2}) = -\delta(\gamma_2 J_2, \gamma_2' J_2') N_2 [j_1] \begin{pmatrix} j_1 & k & j_2 \\ -\frac{1}{2} & 0 & \frac{1}{2} \end{pmatrix}^2 . \tag{1.198}$$

Consequently, the level energies are defined in the relativistic theory by the Coulomb and Breit interactions between and within open subshells. The Breit energy is merely a correction to the electrostatic energy and, therefore, is not considered in the average energy of a subconfiguration having a specific electron distribution in the subshells (the one-configuration *jj*-average energy) that can be found by the same method as in Eq. (1.144):

$$\bar{E}(K) = \sum_i N_i I(\lambda_i) +$$
$$+ \sum_i \frac{N_i(N_i-1)}{2} \left[F^0(\lambda_i, \lambda_i) - \frac{2j_i+1}{2j_i} \sum_{k>0} \begin{pmatrix} j_i & k & j_i \\ -\frac{1}{2} & 0 & \frac{1}{2} \end{pmatrix}^2 F^k(\lambda_i, \lambda_i) \right] +$$
$$+ \sum_{i<p} N_i N_p \left[F^0(\lambda_i, \lambda_p) - \sum_k \begin{pmatrix} j_i & k & j_p \\ -\frac{1}{2} & 0 & \frac{1}{2} \end{pmatrix}^2 G^k(\lambda_i, \lambda_p) \right] . \tag{1.199}$$

If in this equation the functions P and Q are varied, with due regard for the condition

$$\int [P_{nlj}(r) P_{n'lj}(r) + Q_{n\bar{l}j}(r) Q_{n'\bar{l}j}(r)] \, dr = \delta(n, n'), \tag{1.200}$$

that follows from the requirement that one-electron wave functions be orthonormal, then relativistic Dirac–Fock equations for calculating the $P_\lambda(r)$ and $Q_{\bar{\lambda}}(r)$ are obtained [67]:

$$\left(\frac{d}{dr} + \frac{\varkappa}{r} \right) P_\lambda(r) = \frac{1}{c} [2c^2 - \varepsilon_{\lambda, \lambda} - Y_\lambda(r)] Q_{\bar{\lambda}}(r) -$$
$$- X_\lambda^P(r) - \frac{1}{c} \sum_{\lambda'(\neq\lambda)} N_{\lambda'} \varepsilon_{\lambda, \lambda'} Q_{\bar{\lambda}'}(r) \delta(\varkappa, \varkappa'), \tag{1.201*}$$

$$\left(\frac{d}{dr} - \frac{\varkappa}{r} \right) Q_{\bar{\lambda}}(r) = \frac{1}{c} [\varepsilon_{\lambda, \lambda} + Y_\lambda(r)] P_\lambda(r) +$$
$$+ X_\lambda^Q(r) + \frac{1}{c} \sum_{\lambda'(\neq\lambda)} N_{\lambda'} \varepsilon_{\lambda, \lambda'} P_{\lambda'}(r) \delta(\varkappa, \varkappa'). \tag{1.202*}$$

Here, $\lambda \equiv nlj$, $\bar{\lambda} = n\bar{l}j$, and the potential and exchange functions are defined as follows:

$$Y_\lambda(r) = \frac{1}{r}\Big\{ -Z + \sum_{\lambda'} [N_{\lambda'} - \delta(\lambda, \lambda')][Y^0_{\lambda', \lambda'}(r) + Y^0_{\bar{\lambda}', \bar{\lambda}'}(r)] -$$
$$- \sum_{k>0} (N_\lambda - 1) \frac{2j+1}{2j} \begin{pmatrix} j & j & k \\ \frac{1}{2} & -\frac{1}{2} & 0 \end{pmatrix}^2 (jjk)[Y^k_{\lambda, \lambda}(r) + Y^k_{\bar{\lambda}, \bar{\lambda}}(r)] \Big\}, \tag{1.203*}$$

$$X^P_\lambda(r) = -\frac{1}{rc} \sum_{\lambda'(\neq\lambda)} N_{\lambda'} \sum_k \begin{pmatrix} j & j' & k \\ \frac{1}{2} & -\frac{1}{2} & 0 \end{pmatrix}^2 \times$$
$$\times (jj'k)[Y^k_{\lambda, \lambda'}(r) + Y^k_{\bar{\lambda}, \bar{\lambda}'}(r)] Q_{\bar{\lambda}'}(r), \tag{1.204*}$$

$$X^Q_\lambda(r) = -\frac{1}{rc} \sum_{\lambda'(\neq\lambda)} N_{\lambda'} \sum_k \begin{pmatrix} j & j' & k \\ \frac{1}{2} & -\frac{1}{2} & 0 \end{pmatrix}^2 \times$$
$$\times (jj'k)[Y^k_{\lambda, \lambda'}(r) + Y^k_{\bar{\lambda}, \bar{\lambda}'}(r)] P_{\lambda'}(r). \tag{1.205*}$$

$Y_{\lambda,\lambda'}{}^k(r)$ is an integral function of the type given by Eq. (1.110) and $Y_{\bar{\lambda},\bar{\lambda}'}{}^k(r)$ is the same type of function, but for the radial functions $Q_{n\bar{l}j}(r)$; $\varepsilon_{\lambda,\lambda'}$ is a Lagrangian multiplier ($-\varepsilon_{\lambda,\lambda} \equiv -\varepsilon_\lambda$ has the sense of the one-electron energy from which the electron's rest energy has been excluded).

Along with Eqs. (1.201) and (1.202), the Dirac–Fock equations for the average energy of a system of relativistic subconfigurations [70] obtained by varying the energy

$$\bar{E}(K) = g(K)^{-1} \sum_{K_j \in K} g(K_j) \bar{E}(K_j) \quad \Big(g(K) = \sum_{K_j \in K} g(K_j)\Big), \tag{1.206}$$

are used: (K_j is a subconfiguration that corresponds to the nonrelativistic configuration K and $g(K_j)$ is its statistical weight), as well as equations for a specific value of the total momentum J [71].

It must be noted that the use of the variational principle on the functionals given by Eqs. (1.199) or (1.206) is not rigorous [68, 72, 73]. This is associated with the existence of not only a discrete spectrum of eigenvalues for the one-electron energy ($-2mc^2 < \varepsilon_\lambda < 0$) and a continuous spectrum ($\varepsilon_\lambda > 0$) for the Dirac equation and its generalization, the Dirac—Fock equation, but a second continuum ($\varepsilon_\lambda < -2mc^2$) as well.

In order that a spontaneous transition of interacting electrons into a negative energy state, $\varepsilon_\lambda < -2mc^2$, be excluded, Dirac assumed that all these states are ordinarily occupied by electrons and that their continuum is an unobservable vacuum; only a vacancy – a positron – is observed. Because such forbidden transitions were not considered when the Dirac–Fock equations were derived, the equations contain nonphysical solutions that describe an atom as an unstable system. A more rigorous approach requires that the two-electron operator H^e in the functional $\int \Psi^* H \Psi d\tau$ be replaced by the operator $\Lambda_+ H^e \Lambda_+$, where Λ_+ is the operator for the projection into one-electron states having a positive total energy of $\varepsilon_\lambda + mc^2 > 0$, although this significantly complicates the equations. In practice, nonphysical solutions are excluded by the same iteration method used for solving the ordinary Dirac–Fock equations, since the initial wave functions for each iteration are generated from the correct solutions of the previous iteration [68].

The one-configuration relativistic theory is used mainly to describe the properties of inner electrons for a single inner vacancy, when mixing of the relativistic configurations that correspond to the same nonrelativistic configuration does not occur.

1.7. An Approximate Description of Relativistic Effects

Because the relativistic method is so much more complex than the nonrelativistic approach, especially for configurations with open shells when we must consider the mixing of quasidegenerate relativistic configurations, and because of the need to use *jj*-coupling in the inner shells in the relativistic method, even when it is often unrealistic to do so, it is worthwhile to consider relativistic effects when we can, in an approximation, starting from the nonrelativistic model.

The nonrelativistic theory can be combined with relativistic theory, using the advantages of each, by the method of effective or equivalent operators [74, 75]. An effective operator O^e whose matrix element for the nonrelativistic wave functions is equal to the matrix element of O^r for the relativistic functions,

$$(\gamma | O^r | \gamma) = \langle \gamma | O^e | \gamma \rangle \tag{1.207}$$

is found for the relativistic operator O^r. An expression was obtained in [74] for the effective Breit interaction operator which contains two-electron operators of the ($[a^\dagger \times$

$\tilde{a}]^{(k_1 k_2 k)} \cdot [a^\dagger \times \tilde{a}]^{(k_1' k_2' k)})$ type (the orbital and spin ranks are combined into a common rank), 6*j*- and 9*j*-symbols, and relativistic radial integrals.

If only the term with $k_2 = k_{2'} = 0$ that corresponds to classical Coulomb repulsion is left and the number of relativistic integrals is maintained by means of the approximate relations that exist between them, we can derive simple formulas for the effective integrals — linear combinations of relativistic integrals that correspond to nonrelativistic integrals [76]:

$$X^k(nl,\ n'l') = \frac{\sum\limits_{jj'} g(j,\ j')\, X^k(nlj,\ n'l'j')}{\sum\limits_{jj'} g(j,\ j')}, \quad X^k = F^k,\ G^k. \tag{1.208}$$

When $nl = n'l'$ the summation over j and j' is restricted by the condition that $j \le j$; $g(j, j')$ is the statistical factor:

$$g(j,\ j') = \begin{cases} j(2j+1), & \text{if} \quad n = n',\ l = l',\ j = j', \\ (2j+1)(2j'+1) & \text{otherwise.} \end{cases} \tag{1.209}$$

Equation (1.208) supplements the analogous expressions for the integral $I(nl)$ and the constant ζ_{nl} [76, 77]:

$$I(nl) = (4l+2)^{-1} \sum_j (2j+1)\, I(nlj), \tag{1.210}$$

$$\zeta_{nl} = \frac{2}{2l+1}\, \frac{1}{\mathcal{N}}\, [\bar{E}^{rel}(K_h) - \bar{E}^{rel}(K_1)], \tag{1.211}$$

where K_1 and K_h denote the lowest and highest subconfigurations, and

$$\mathcal{N} = \begin{cases} N & \text{for} \quad 1 \leqslant N \leqslant 2l, \\ 2l & \text{for} \quad 2l+1 \leqslant N \leqslant 2l+2, \\ 4l+2-N & \text{for} \quad 2l+1 \leqslant N \leqslant 4l+2. \end{cases} \tag{1.212}$$

All of the relativistic integrals in Eqs. (1.208), (1.210), and (1.211) must be calculated with the same wave functions. If there are other open shells the E^{rel} in Eq. (1.211) are averaged over them. Relations similar to Eqs. (1.208)–(1.210) were also derived in [77] by comparing the average values of the nonrelativistic and relativistic matrix elements.

The relations presented above enable us to make nonrelativistic calculations more exact by replacing the nonrelativistic integrals in the matrix elements by relativistic integrals.

A more consistent method of describing relativistic corrections is to use first-order perturbation theory to calculate them for the nonrelativistic wave functions. The correction operators are obtained from the relativistic equation for a two-electron system,

$$E = \int \Psi^* (H^D + H^e + H^B) \Psi d\tau, \tag{1.213}$$

if the small component of the one-electron four-component function can be approximated by a large component and terms on the order of $(v/c)^2$ be left inclusively [78]. [For an electron in a hydrogenic atom $\langle v^2/c^2 \rangle = (Z\alpha/n)^2$; therefore, relativistic effects will be described as in the initial operator, Eq. (1.179), with an accuracy of $(Z\alpha)^2$.] After some transformations, from Eq. (1.213) we derive, in addition to the nonrelativistic Hamiltonian, the following operators: H^m, which takes the dependence of the electron mass on velocity into account; H^{so}, the operator for spin–orbit interaction in the field of the nucleus, Eq. (1.132); and H^d, the so-called Darwin term:

$$H^m = -\frac{1}{8m^3c^2} \sum_i \mathbf{p}_i^4; \quad H^d = \frac{Z\pi e^2 \hbar^2}{2m^2c^2} \sum_i \delta(\mathbf{r}_i). \tag{1.214}$$

In this approximation the Breit operator H^B is transformed to the sum

$$H^B \rightarrow H^{soo} + H^{oo} + H^{ss} + H^{sc} + H^c \tag{1.215}$$

for operators of the two-electron type, which describe the magnetic interactions and retarded electromagnetic field. We have the following operators: H^{soo} – the "spin–other orbit"; H^{oo} – the "orbit–orbit"; H^{ss} – the spin–spin interaction; H^c and H^{sc} – the contact and spin–contact interactions. They are expressed in terms of the angular momenta of the electrons, their spin, and the distance vector between the electrons [54, 39]. These operators were given, e.g., in [48], as irreducible tensors. The method of calculating the matrix elements for these operators, as well as H^d and H^m in the energy matrix for Hartree–Fock functions, is called the Hartree–Fock–Pauli method.

Among the relativistic corrections, the matrix elements for the operators H^m and H^d are singled out for magnitude; however, they are independent of the many-electron quantum numbers and alter only the total energy of a configuration (the contributions from the individual electrons are additive):

$$\langle nl^N\gamma | H^m | nl^N\gamma' \rangle = \delta(\gamma, \gamma') \frac{\hbar^4}{8m^3c^2} N\{ I''(nl|0) +$$
$$+ 2l(l+1)I'(nl|-2) + l(l+1)[l(l+1)-6]I(nl|-4)\}, \tag{1.216}$$

$$\langle nl^N\gamma | H^d | nl^N\gamma' \rangle = \delta(\gamma, \gamma') \frac{Z\hbar^2\pi e^2}{2m^2c^2} \rho(nl^N|0). \tag{1.217}$$

The notations

$$I(nl|\beta) = \int_0^\infty r^\beta P_{nl}^2(r)\,dr, \quad I'(nl|\beta) = \int_0^\infty r^\beta \left(\frac{dP_{nl}(r)}{dr}\right)^2 dr,$$
$$I''(nl|0) = \int_0^\infty \left(\frac{d^2P_{nl}(r)}{dr^2}\right)^2 dr. \tag{1.218}$$

are used here, and ρ is the probability density of finding electrons at the nucleus. This density is nonzero only for s-electrons in the nonrelativistic theory

$$\rho(ns^N|0) = \frac{N}{4\pi}\left(\frac{P_{ns}(r)}{r}\right)^2_{r=0}. \tag{1.219}$$

The eigenvalue for the H^d operator is positive and for the H^m operator it is negative; they often compensate each other for s-electrons.

The matrix elements for the H^c, H^{sc}, and H^{oo} operators are independent of the total momentum J, just as is the electrostatic interaction operator, and make the positions of the terms more accurate. The matrix elements of the operators for both contact [containing $\delta(\mathbf{r}_{ij})$] interactions differ by only a constant [48]. Their spin-angular part is the same as for the electrostatic interaction operator; therefore, these corrections are taken into account by merely changing the F^k and G^k integrals:

$$F^k(nl, nl) \to F^k(nl, nl) + \delta(\upsilon, \upsilon')[k]R_2(nl, nl), \tag{1.220}$$

$$X^k(n_1l_1, n_2l_2) \to X^k(n_1l_1, n_2l_2) + [k]R_2(n_1l_1, n_2l_2), \quad X = F, G, \tag{1.221}$$

where

$$R_2(n_1l_1, n_2l_2) = \frac{\alpha^2}{4}\int_0^\infty r^{-2}P_{n_1l_1}^2(r)P_{n_2l_2}^2(r)\,dr. \tag{1.222*}$$

The one-shell matrix element for the H^{oo} operator is

$$\langle nl^N\gamma\upsilon LS \| H^{oo} \| nl^N\gamma'\upsilon' LS\rangle = -\delta(\upsilon,\ \upsilon')\, l(l+1)(2l+1)\times$$
$$\times\sum_{k=0}^{2l-2}\frac{4\,(2k+1)\,(2k+3)}{k+2}\begin{Bmatrix} k & 1 & k+1 \\ l & l & l \end{Bmatrix}^2 \langle l\| C^{(k)}\| l\rangle^2 M^k(nl, nl)\times$$
$$\times\Big[[L,\ S]^{-1}\sum_{\gamma''\upsilon''L''}\langle l^N\gamma\upsilon LS\| U^{(k+1)}\| l^N\gamma''\upsilon''L''S\rangle\times$$
$$\times\langle l^N\gamma'\upsilon' LS\| U^{(k+1)}\| l^N\gamma''\upsilon''L''S\rangle - \delta(\gamma\gamma')\,\delta(\upsilon,\ \upsilon')\, N\,[l]^{-1}\Big], \tag{1.223}$$

$$M^k(nl,\ n'l') = \frac{\alpha^2}{4}\int_0^\infty r^{-k-3}P_{nl}^2(r)\int_0^r r_1^k P_{n'l'}^2(r_1)\,dr_1. \tag{1.224*}$$

The "orbit–orbit" interaction between shells is weaker than the interaction within a shell and is usually omitted from the calculations.

The matrix elements for the other two terms H^{soo} and H^{ss} in the expansion of the Breit operator are a function of J and more accurately define the positions of the energy levels. The fundamental part of the "spin–other orbit" interaction (the interaction between an open shell and a closed shell and part of the interaction within an open shell) has the same spin dependence as does a simple spin–orbit interaction and can be described when a refined expression for the spin–orbit coupling constant is used [52]:

$$\zeta_{nl} = \frac{Z\alpha^2}{2}\langle nl\,|\,r^{-3}\,|\,nl\rangle + \sum_{n'l'}\zeta_{nl,\,n'l'} + \zeta'_{nl}. \tag{1.225*}$$

$\zeta_{nl,n'l'}$ is the contribution from an interaction with a closed $n'l'$ shell; therefore, summation is carried out only over the closed shells. Thus,

$$\zeta_{nl,\,n'l'} = -4\,(2l'+1)\,M^0(nl,\ n'l') + 6\sum_k\langle l\| C^{(k)}\| l'\rangle^2\times$$
$$\times\Big\{2\,(2k+1)\Big[(2k+3)\begin{Bmatrix} l & k+1 & l' \\ k & l & 1 \end{Bmatrix}^2 N^k(nl,\ n'l') - (2k-1)\times$$

$$\times\begin{Bmatrix} l & k-1 & l' \\ k & l & 1 \end{Bmatrix}^2 N^{k-2}(nl, n'l')\Bigg] + \left[\frac{k(k+1)(2k+1)}{l(l+1)(2l+1)}\right]^{1/2} \begin{Bmatrix} l & k & l' \\ k & l & 1 \end{Bmatrix} \times$$

$$\times\big[(k-1)N^{k-2}(nl, n'l') - (k+2)N^k(nl, n'l') + 2K^k(nl, n'l') +$$

$$+ (l+l'+1)(l-l')\left(k^{-1}N^{k-2}(nl, n'l') - (k+1)^{-1}N^k(nl, n'l')\right)\big]\Bigg\}. \quad (1.226)$$

Here M^0 is the integral for a direct magnetic interaction, Eq. (1.224), and N^k and K^k are exchange integrals:

$$N^k(nl, n'l') = \frac{\alpha^2}{4}\int_0^\infty r^{-k-3}P_{nl}(r)P_{n'l'}(r)\,dr\int_0^r r_1^k P_{nl}(r_1)P_{n'l'}(r_1)\,dr_1, \quad (1.227)^*$$

$$K^k(nl, n'l') = \frac{\alpha^2}{4}\int_0^\infty r^{-2}\,\frac{dP_{nl}(r)}{dr}\,P_{n'l'}(r)\,dr \times$$

$$+\left[r^{-k}\int_0^r r_1^k P_{nl}(r_1)P_{n'l'}(r_1)\,dr_1 + r^{k+1}\int_r^\infty r_1^{-k-1}P_{nl}(r_1)P_{n'l'}(r_1)\,dr_1\right]. \quad (1.228)^*$$

Coefficients for the integrals in $\zeta_{nl,n\ l}$ have been tabulated [52]; ζ'_{nl} corresponds to the one-electron part of the "spin–other orbit" interaction within an nl^N shell:

$$\zeta'_{np} = (3-2N)\,M^0(np, np), \quad (1.229)$$

$$\zeta'_{nd} = (3-2N)\,M^0(nd, nd) + \frac{6}{7}\,M^2(nd, nd), \quad (1.230)$$

$$\zeta'_{nf} = (3-2N)\,M^0(nf, nf) + M^2(nf, nf) + \frac{5}{11}\,M^4(nf, nf). \quad (1.231)$$

The remaining part of the "spin–other orbit" interaction is neglected in energy calculations, just as it is in a relatively weak spin–spin interaction.

The Hartree–Fock–Pauli method lets us examine rather effectively the relativistic effects for the electron binding energy in the intermediate shells and the energies of transitions to these shells [79]. In heavy atoms the method is inaccurate not only for the deep inner shells, but for the outer *d*- and *f*- electrons. The wave functions of these electrons penetrate the nucleus only slightly and direct relativistic effects for them are small. However, relativistic contraction of the wave functions of the inner electrons, especially the *s*-electrons, reduces the effective nuclear charge for the outer *d*- and *f*-electrons [73, 80]. This secondary relativistic effect can only be accounted for by relativistic wave functions.

Relativistic effects that are not accounted for by the Breit operator – the interaction between an electron and an intrinsic radiation field and the vacuum polarization (both are often called a Lamb shift) – are significant for the deepest shells in heavy atoms. An electron emits and absorbs virtual photons, changing its energy (the correction is called the electron's self-energy), and causes an anomalous magnetic moment. Also, virtual electron–positron pairs arise around an electron in vacuum. The nucleus of an atom attracts electrons and repulses virtual positrons, and the so-called polarization of a vacuum occurs, all of which causes the nuclear field to be screened.

These corrections to the energy of the electron are presented as a series in powers of $Z\alpha$ for the one-electron model that is ordinarily used. For one *nlj*-electron, an overall correction that contains terms of lower order is the expression [81, 82]

$$\Delta E_{nlj} = \frac{4Z^4\alpha^5 mc^2}{3\pi n^3}\Big\{A_{40} + A_{41}\ln(Z\alpha)^{-2} + A_{50}Z\alpha +$$
$$+ (Z\alpha)^2[A_{62}\ln^2(Z\alpha)^{-2} + A_{61}\ln(Z\alpha)^{-2} + A_{60} + O(Z\alpha)] +$$
$$+ \frac{\alpha}{\pi}[B_{40} + O(Z\alpha)] + O\left(\frac{\alpha^2}{\pi^2}\right)\Big\} + \Delta E_m + \Delta E_r. \quad (1.232)$$

The constants A_{mn} consist of terms that are determined by the self-energy of the electron (*SE*), the magnetic moment (*MM*), and the vacuum polarization (*VP*):

$$A_{mn} = A_{mn}^{SE} + A_{mn}^{VP} + A_{mn}^{MM}. \quad (1.233)$$

The lower-order terms in $Z\alpha$ are

$$A_{40}^{SE} = \ln\left[\frac{(Z\alpha)^2\,\mathrm{Ry}}{\Delta\varepsilon_{nl}}\right] + \frac{11}{24}\,\delta(l,\,0),\quad A_{40}^{VP} = -\frac{1}{5}\,\delta(l,\,0), \tag{1.234}$$

$$A_{40}^{MM} = -\frac{3}{8\varkappa\,(2l+1)},\quad B_{40} = 0{,}6505\,\delta(l,\,0) - 0{,}2464\,\frac{1}{\varkappa\,(2l+1)}, \tag{1.235}$$

where Ry – the Rydberg – is equal to half an atomic energy unit: κ is the quantum number, Eq. (1.175); $\Delta\varepsilon_{nl}$ is the average Bethe excitation energy that arises from the renormalized nonrelativistic self-energy of an electron [54, 82]; ΔE_m and ΔE_r are corrections that are governed by the finite size and mass of the nucleus. Expressions for these corrections and for the A_{m1} are given, for example, in [82].

For large Z the convergence of the series for ΔE_{nlj} becomes slower; therefore, the summation must be carried out for an infinite number of corrections [83].

Results from calculating the self-energy of electrons for $n \leq 2$ and $10 \leq Z \leq 100$ with hydrogenlike functions have been tabulated in [83]. These are used for many-electron atoms by making the replacement $Z \to Z^*$ and summing over the contributions from the individual electrons [84, 85]. The correction due to vacuum polarization is roughly 20% of the self-energy [84, 86], but has an inverse – negative – sign and partially compensates the radiative correction. By way of example, we can show that the entire Lamb shift for Hg is -157 eV for the binding energy of a $1s$-electron, -22 eV for a $2s$-electron, and -2 eV for the $2p_{1/2}$ electron.

When refined relativistic calculations are done for the energies of the inner electrons of heavy atoms, the dependence of energy on the frequency of a virtual photon must be considered in first-order perturbation theory. This dependence is found from the difference between the interactions given by Eqs. (1.178) and (1.179). (The matrix elements for the operator H^{rel} have the same spin-angular parts as for the operator H^B; only the radial integrals are different [87,48].) In Hg this correction is 6 eV for I_{1s}, -0.5 eV for I_{2s}, and -0.08 eV for $I_{2p_{1/2}}$ [86].

1.8. Interaction between an Atom and an Electromagnetic Field. The Radiative Transition Operators

The Hamiltonian for a system atom + radiation comprises terms that correspond to the Hamiltonians of an atom, a free electromagnetic field, and the interaction between them. The operator for the interaction between an N-electron system and a field when Coulomb gauge of the field (div $\mathbf{A}$ = 0. φ = 0) is used [50, 78] is

$$H' = -\frac{e}{c}\sum_{i=1}^{N}\int \mathbf{j}_i(\mathbf{r})\cdot\mathbf{A}(\mathbf{r})\,d\mathbf{r}, \tag{1.236}$$

where $\mathbf{A}(\mathbf{r})$ is the vector potential and $\varphi(\mathbf{r})$ the scalar potential of the electromagnetic field. The Dirac flux density of electron i at a point $\mathbf{r}$ is

$$\mathbf{j}_i(\mathbf{r}) = c\,\boldsymbol{\alpha}_i\,\delta(\mathbf{r}_i - \mathbf{r}), \tag{1.237}$$

where α_i is the Dirac matrix of Eq. (1.173). The summation in Eq. (1.236) is carried out over all the electrons in the atom. Substituting Eq. (1.236) into (1.235) gives

$$H' = -e\sum_i \boldsymbol{\alpha}_i \cdot \mathbf{A}(\mathbf{r}_i). \tag{1.238}$$

In the nonrelativistic theory the interaction operator is obtained by replacing the nonrelativistic expression for $\mathbf{j}_i(\mathbf{r})$ [50] in Eq. (1.236) or, what is easier, by replacing the electron's momentum, $\mathbf{p}_i$, in the Hamiltonian for the atom, Eq. (1.94), with $\vec{p}_i - (e/c)\mathbf{A}(\mathbf{r}_i)$ and keeping the H term that describes the interaction with the field:

$$H' = -\frac{e}{2mc}\sum_i [\mathbf{p}_i \cdot \mathbf{A}(\mathbf{r}_i) + \mathbf{A}(\mathbf{r}_i)\cdot \mathbf{p}_i] + \frac{e^2}{2mc^2}\sum_i \mathbf{A}(\mathbf{r}_i)^2. \tag{1.239}$$

With a Coulomb gauge, the operators $\mathbf{p}$ and $\mathbf{A}$ commutate between themselves; therefore,

$$H' = H_1' + H_2', \quad H_1' = -\frac{e}{mc}\sum_i \mathbf{p}_i \cdot \mathbf{A}(\mathbf{r}_i), \quad H_2' = \frac{e^2}{2mc^2}\sum_i \mathbf{A}(\mathbf{r}_i)^2. \tag{1.240}$$

In atomic units, the quantity e/mc is equal to the small parameter α. This allows us to use perturbation theory to examine the processes that occur when an atom interacts with an electromagnetic field.

The vector potential $\mathbf{A}(\mathbf{r}, t)$ can be written as an expansion in plane waves:

$$\mathbf{A}(\mathbf{r}, t) = \sum_{\mathbf{k}} \sum_{\rho=1}^{2} \hat{e}_{\mathbf{k}\rho}\{a_{\mathbf{k}\rho}e^{i(\mathbf{k}\cdot\mathbf{r}-\omega t)} + a^*_{\mathbf{k}\rho}e^{-i(\mathbf{k}\cdot\mathbf{r}-\omega t)}\}, \tag{1.241}$$

where ω is the angular frequency and $\mathbf{k}$ is a wave vector whose components for the boundary conditions of field periodicity (in the "simple" variant being used — at the edges of a cube of size V) become a discrete series; $\hat{e}_{\mathbf{k}1}$ and $\hat{e}_{\mathbf{k}2}$ are unit photon polarization vectors ($\hat{e}_{\mathbf{k}1} \perp \hat{e}_{\mathbf{k}2}$ and both are perpendicular to the wave vector $\mathbf{k}$); $a_{\mathbf{k}\rho}{}^*$

and $a_{\mathbf{k}\rho}$ are, respectively, within a constant, equal to the creation and annihilation operators of a photon having a wave vector **k** and polarization ρ. Their matrix elements are nonzero only between field states in which the number of quanta $n_{\mathbf{k}\rho}$ is different by one (the initial state is indicated on the right),

$$\langle n_{\mathbf{k}\rho}-1 \mid a_{\mathbf{k}\rho} \mid n_{\mathbf{k}\rho}\rangle = \left[\frac{2\pi c^2 \hbar\, n_{\mathbf{k}\rho}}{\omega V}\right]^{1/2}, \tag{1.242}$$

$$\langle n_{\mathbf{k}\rho}+1 \mid a^*_{\mathbf{k}\rho} \; n_{\mathbf{k}\rho}\rangle = \left[\frac{2\pi c^2 \hbar\, (n_{\mathbf{k}\rho}+1)}{\omega V}\right]^{1/2}. \tag{1.243}$$

Thus, in first-order perturbations theory, an operator H_1' that is linear in **A** describes one-photon transitions and an operator H_2' that contains **A** and small parameter in second power describes two-photon transitions (these are also defined by the operator H_1', but in second-order perturbation theory).

The probability that a system atom + radiation will make a transition from its initial state i into the final states of the continuous spectrum f, $f + df$ in unit time is given by

$$dW_{i\to f} = \frac{2\pi}{\hbar} \mid M_{fi} \mid^2 \delta (E_i - E_f)\, df. \tag{1.244}$$

in first-order perturbation theory [56], where E_i and E_f are the total energies of the entire system in the initial and final states, and M_{fi} is the amplitude of the process and is equal to the matrix element for the transitions operator at $t = 0$:

$$M_{fi} = \langle f \mid H'(0) \mid i\rangle . \tag{1.245}$$

The wave functions for the final state in Eq. (1.244) must be normalized to a δ-function in the f-space. Equations (1.244) and (1.245) remain valid in the relativistic theory.

If we are interested in only the system energy in the final state, Eq. (1.244) can be integrated over the values of the remaining parameters df' for this state, after which it is transformed into

$$dW_{i\to E} = \frac{2\pi}{\hbar} \mid M_{Ei} \mid^2 \delta (E_i - E)\, \rho (E)\, dE, \quad E \equiv E_f. \tag{1.246}$$

Here we have introduced the notation

$$\int \mid M_{fi} \mid^2 df' = \mid M_{Ei} \mid^2 \rho (E), \quad df = df'\, dE, \tag{1.247}$$

where $\rho(E)$ is the density of the final state.

Let a free electron appear during the process (e.g., photoionization of the atom occurs). Then,

$$df \equiv d\mathbf{q} = dq_x dq_y dq_z = q^2 dq \, d\Omega_q. \qquad (1.248)$$

$\mathbf{p} = \hbar\mathbf{q}$ is the momentum, and $\mathbf{q}$ is the electron's wave vector. The electron's wave function is normalized to $\delta(\mathbf{q} - \mathbf{q}')$.

If in the final state a photon should appear, then, keeping in mind the fact that the electromagnetic field is thought of as being in a cube of volume V, we have

$$df = \frac{V}{(2\pi)^3} \, d\mathbf{k}. \qquad (1.249)$$

When a photon and a free electron occur in the final state (e.g., the Compton effect) df is equal to the product of Eqs. (1.248) and (1.249). Similarly, we can find df when two free electrons appear during the process.

Equation (1.246) is also useful for describing the photoexcitation of an atom from its ground state into a discrete spectrum state having a natural width defined by its radiation lifetime. Then the density of states $\rho(E)$ is described by a Lorentz function (Sec. 9.1). [If the initial level has a nonzero radiation width, Eq. (1.246) must be multiplied by the density of the initial states $\rho(E')$ and dE'.] An equation for the total probability of exciting an atom from a state γ into a state γ' is obtained when we integrate Eq. (1.246) with respect to energy:

$$W^{ex}_{\gamma\to\gamma'} = \frac{2\pi}{\hbar} \left| M_{\gamma' n_{\mathbf{k}\rho}-1,\, \gamma n_{\mathbf{k}\rho}} \right|^2. \qquad (1.250)$$

In the inverse process – the emission of a photon when an atom makes a transition between discrete spectrum states – the probability will contain the density of photon states for the final state, Eq. (1.241). The probability that an atom will make the transition from a state γ to a state γ', and be accompanied by the emission of a photon having polarization $\hat{e}_{\mathbf{k}\rho}$ and a wave vector in the interval $\mathbf{k}$, $\mathbf{k} + d\mathbf{k}$, is

$$dW^{e}_{\gamma\to\gamma'} = \frac{2\pi}{\hbar} \left| M_{\gamma' n_{\mathbf{k}\rho}+1,\, \gamma n_{\mathbf{k}\rho}} \right|^2 \delta(E_\gamma - E_{\gamma'} - \hbar\omega) \, \frac{V d\mathbf{k}}{(2\pi)^3}. \qquad (1.251)$$

The matrix element $\langle f \,|\, H' \,|\, i \rangle$ is zero for some processes. Then, in second-order perturbations theory the probability of a transition is given by Eq. (1.244), but with an amplitude of

$$M_{fi} = \sum_j \frac{\langle f | H' | j \rangle \langle j | H' | i \rangle}{E_i - E_j}. \tag{1.252}$$

The summation over intermediate states j will also contain an integration over the continuous spectrum.

Instead of the process probability, another characteristic – the process cross section σ – is often used. Let a particle flux having density j impact an atom. Then the process probability and cross section are related by

$$d\sigma = dW/j. \tag{1.253}$$

Their numerical values coincide when the flux of incident particles has unit density.

When the interaction between an electromagnetic field and an atom is considered, it is convenient to represent the field as a superposition of the various multipole fields. This can be done by expanding the exponents in the equation for the vector potential over the vector spherical functions [39, 78]:

$$\mathbf{Y}_{tp}^{(e)}(\hat{\mathbf{r}}) = [t(t+1)]^{-1/2} r \nabla Y_{tp}(\hat{\mathbf{r}}), \tag{1.254}$$

$$\mathbf{Y}_{tp}^{(m)}(\hat{\mathbf{r}}) = [t(t+1)]^{-1/2} \mathbf{t}\, Y_{tp}(\hat{\mathbf{r}}). \tag{1.255}$$

Here, $\mathbf{t}$ is an operator $-i[\mathbf{r} \times \nabla]$, Y_{tp} is a spherical harmonic, and $\hat{\mathbf{r}}$ is a unit vector defined by the angles θ, φ.

The radiation field is represented as a superposition of vector spherical harmonics [88]:

$$\hat{\epsilon}\, e^{i\mathbf{k}\cdot\mathbf{r}} = 4\pi \sum_{tp} [\hat{\epsilon}\mathbf{Y}_{tp}^{(m)}(\hat{\mathbf{k}})^* \cdot \mathbf{A}_{tp}^{(m)}(k, \mathbf{r}) - \hat{\epsilon}\,\mathbf{Y}_{tp}^{(e)}(\hat{\mathbf{k}})^* \cdot \mathbf{A}_{tp}^{(e)}(k, \mathbf{r})], \tag{1.256}$$

$$\mathbf{A}_{tp}^{(m)}(k, \mathbf{r}) = i^t [t(t+1)]^{-1/2} j_t(kr)\, \mathbf{t}\, Y_{tp}(\hat{\mathbf{r}}), \tag{1.257}$$

$$\mathbf{A}_{tp}^{(e)}(k, \mathbf{r}) = i^t [t(t+1)]^{-1/2} k^{-1} [\nabla \times \mathbf{t}] j_t(kr)\, Y_{tp}(\hat{\mathbf{r}}), \tag{1.258}$$

where $\mathbf{A}^{(e)}$ and $\mathbf{A}^{(m)}$ are solutions of the wave equation $(\Delta + \mathbf{k})\mathbf{A} = 0$ which, respectively, have parity $(-1)^t$ and $(-1)^{t+1}$; $j_t(kr)$ is a spherical Bessel function, and $\hat{\epsilon}$ is the unit photon polarization vector.

Terms in the expansion of Eq. (1.256) which have the superscript e correspond to the radiation field from an electric multipole and m corresponds to a magnetic multipole, all of order t.

If the fundamental contribution to the process amplitude is given only by values of $\mathbf{k}\cdot\mathbf{r} << 1$, which corresponds to the condition that $\lambda >> a$, where λ is the wavelength and a is the radius of that area of the atom that is defined by the radial integral for the transition, then the magnitude of the function $j_t(kr)$ [and the terms in the expansion, Eq. (1.256), that contain it] falls off rapidly as t increases. The condition $\lambda >> a$ is satisfied not only for the optical but for most x-ray transitions in the atom as well. The fact of the matter is that not only does λ decrease for them, but a decreases also: transitions occur in an atom's inner shells and, because of the rapid asymptotic decay of their wave functions, a actually corresponds to the area in which the wave function of the deeper of the two shells taking part in the radiative transition has significant amplitude. Even for U the total probability of radiation decay of a K-vacancy is determined mainly by dipole transitions (they contribute 98.8%, electric quadrupoles contribute 0.6%, magnetic dipoles contribute 0.2%, and the remaining multipole transitions contribute only 0.4% [89]). Thus, in the nonrelativistic theory we can ignore the correction multipole terms and consider only the fundamental term in Eq. (1.256) – the electric dipole – when the transition is between configurations of different parity, and consider the electric quadrupole when the transition is between configurations having the same parity.

Expanding Eq. (1.256) allows us to write the Hamiltonian H' for the interaction between an atom and the radiation as the sum of the multipole transition operators. They can conveniently be defined so that they represent the corresponding multipole angular momenta in the nonrelativistic long-wave approximation ($\mathbf{k}\cdot\mathbf{r} << 1$, $j_t(kr) \sim (kr)^t/(2t + 1)!!$). Since in the nonrelativistic theory magnetic dipole transitions are only possible between levels of the same configuration [39], we will consider only the electric multipole (dipole and quadrupole) transition operators in this approximation:

$$O_p^{(t)} = -e \sum_i r_i^t C_p^{(t)}(\theta_i, \varphi_i). \tag{1.259}$$

The subscript i covers the coordinates of all of an atom's electrons; $O^{(1)} \equiv D^{(1)}$ is the dipole moment and $O^{(2)}$ is the quadrupole moment of an atom's electron shells.

The rank of a transition operator is not zero; therefore, its matrix element, according to Eq. (1.56), depends on the projections of the total momentum:

$$\langle \gamma JM \,|\, O_p^{(t)} \,|\, \gamma' J' M' \rangle = (-1)^{J-M} \begin{pmatrix} J & t & J' \\ -M & p & M' \end{pmatrix} \langle \gamma J \,\|\, O^{(t)} \,\|\, \gamma' J' \rangle . \tag{1.260}$$

Equation (1.260) defines the selection rule for the quantum number J. For example, for dipole transitions

$$\Delta J = 0, \ \pm 1; \quad J + J' \geqslant 1 . \tag{1.261}$$

The operator $D^{(1)}$ acts on orbital, but not on spin variables. Therefore, in the LS-coupling

$$\Delta S = 0; \quad \Delta L = 0, \ \pm 1; \quad L + L' \geqslant 1 . \tag{1.262}$$

Since the dipole moment is odd, the initial and final states of the atom must have opposite parity P ($\Delta P \neq 0$).

The following relationships [46] are usually satisfied for a coupling that is nearly an LS-coupling: the matrix element for $D^{(1)}$, especially for large L and L', reaches a maximum in absolute value when $\Delta J = \Delta L$ and tends to decrease as J decreases. If $\Delta J \neq \Delta L$ the matrix element has an extremum for intermediate values of J. A general rule, valid for other types of coupling, is that transitions are stronger if the many-electron angular momenta change in the same manner — either increasing or decreasing.

The radiative transition of an electron changes the field in which the atom's other electrons are moving. When describing them in both configurations by different one-electron wave functions the matrix element for the transition operator will contain overlap integrals along with small exchange terms. These corrections partially compensate each other [46]; therefore, they can usually be neglected in approximate one-configuration calculations (but the radial wave functions for both configurations are used in the radial integral and in the transition energy). About the same accuracy is achieved when the wave functions for the initial and final states are optimized simultaneously, e.g., by the transition operator method [90]. In some cases taking the exchange and wave function overlap corrections into account can eliminate systematic differences between theoretical and experimental results [91].

If these corrections are ignored, the initial and final transition states described by a one-electron operator $O^{(t)}$ in a one-configuration approximation may differ in quantum numbers for only a single electron. The simplest form of a submatrix element is found in the LS-coupling, since the operator does not act on spin variables:

$$\begin{aligned}&\langle n_1 l_1^{N_1}\gamma_1 L_1 S_1 n_2 l_2^{N_2}\gamma_2 L_2 S_2 LSJ \| O^{(t)} \| n_1 l_1^{N_1-1}\gamma_1' L_1' S_1' n_2 l_2^{N_2+1}\gamma_2' L_2' S_2' L' S' J'\rangle = \\ &= \delta(S, S')(-1)^{l_1+S_1'+L_2'+S_2'+L_2+L-J'+N_2+1}[N_1(N_2+1)]^{1/2} \times \\ &\times [L_1, S_1, L_2', S_2', L, L', J, J']^{1/2}(l_1^{N_1}\gamma_1 L_1 S_1 \| l_1^{N_1-1}\gamma_1' L_1' S_1' l_1) \times \\ &\times (l_2^{N_2}\gamma_2 L_2 S_2 l_2 \| l_2^{N_2+1}\gamma_2' L_2' S_2') \begin{Bmatrix} L & J & S \\ J' & L' & t \end{Bmatrix} \begin{Bmatrix} S_1 & S_1' & 1/2 \\ S_2' & S_2 & S \end{Bmatrix} \times \\ &\times \begin{Bmatrix} L_1 & L_1' & l_1 \\ L_2 & L_2' & l_2 \\ L & L' & t \end{Bmatrix} s^{(t)}_{n_1 l_1, n_2 l_2}. \end{aligned} \tag{1.263}$$

The factor $(-1)^{N_2}[N_1(N_2+1)]^{1/2}$ is obtained by antisymmetrizing the wave functions of the electrons in different shells, which must be considered in an interconfiguration matrix element; it also takes into account the difference of the normalization factors for wave functions of both states (Sec. 2.2). The other factors in Eq. (1.263) appear when a single electron is separated out from a shell of equivalent electrons, when the angular momenta are recoupled, and when the submatrix element is written in terms of the one-electron quantity:

$$s^{(t)}_{n_1 l_1, n_2 l_2} = e\langle l_1 \| C^{(t)} \| l_2\rangle \langle n_1 l_1 | r^t | n_2 l_2\rangle, \tag{1.264}$$

where e is the absolute value of the electron charge.

If there are other open shells in the atom and they do not participate in the transition, then the dependence of the transition operator matrix element on the quantum numbers is found from Eqs. (1.68) and (1.69) (and the transformation matrices when the angular momenta of the shells on which the operator acts are not directly coupled among themselves).

The submatrix element for the transition operator is transposed according to

$$\begin{aligned}&\langle\gamma J \| O^{(t)} \| \gamma' J'\rangle = (-1)^{J-J'}\langle\gamma' J' \| O^{(t)} \| \gamma J\rangle, \ \langle\gamma LS \| O^{(t)} \| \gamma' L' S\rangle = \\ &= (-1)^{L-L'}\langle\gamma' L' S \| O^{(t)} \| \gamma LS\rangle. \end{aligned} \tag{1.265}$$

We will give the equations for the submatrix elements of the $O^{(t)}$ operator for some practically important cases (the quantum numbers n_i are omitted):

$$\langle l_1^{4l_1+2} l_2^{N_2}\gamma_2 L_2 S_2 J \| O^{(t)} \| l_1^{4l_1+1} l_2^{N_2+1}\gamma_2' L_2' S_2' L' S J'\rangle =$$

$$= \delta(S_2, S)(-1)^{L_2+S+J'+L_2'+L'+l_2+N_2+1}\sqrt{N_2+1}\,[S_2]^{-1/2}\times$$

$$\times [L', L_2', S_2', J, J']^{1/2}\begin{Bmatrix} L_2 & J & S \\ J' & L' & t \end{Bmatrix}\begin{Bmatrix} t & L' & L_2 \\ L_2' & l_2 & l_1 \end{Bmatrix}\times$$

$$\times (l_2^{N_2+1}\gamma_2' L_2' S_2' \| l_2^{N_2}\gamma_2 L_2 S_2 l_2)\, s_{l_1, l_2}^{(t)}, \tag{1.266}$$

$$\langle l_1^{4l_1+2} l_2^{4l_2+1} j_2 l_3^{N_3}\gamma_3 J_3 J \| O^{(t)} \| l_1^{4l_1+1} j_1 l_2^{4l_2+2} l_3^{N_3}\gamma_3' J_3' J'\rangle =$$

$$= \delta(\gamma_3 J_3, \gamma_3' J_3')(-1)^{J_3+J'+j_1+1}[J, J']^{1/2}\begin{Bmatrix} j_1 & J' & J_3 \\ J & j_2 & t \end{Bmatrix}\langle l_1 j_1 \| o^{(t)} \| l_2 j_2\rangle,$$

$$\langle l_1^{4l_1+2} l_2^{N_2}\gamma_2 J \| O^{(t)} \| l_1^{4l_1+1} j_1 l_2^{N_2}\gamma_2' J_2'(J'') l_3 j_3 J'\rangle =$$

$$= \delta(\gamma_2 J, \gamma_2' J_2')(-1)^{J+J''+j_3+t}[J', J'']^{1/2}\begin{Bmatrix} j_1 & J & J'' \\ J' & j_3 & t \end{Bmatrix}\langle l_1 j_1 \| o^{(t)} \| l_3 j_3\rangle, \tag{1.267}$$

$$\langle l_1 j_1 \| o^{(t)} \| l_2 j_2\rangle = (-1)^{l_1-1/2+j_2+t}[j_1, j_2]^{1/2}\begin{Bmatrix} l_1 & j_1 & 1/2 \\ j_2 & l_2 & t \end{Bmatrix} s_{l_1, l_2}^{(t)} =$$

$$= e(-1)^{j_2+1/2+t}[j_1, j_2]^{1/2}\begin{pmatrix} j_1 & t & j_2 \\ -\frac{1}{2} & 0 & \frac{1}{2} \end{pmatrix}(l_1 \| C^{(t)} \| l_2)\langle n_1 l_1 | r^t | n_2 l_2\rangle, \tag{1.268}$$

$$\langle l_1^{N_1}\gamma_1 L_1 S_1 j_1 \| O^{(t)} \| l_1^{N_1-1}\gamma_1' L_1' S_1' l_2 LSj\rangle$$

$$= \delta(S_1, S)(-1)^{L_1'+l_1+S_1+j+t+1}\sqrt{N_1}\,[L_1, L, j_1, j]^{1/2} \tag{1.269}$$

$$\times \begin{Bmatrix} l_1 & L_1 & L_1' \\ L & l_2 & t \end{Bmatrix}\begin{Bmatrix} j_1 & j & t \\ L & L_1 & S \end{Bmatrix} s_{l_1, l_2}^{(t)}. \tag{1.270}$$

$$\langle l_1^{4l_1+2} l_2^{4l_2+1} l_2 j_2 \| O^{(t)} \| l_1^{4l_1+1} l_2^{4l_2+2} l_1 j_1\rangle = (-1)^{j_1+j_2+t}\langle l_1 j_1 \| o^{(t)} \| l_2 j_2\rangle, \tag{1.271}$$

$$\langle l_1^{4l_1+2}\,{}^1S_0 \| O^{(t)} \| l_1^{4l_1+1} l_2 LSJ\rangle = -\delta(L, t)\delta(S, 0)\delta(J, t)\sqrt{2}\, s_{l_1, l_2}^{(t)}, \tag{1.272}$$

$$\langle l_1^{4l_1+2} \| O^{(t)} \| l_1^{4l_1+1} j_1 l_2 j_2 J\rangle = \delta(J, t)\langle l_1 j_1 \| o^{(t)} \| l_2 j_2\rangle. \tag{1.273}$$

The radial transition integral $\langle nl \mid r^t \mid n'l' \rangle$ for one-electron functions depends on Z as Z^{-t} [54, 46]. When $n \neq n'$ the dipole integral in the basis of these functions is always positive [26, 46]. If more realistic functions are used, the dipole integral as a function of n' or of the energy of the respective continuum ε can change sign and become negative. According to results of calculating via the HF-av. method with the Slater potential [92], the integral $\langle nl \mid r \mid n'(\varepsilon)l + 1 \rangle$, corresponding to an atom that has been photoexcited from the ground state, has, with the exception of the $2s$-shell, in the subthreshold energies at least one zero for all nl-shells whose radial wave functions contain nodes. After an nl-shell has appeared in an atom's ground configuration, as a rule, a zero is found in the continuum and, as Z increases, it moves out, away from threshold; increasing Z further, this trend reverses and zero moves to lower photoelectron energies and finally into the discrete spectrum. Near the point where the integral passes through zero it has small values, which causes some lines in the Rydberg series to disappear or the so-called Cooper minimum to appear in the photoionization cross section [26].

The radiative transition operator can be written in another form. Using the expression for the commutator of a Hamiltonian with a radius-vector

$$[H, \mathbf{r}] = -i\hbar \frac{\mathbf{p}}{m}, \tag{1.274}$$

we find

$$\left\langle \gamma \middle| \sum_j \mathbf{r}_j \middle| \gamma' \right\rangle = -\frac{i\hbar}{m(E_\gamma - E_{\gamma'})} \left\langle \gamma \middle| \sum_j \mathbf{p}_j \middle| \gamma' \right\rangle. \tag{1.275}$$

A dipole operator in the form given by Eq. (1.259) is called its r-form, and that corresponding to the right-hand side of Eq. (1.275) is called its p- or v-form.

Expressions for the matrix elements of the dipole transition operator $nl \to n'l'$ in the v-form are obtained from those given earlier by the substitution

$$\langle nl | r | n'l' \rangle \to (\Delta E)^{-1} \left\langle nl \middle| \left(\frac{d}{dr} \mp \frac{l_>}{r} \right) \middle| n'l' \right\rangle, \tag{1.276*}$$

where $l_> = \max(l, l')$ and ΔE is the transition energy, equal to the difference in the total energies of the initial and final states $E(\gamma) - E(\gamma')$ (the average energy, if the radial wave functions for the HF-av. are used, or the difference in the energy of the terms for the case of HF-t. functions). The "+" sign in Eq. (1.276) corresponds to a $l \to l+1$ transition, and the "$-$" sign corresponds to a $l \to l-1$ transition.

The r- and v-forms are equivalent for exact wave functions. The results coincide even when calculating with approximate functions if they are solutions of a homogeneous equation, but differ for radial Hartree–Fock functions because of an exchange term in the equations. For inner shell electrons this term is usually small; therefore, the probabilities of x-ray transitions in the r- and v-forms are in good agreement for most transitions. In practice, the r-form is more widely used because it is simpler and does not contain a derivative of the approximate function.

The relativistic operator for electric multipole radiation in the form of irreducible tensors becomes

$$ {}_eO_p^{(t)} = -\frac{(2t-1)!!}{k^{t-1}}\, er \left\{ C_p^{(t)} \left[j_{t-1}(kr) - \frac{t}{t+1}\, j_{t+1}(kr) \right] + \frac{i}{t+1}\, j_t(kr) \times \right. $$
$$ \left. \times \left(\sqrt{t(2t-1)}\,[C^{(t-1)} \times \alpha^{(1)}]_p^{(t)} - \sqrt{(t+1)(2t+3)}\,[C^{(t+1)} \times \alpha^{(1)}]_p^{(t)} \right) \right\}, \qquad (1.277) $$

where $k!!$ is equal to $k(k-2)\ldots 2$ when k is even, and $k(k-2)\ldots 1$ when k is odd. The operator given by Eq. (1.277) corresponds to the Coulomb gauge of an electromagnetic field with a gauge constant of $\mathcal{K} = 0$ and becomes Eq. (1.259) in the nonrelativistic long-wave limit. A general equation for $O^{(t)}$ [93, 48] contains additional terms with constant $\mathcal{K}$. These go to zero if exact wave functions are used, but they make a nonzero contribution when fairly approximate functions are used, which makes the results dependent upon the choice of $\mathcal{K}$ (the constant can be used as a semiempirical parameter [94]). When the transition is made to the nonrelativistic limit in this operator, we can also derive a nonrelativistic transition operator that contains the gauge constant [93]; the r- and v-forms are its special cases.

The relativistic magnetic multipole radiation operator is independent of the gauge constant $\mathcal{K}$ [48]:

$$ {}_mO_p^{(t)} = -ie\sqrt{\frac{t}{t+1}}\,(2t+1)!!\,k^{-t}\,[C^{(t)} \times \alpha^{(1)}]_p^{(t)}\, j_t(kr). \qquad (1.278) $$

Its radial component does not disappear even when $t = 1$. Thus, a relativistic magnetic dipole operator, in contrast to a nonrelativistic operator, also describes the transitions between levels of different configurations.

The dependence of a one-electron matrix element on the many-electron quantum numbers is determined by only the rank of the operator. Therefore, submatrix elements for the ${}_eO^{(t)}$ and ${}_mO^{(t)}$ operators are uniquely defined in terms of the one-electron submatrix element:

$$(n_1 l_1 j_1^{N_1} \gamma_1 J_1 n_2 l_2 j_2^{N_2} \gamma_2 J_2 J \| O^{(t)} \| n_1 l_1 j_1^{N_1-1} \gamma_1' J_1' n_2 l_2 j_2^{N_2+1} \gamma_2' J_2' J') =$$

$$= (-1)^{N_2 + j_2 + J_2 - J_2'} \sqrt{N_1 (N_2+1)} (j_1^{N_1} \gamma_1 J_1 \| j_1^{N_1-1} \gamma_1' J_1' j_1) \times$$

$$\times (j_2^{N_2} \gamma_2 J_2 j_2 \| j_2^{N_2+1} \gamma_2' J_2') [J_1, J_2', J, J']^{1/2} \begin{Bmatrix} j_1 & j_2 & t \\ J_1 & J_2 & J \\ J_1' & J_2' & J' \end{Bmatrix} \times$$

$$\times (n_1 l_1 j_1 \| o^{(t)} \| n_2 l_2 j_2). \tag{1.279}$$

The one-electron submatrix elements for the electric and magnetic operators are [93]

$$(n_1 l_1 j_1 \| {}_e o^{(t)} \| n_2 l_2 j_2) = (-1)^{j_1 + 1/2 + t} (2t-1)!! [j_1, j_2]^{1/2} \times$$

$$\times \begin{pmatrix} j_1 & t & j_2 \\ -\frac{1}{2} & 0 & \frac{1}{2} \end{pmatrix} R_t(e), \tag{1.280}$$

$$(n_1 l_1 j_1 \| {}_m o^{(t)} \| n_2 l_2 j_2) = (-1)^{l_1} (2t+1)!! \sqrt{\frac{t}{t+1}} [j_1, j_2]^{1/2} \times$$

$$\times \begin{pmatrix} j_1 & t & j_2 \\ -\frac{1}{2} & 1 & -\frac{1}{2} \end{pmatrix} R_t(m). \tag{1.281}$$

The radial transition integrals are defined as follows:

$$R_t(e) = \frac{e}{k^{t-1}} \int_0^\infty \Big\{ (l_1 l_2 t) \Big[P_1 P_2 j_{t-1} - \frac{t}{t+1} P_1 P_2 j_{t+1} - \frac{2t+1}{t+1} Q_1 P_2 j_t \Big] +$$

$$+ (\bar{l}_1 \bar{l}_2 t) \Big[Q_1 Q_2 j_{t-1} - \frac{t}{t+1} Q_1 Q_2 j_{t+1} + \frac{2t+1}{t+1} P_1 Q_2 j_t \Big] r dr, \tag{1.282}$$

$$R_t(m) = \frac{e}{k^t} \int_0^\infty [(l_1 \bar{l}_2 t) P_1 Q_2 + (\bar{l}_1 l_2 t) Q_1 P_2] j_t \, dr. \tag{1.283}$$

Here, $j_t \equiv j_t(kr)$, $P_i \equiv P_{n_i l_i j_i}(r)$; $Q_i \equiv Q_{n_i \bar{l}_i j_i}(r)$ and $(l_1 l_2 t)$ is a triad condition having the additional requirement that the sum $l_1 + l_2 + t$ be even.

When there is one vacancy making a transition the submatrix element for the ${}_eO^{(t)}$ or ${}_mO^{(t)}$ operator, similar to Eq. (1.271), is, to within the accuracy of the phase factor, equal to the one-electron submatrix element:

$$j_1^{2j_1+1} j_2^{2j_2} j_2 \| O^{(t)} \| j_1^{2j_1} j_2^{2j_2+1} j_1) = (-1)^{j_1+j_2+t} (j_1 \| o^{(t)} \| j_2). \tag{1.284}$$

The probability of a radiative transition is given by the square of the absolute value of the transition operator matrix element. When the radiation is unpolarized, it is convenient to introduce a quantity that is independent of the projections — the so-called line strength:

$$S_t(\gamma J, \gamma' J') = \sum_{MM'p} |\langle \gamma JM | O_p^{(t)} | \gamma' J' M' \rangle|^2 = |\langle \gamma J \| O^{(t)} \| \gamma' J' \rangle|^2. \tag{1.285}$$

This line strength is symmetric to the exchange of quantum numbers for the initial and final levels:

$$S_t(\gamma J_1, \gamma' J') = S_t(\gamma' J', \gamma J). \tag{1.286}$$

The line strength, summed over all quantum numbers except those that designate both configurations (these we pick out from the γ and γ'), is the total line strength. It describes the transition strength between all the states in the configurations being examined and is independent of the coupling type. A summation can be done for the general case, in which a simple algebraic expression is obtained. In the nonrelativistic theory [38]

$$S_t(K_0 l_i^{N_i} l_k^{N_k}, K_0 l_i^{N_i-1} l_k^{N_k+1}) =$$

$$= \binom{4l_i+1}{N_i-1} \binom{4l_k+1}{N_k} \prod_{p(\neq i,k)} \binom{4l_p+2}{N_p} 2 |s_{l_i, l_k}^{(t)}|^2. \tag{1.287}$$

In the relativistic theory

$$S_t(K_0 l_i j_i^{N_i} l_k j_k^{N_k},\ K_0 l_i j_i^{N_i-1} l_k j_k^{N_k+1}) =$$
$$= \binom{2j_i}{N_i-1}\binom{2j_k}{N_k}\prod_{p(\neq i,\,k)}\binom{2j_p+1}{N_p} |(l_i j_i \| o^{(t)} \| l_k j_k)|^2. \tag{1.288}$$

The K_0 denotes passive shells and $s^{(t)}$ is defined by Eq. (1.264).

A dimensionless quantity, the oscillator strength,

$$f(\gamma J,\ \gamma' J') = \frac{2m}{3\hbar^2 e^2 (2J+1)} [E(\gamma' J') - E(\gamma J)]\, S(\gamma J,\ \gamma' J'). \tag{1.289}$$

is also widely used to examine electric dipole transitions. The oscillator strength has the sense of an effective number of harmonic oscillators – classical electrons that would have absorbed radiation in the same way as an atom would during a given transition.

When a transition is made into the continuous spectrum, instead of Eq. (1.289) the oscillator strength density

$$\frac{df(\gamma J,\ \gamma' J' \varepsilon l j J'')}{d\varepsilon} = \frac{2m\,[E(\gamma' J') + \varepsilon - E(\gamma J)]}{3\hbar^2 e^2 (2J+1)}\, S(\gamma J,\ \gamma' J' \varepsilon l j J''), \tag{1.290}$$

is introduced, where ε is the energy of a free electron.

The sum of the oscillator strengths over all possible final configurations and their states, including the continuous spectrum, is the total oscillator strength. It follows from the general commutation relation for **r** and **p** that in the nonrelativistic theory this quantity is equal to the number of electrons in the atom [54, 46]:

$$\sum_{\gamma'} f(\gamma,\ \gamma') + \sum_{\gamma''} \int \frac{df(\gamma,\ \gamma''\varepsilon)}{d\varepsilon}\, d\varepsilon = N. \tag{1.291}$$

There is also a sum rule for excitation from an individual shell:

$$\sum_{\gamma' J' l'} \Big[\sum_{n'} f(K l_i^{N_i} \gamma J,\ K l_i^{N_i-1} n' l' \gamma' J') +$$
$$+ \int \frac{df(K l_i^{N_i} \gamma J,\ K l_i^{N_i-1} \varepsilon l' \gamma' J')}{d\varepsilon}\, d\varepsilon \Big] = N_i. \tag{1.292}$$

This relation, in contrast to Eq. (1.291), is not exact: it is satisfied only in a one-configuration approximation and, additionally, states prohibited by the Pauli exclusion principle appear in the sum over $n'l'$ [54, 46].

Chapter 2

CORRELATION METHODS IN ATOMIC THEORY

With respect to refining methods for recording atomic spectra, studying them near the threshold energies, and examining many-electron transitions and other anomalies, the development of effective, simple, and universal methods of accounting for the correlation between electrons is the most important problem in atomic physics, including the theory of electron and x-ray spectra. This chapter does not presume to be a systematic review of the various correlation methods. Methods of this type and variations of them have been proposed, but most of these are quite complex and are only used in practice for configurations containing either several electrons or closed electron shells. We will only describe the most widely used methods for describing electron and x-ray spectra – semiempirical and simple theoretical methods of accounting for correlations (Sec. 2.1), the many-configuration approximation and its variations, the method of configuration interaction (Sec. 2.2), stationary perturbation theory (Sec. 2.3), and the method that takes into account the interaction between a discrete state and a continuum, which is extremely important for highly excited states (Sec. 2.4). The various correlation methods in atomic theory have been thoroughly reviewed in the monographs [47, 45, 95, 96, etc.].

2.1. The Correlation Energy and the Simplest Methods of Accounting for It

According to Löwdin [97], the correlation energy is the difference between the exact nonrelativistic energy E^{nr}, obtained approximately by subtracting the relativistic correction from the experimental energy, and the Hartree–Fock energy:

$$E^{cor} = E^{nr} - E^{HF} \approx (E^{exp} - E^{rel}) - E^{HF} \tag{2.1}$$

Ordinarily, E^{HF} is understood to be the energy calculated by the Hartree–Fock method for the average energy (HF-av.). Thus, we will call those methods that go beyond the scope of the HF-av. method correlation methods.

The Hartree–Fock method can be made more accurate by removing the constraint that all electrons in a shell have the same radial wave function. If we assume that this wave function also depends on the electron's direction of spin, we obtain the spin-polarized Hartree–Fock method [98–100]. However, when we fix the projections of the electron spins, the resulting spin angular momentum does not have an exact value. In addition dividing a shell into two subshells having upward and downward spins, just as in the relativistic theory, the configuration is divided into several subconfigurations that correspond to the different electron distributions in the subshells – these spin-polarized configurations mix via spin–orbit interaction. It is for these reasons that the spin-polarized method has not been widely adopted.

We can go even further in this direction and assume that the radial wave function is also dependent on the projection of the one-electron orbital angular momentum, i.e., that each electron is described by a different radial function. This corresponds to the unrestricted Hartree–Fock method [101]. A simpler version of this method is the extended method [102], in which a one-electron radial wave function does not belong to a specific spin-angular function: the wave function for a shell is presented as the product of an antisymmetrized spin-angular function and a symmetrized radial function as a permanent from the radial functions. All in all, the extended method is much more complex than the HF-av. method and allows us to calculate only a fraction of the correlation energy [45], i.e., only the radial correlations between equivalent electrons.

With the Hartree–Fock method we can also dispense with the condition that one-electron radial wave functions having the same l be orthogonal [103]. This makes the wave functions and energy more precise for configurations containing two or more open shells that have the same l. Then, the energy calculated by using nonorthogonal wave functions is obtained exactly in the first and second terms of the expansion in powers of Z, whereas only the first term is exact when the condition given by Eq. (1.150) is taken into account [45]. However, when we dispense with the orthogonality condition the equations for the Hartree–Fock energy and potential become much more complex: additional terms containing overlap integrals and integrals $I(nl)$ appear in the expression for the potential.

A comparatively simple method of accounting for correlation methods, although it is neither very strict nor universal, is to use a potential function having one or more semiempirical parameters in the one-electron equation. Different semiempirical potentials are used most successfully to describe one, and less often, two outer electrons besides closed shells [57]. The Klapisch potential [104] is fairly effective for complex configurations.

This potential is obtained by approximating the radial distribution density for each electron in the atom by Slater's nodeless function having a single parameter θ_i:

$$4\pi r^2 \rho_{n_i l_i}(r) = r^{2l_i+2} e^{-\theta_i r}. \tag{2.2}$$

Then the electric field in an N-electron atom is given by

$$V(r) = -\frac{e^2}{r}\{Z - N + 1 + (N-1)e^{-\alpha_1 r} + \alpha_2 r e^{-\alpha_3 r} + \ldots + \\ + \alpha_{n-1} r^n e^{-\alpha_n r}\}. \tag{2.3}$$

The number of unknown parameters α_i is reduced when we consider that the wave functions must be orthogonal and the remaining parameters are determined by the method of least squares with the requirement that the best agreement between the calculated energy levels or binding energies and their experimental values be obtained [57]. The Klapisch potential is also used in the relativistic version [104].

Whether or not the parametric method will be successful will depend in large degree on how realistic the physical model being used is. A one-parameter g-Hartree method [105] has recently been developed from quantum electrodynamics, which gives good agreement between experimental values for the binding energies and the other characteristics of an atom's inner shells [106].

The model for a uniform electron gas can be used to estimate correlation energy, as well as the exchange energy [107]. When the electron charge density is weakly varying (an approximation of the local density) an atom's correlation energy is given in terms of the electron's average correlation energy, which is approximated by a semiempirical formula [107, 46]. The local-density approximation is more accurate for one-electron characteristics, including the binding energy [46, 107].

It is common knowledge that the one-configuration values for the electrostatic interaction integrals F^k and G^k are, as a rule, higher than the semiempirical values. These semiempirical values are found by the method of least squares from the experimental values of the energy levels and the theoretical expressions for the coefficients at the F^k, G^k, and ζ_{nl} integrals, which are thought of as free parameters. The number of levels in a configuration is usually much larger than the number of integrals that define the energy spectrum in a one-configuration approximation; therefore, additional parameters, such as, for example, the known correction $\alpha L(L+1)$ or parameters that describe an interaction with lower-lying configurations may be included in the energy equation. The method is used to define and classify the levels if only a few of them are known [108, 46]. The number of parameters increases significantly in the relativistic theory, which lowers its effectiveness.

The reduction in the semiempirical values of the F^k and G^k integrals in comparison with the Hartree–Fock values, as well as the need to introduce additional parameters (α, β, etc.), is caused by the interaction between the configuration being examined and the many configurations that are energetically removed from it. This follows from an examination of the energy correction in second-order perturbation theory [109, 40]:

$$\Delta E(K\gamma) = \sum_{K'(\neq K)} \sum_{\gamma'} \frac{\langle K\gamma | H | K'\gamma' \rangle \langle K'\gamma' | H | K\gamma \rangle}{E(K\gamma) - E(K'\gamma')}. \quad (2.4)$$

If we take the average value of the energy denominator $E(K\gamma) - \bar{E}(K')$ outside the summation over γ', then the matrix elements can be summed over γ in algebraic form. The correction to the energy of a level $K\gamma$ can, because of its interaction with the levels of the K', be given as the matrix element of an effective operator H^{ef}:

$$\Delta E(K\gamma,\ K') = [E(K\gamma) - \bar{E}(K')]^{-1} \times$$
$$\times \sum_{\gamma'} \langle K\gamma | H | K'\gamma' \rangle \langle K'\gamma' | H | K\gamma \rangle = \langle K\gamma | H^{ef} | K\gamma \rangle. \quad (2.5)$$

This operator can most easily be found by the method of second quantization. If in Eq. (2.5) we use the Hamiltonian H, expressed in terms of electron creation and annihilation operators (which transform the wave function for the K configuration into a function for the K' configuration and vice-versa), we can perform the summation by means of the condition for the completeness of a basis functions, since terms absent from the summation contribute nothing. Then, using the rules of anticommutation, Eq. (1.84), all operators $a^\dagger$ are carried to the left, all operators a are carried to the right, and when the formulas for summing Wigner symbols and $3nj$-symbols are used, the separate terms are written as matrix elements for p-electron operators:

$$H^{ef} = \sum_{p} H_p^{ef}, \quad H_p^{ef} = \sum_{i<k<\ldots<l} h_{ik\ldots l}^{ef} = \frac{1}{p!} \sum_{\substack{\alpha\ldots\beta \\ \vartheta\ldots\gamma}} \underbrace{a_\alpha^\dagger \ldots a_\beta^\dagger}_{p} \times$$
$$\times \langle \alpha \ldots \beta | h_{1\ldots p}^{ef} | \gamma \ldots \vartheta \rangle \underbrace{a_\vartheta \ldots a_\gamma}_{p}. \quad (2.6)$$

Operators more complex than four-electron operator cannot be contained in a sum over p because H includes only one- and two-electron operators. An effective operator as an energy operator must be a scalar for the resulting angular momenta; therefore, its one-electron term, as well as a term with $p = 0$, causes the same shift for all levels. Some of the two-electron effective operator turns out to be proportional to the electrostatic interaction operator and, because the spin-angular parts of their matrix elements are the same, gives only a correction to the radial integrals F^k and G^k. An effective operator $(U^{(1)} \cdot U^{(1)})$ leads to the appearance of an $\alpha L(L + 1)$ term. In general, H^{ef} also contains three- and four-electron terms whose contribution can be approximated by additional semiempirical terms.

When the method of effective operators is used in theoretical form, it converges slowly. This convergence is significantly improved if transformed functions of the configuration being examined are used as radial functions of the admixed configurations [110].

The higher value of the Hartree–Fock integrals for an electrostatic interaction in comparison with the semiempirical values leads to a systematic expansion of the energy spectrum. This is sometimes corrected by reducing the values of the F^k and G^k integrals by a factor of 1.2–1.5 [111–113].

In studying the levels in a Rydberg series, including the excitation of electrons to these levels by soft x-ray absorption, the semiempirical quantum defect method [114–115] is widely used. In its simplest – the one-channel – form (a series is not perturbed by the levels of other Rydberg series) the binding energy of an excited electron is approximated by a one-electron expression with an effective quantum number n^*:

$$I_{nl} = \frac{(Z^*)^2}{2\,(n^*)^2} \approx \frac{(Z-N+1)^2}{2\,(n-\mu_l)^2}\,, \tag{2.7*}$$

where μ_l is a quantum defect that is weakly dependent on n, which lets us interpolate or extrapolate the binding energy. Additional parameters, which also are weakly dependent on energy, are introduced in the more general multichannel quantum defect method. These parameters take the interaction between series into consideration. This method has been formulated for the relativistic case [116].

2.2. The Many-Configuration Approximation

A fundamental method for taking the correlations between electrons into account is the many-configuration (MC) method in which the atom's wave function is found as an expansion in not only the basis functions of that very configuration, as is done in an intermediate coupling, but in other configurations having the same parity:

$$\Phi(\Gamma J) = \sum_{K'\gamma'} C^{\Gamma}_{K'\gamma' J}\,\Psi(K'\gamma' J). \tag{2.8}$$

Here $C_{K'\gamma' J}{}^{\Gamma}$ is the expansion coefficient; Ψ is the one-configuration function defined in a one- or many-configuration approximation; Φ is the many-configuration function that corresponds to a mixture of configurations, although in terms of greatest weight or origin it is often designated approximately by the quantum numbers for one configuration.

If the functions Ψ are calculated in a one-configuration approximation, and the expansion coefficients are defined by diagonalizing the energy matrix in the basis of these functions, we then have a simpler version of the many-configuration method, called the method of configuration interaction or superposition.

In an orthonormalized basis, the many-configuration matrix elements go to zero if the configurations are distinguished from one another by the quantum numbers of more than two electrons. Configurations such as these affect one another only through an interaction with other configurations. If the configurations are distinguished by the states of two electrons or one electron that changes the orbital quantum number, then the many-configuration matrix elements for one-electron operators of the H^p, H^k, and H^{so} type are equal to zero (in the second case because of the scalar nature of H^p and H^k and the diagonality of the H^{so} matrix elements relative to l). With the exception of the p-shells in heavy atoms, the comparatively small interconfiguration matrix elements H^{so} are usually omitted for open valent and subvalent shells. When the interaction between configurations cannot be neglected, we can use the approximation

$$\zeta_{nl,\,n'l} \approx [\zeta_{nl}\,\zeta_{n'l}]^{1/2}. \tag{2.9}$$

to calculate the interconfiguration constant of spin–orbit interaction $\zeta_{nl,n'l'}$. This approximation is obtained from the Schwarz inequality, which, because of the similarity of the wave functions $P_{nl}(r)$ and $P_{n'l'}(r)$ for the small r that makes the fundamental contribution to $\zeta_{nl,n\ l}$, can, for all practical purposes, be replaced by an equality [46].

If the one-configuration wave functions are calculated separately for each configuration, they will be only approximately orthogonal to one another. A complete orthonormalized basis can be obtained by freezing the wave functions of the core and calculating the functions of excited electrons for the same local potential. Freezing, however, significantly deteriorates the accuracy of the basis wave functions and can only be justified when the characteristics of the one configuration are important and the others are treated as admixed configurations.

On the other hand, the radial functions for configurations having the same l are orthogonal to within 10^{-2} to 10^{-3}. It was shown in [46] that the errors caused by omission of the overlap integrals in different elements of the energy matrix compensate one another when functions that are roughly orthogonal are used. Consequently, the configuration interaction method wins in simplicity and loses only slightly in accuracy if true one-configuration wave functions are used in the expansion given by Eq. (2.8), but the energy matrix and transition probabilities are calculated on the assumption that the functions are orthogonal.

The interconfiguration matrix elements for the Hamiltonian are given in terms of one- and two-electron matrix elements when the technique of irreducible tensors and fractional parentage coefficients is used, i.e., as in the one-configuration case. The fundamental difference is in the factor that appears when the transition is made from the matrix element of a T operator that contains a sum over the coordinates of all electrons in the system to the matrix element for a one- or two-electron operator t that acts only on the coordinates of electrons that have separated out by the coefficients of fractional parentage ($\|$):

$$\langle K|T|K'\rangle = A(\|)\langle K|t|K'\rangle. \tag{2.10}$$

The factor A takes into account the different normalization of bra and ket wave functions that contain different numbers of equivalent and nonequivalent electrons, the equality of the contributions of different terms in the operator, and the phase that appears when particles are transposed so that the electrons upon which the operator is acting would occupy the same position in both functions [117]. For a one-electron operator we have

$$A_1(K,\ K') = (-1)^{\nu_p - \nu'_r} \prod_{i=1}^{q} \binom{N_i}{m_i}^{1/2} \binom{N'_i}{m'_i}^{1/2}, \tag{2.11}$$

where $K = n_1 l_1^{N_1} n_2 l_2^{N_2}, \ldots, n_q l_q^{N_q}$; $K' = n_1 l_1^{N_1'} n_2 l_2^{N_2'}, \ldots, n_q l_q^{N_q'}$; p and r are the numbers of the shells in which the numbers of electrons N_i and N_i' differ by one (they must be equal for the remaining $q - 2$ shells); m_i and m_i are the numbers of electrons in the $l_i^{N_i}$ and $l_i^{N_i'}$ shells that the operator is acting on; ν_p and ν_r' in Eq. (2.11) are

$$\nu_p = \sum_{j=1}^{p-1} N_j + (N_p - 1), \quad \nu'_r = \sum_{j=1}^{r-1} N'_j + (N'_r - 1), \quad N_p > N_{p'},\ N_r < N_{r'}. \tag{2.12}$$

In the two-electron matrix element that is nondiagonal relative to the configurations, the analogous factor is

$$A_2(K,\ K') = (-1)^{\nu_p + \nu_t - \delta_{pt} - \nu'_r - \nu'_s + \delta_{rs}} \prod_{i=1}^{q} \binom{N_i}{m_i}^{1/2} \binom{N'_i}{m'_i}^{1/2}. \tag{2.13}$$

Here, p, r, s, and t are the numbers of the shells having different numbers of electrons N_i and N_i' in both configurations, and ν_i, ν_i', m_i, and m_i' mean the same thing as in Eq. (2.11).

The corresponding expressions in the relativistic theory are obtained by replacing the shells with subshells.

After the factor A_2 has been defined and the electrons from the p, t, r, and s shells have been separated out by means of coefficients of fractional parentage, the angular momenta of those electrons upon which the operator is acting are combined via transformation matrices into common resultant angular momenta and the interconfiguration matrix element is then expressed in terms of a two-electron element, in keeping with Eqs. (1.68) and (1.69).

In the nonrelativistic theory the common two-electron matrix element for an electrostatic interaction operator is

$$\langle n_1 l_1 n_2 l_2 LS | h^e | n_1' l_1' n_2' l_2' LS\rangle = 2NN' (-1)^{l_1'+l_1} \sum_k \Big[(-1)^L \times$$

$$\times \begin{Bmatrix} l_1 & l_2 & L \\ l_2' & l_1' & k \end{Bmatrix} \langle l_1 \| C^{(k)} \| l_1'\rangle \langle l_2 \| C^{(k)} \| l_2'\rangle R^k (n_1 l_1 n_2 l_2,\ n_1' l_1' n_2' l_2') +$$

$$+ (-1)^S \begin{Bmatrix} l_1 & l_2 & L \\ l_1' & l_2' & k \end{Bmatrix} \langle l_1 \| C^{(k)} \| l_2'\rangle \langle l_2 \| C^{(k)} \| l_1'\rangle \times$$

$$\times R^k (n_1 l_1 n_2 l_2,\ n_2' l_2' n_1' l_1') \Big], \tag{2.14}$$

where N and N' are normalization factors for the electron wave functions $n_1 l_1$, $n_2 l_2$, and $n_1' l_1'$, $n_2' l_2'$ [Eq. (1.107)], and R^k is the integral for an electrostatic interaction:

$$R^k (\alpha\beta,\ \alpha'\beta') = e^2 \int_0^\infty\!\!\int \frac{r_<^k}{r_>^{k+1}} P_\alpha (r_1) P_\beta (r_2) P_{\alpha'} (r_1) P_{\beta'} (r_2)\, dr_1 dr_2 . \tag{2.15}$$

which does not change when the following sets of quantum numbers are transposed:

$$R^k (\alpha\beta,\ \alpha'\beta') = R^k (\alpha'\beta',\ \alpha\beta) = R^k (\alpha\beta',\ \alpha'\beta) = R^k (\alpha'\beta,\ \alpha\beta') = \\ = R^k (\beta\alpha,\ \beta'\alpha'). \tag{2.16}$$

In special cases the integral R^k equals the one-configuration integrals F^k and G^k:

$$R^k (\alpha\beta,\ \alpha\beta) = F^k (\alpha,\ \beta), \quad R^k (\alpha\beta,\ \beta\alpha) = G^k (\alpha,\ \beta). \tag{2.17}$$

In contrast to the F^k and G^k, which are always positive, the R^k integrals, which contain various radial functions, may be positive or negative.

The two-electron matrix element for the h^e operator for relativistic wave functions has the form [71, 48]

$$(\lambda_1\lambda_2 J\,|\,h^e\,|\,\lambda_3\lambda_4 J)^a = 2N_{\lambda_1,\lambda_2}N_{\lambda_3,\lambda_4}[(\lambda_1\lambda_2 J\,|\,h^e_{12}\,|\,\lambda_3\lambda_4 J) - \\ -(-1)^{j_3+j_4-J}(\lambda_1\lambda_2 J\,|\,h^e_{12}\,|\,\lambda_4\lambda_3 J)], \quad \lambda_i \equiv n_i l_i j_i, \tag{2.18}$$

$$(\lambda_1\lambda_2 J\,|\,h^e_{12}\,|\,\lambda_3\lambda_4 J) = (-1)^{j_3+j_4+J}[j_1, j_2, j_3, j_4]^{1/2} \sum_k \begin{pmatrix} j_1 & k & j_3 \\ -\frac{1}{2} & 0 & \frac{1}{2} \end{pmatrix} \times \\ \times \begin{pmatrix} j_2 & k & j_4 \\ -\frac{1}{2} & 0 & \frac{1}{2} \end{pmatrix} \begin{Bmatrix} j_1 & j_2 & J \\ j_4 & j_3 & k \end{Bmatrix} [R^k(\lambda_1\lambda_2, \lambda_3\lambda_4) + R^k(\lambda_1\bar{\lambda}_2, \lambda_3\bar{\lambda}_4) + \\ + R^k(\bar{\lambda}_1\lambda_2, \bar{\lambda}_3\lambda_4) + R^k(\bar{\lambda}_1\bar{\lambda}_2, \bar{\lambda}_3\bar{\lambda}_4)]. \tag{2.19}$$

Here the R^k integral is defined according to Eq. (1.187), and the $N_{\lambda,\lambda'}$ is defined according to Eq. (1.107). The matrix elements for a retarded interaction operator that are nondiagonal to the configurations [the second term in Eq. (1.179b)] are zero [71, 48].

General expressions for the interconfiguration matrix elements of a retarded interaction operator are given in [117, 43, 46] for the nonrelativistic theory, and of the Breit operator for the relativistic theory is given in [71, 48].

If the configurations differ by only the quantum numbers of a single electron, then the interconfiguration matrix element for the H^e operator is the sum of the terms that correspond to the interaction between this electron and the electrons in different shells, including closed shells. For the case of K and K' configurations, which differ by the principal quantum number of a single electron (which correspond to the one-electron excitation $nl \to n'l$), when using the K' configuration for all electrons except of the $n'l$ frozen wave functions of the K configuration, the interconfiguration matrix elements for the H^e operator are completely or partially compensated by the matrix elements for the H^k and H^p operators (they are given by the Hartree–Fock equations in terms of the same radial integrals R^k). Total compensation is formulated as the Brillouin theorem:

$$\langle K\gamma\,|\,(H^k + H^p + H^e)\,|\,K'\gamma'\rangle = 0. \tag{2.20}$$

which is satisfied for all configurations of this type only in a basis of one-determinant wave functions, and only for some configurations in a basis of wave functions for coupled angular momenta. From general relationships between the interconfiguration matrix elements that were derived on the basis of the variational principle [118], it follows [119] that in HF-t approximation or the isospin basis the Brillouin theorem is satisfied for all many-electron states of the following (and only these) configurations:

$$K_0' nl - K_0' n' l, \quad K_0 nl^2 - K_0 nl\, n' l, \quad K_0 nl^{4l+2} - K_0 nl^{4l+1} n' l,$$

$$K_0' nl^{4l+2} n' l^{4l+1} - K_0' nl^{4l+1} n' l^{4l+2}, \qquad K_0 nl^{4l+2} n' l^{4l} - K_0 nl^{4l+1} n' l^{4l+1}.$$

Here the K_0 are closed shells, and the K_0' may contain a single open shell of the $n_0 l_0$ or $n_0 l_0^{4l_0+1}$ type in addition to closed shells. Both configurations enter into these pairs in an unequal manner (frozen wave functions from the first configuration are used) and, in general, they cannot be transposed.

In the HF-av. approximation the Brillouin theorem can only be satisfied for the configurations

$$K_0 nl - K_0 n' l, \quad K_0 nl^{4l+2} - K_0 nl^{4l+1} n' l, \quad K_0 nl^{4l+2} n' l^{4l+1} - K_0 nl^{4l+1} n' l^{4l+2}.$$

If the interconfiguration matrix elements go to zero in the HF-t approximation but are nonzero in the HF-av. approximation, the wave function for the HF-t approximation can be given approximately (in first-order perturbation theory) as the superposition of the wave functions for the respective configurations in the basis of the HF-av. approximation, e.g.,

$$\Psi (K_0 n_0 l_0^{4l_0+1} nlLSJ)_{\text{HF-t}} = \sum_{n'} c_{n'} \Psi (K_0 n_0 l_0^{4l_0+1} n' lLSJ)_{\text{HF-av.}} \tag{2.21}$$

Summation over n' also includes integration over the $n_0 l_0^{4l_0+1} \varepsilon l$ continuum.

If Brillouin's theorem is not satisfied, some terms in the interconfiguration matrix element that correspond to one-electron excitation will go to zero. In the matrix element for a nonrelativistic Hamiltonian between $Kn_1l_1^{N_1}n_2l_2^{N_2}n_3l_3^{N_3}$ and $Kn_1l_1^{N_1}n_2l_2^{N_2-1}n_3l_3^{N_3+1}$ configurations (the K are any shells) the following terms are either compensated ($l_2 = l_3$) or are equal to zero ($l_2 \neq l_3$) [119]: 1) for any N_1, N_2, or N_3 that contain integrals $R^0(n_2l_2n_2l_2, n_2l_2n_3l_3)$, $R^0(n_3l_3n_3l_3, n_3l_3n_2l_2)$, and $R^0(n_2l_2n_1l_1, n_3l_3n_1l_1)$; 2) with the integrals $R^k(n_2l_2n_1l_1, n_3l_3n_1l_1)$ and $R^k(n_2l_2n_1l_1, n_1l_1n_3l_3)$ when $N_1 = 4l_1 + 2$; and 3) with the integrals $R^k(n_2l_2n_2l_2, n_2l_2n_3l_3)$ and $R^k(n_3l_3n_3l_3, n_3l_3n_2l_2)$, respectively, when $N_2 = 4l_2 + 2$, or $N_3 = 4l_3 + 1$.

According to the energy correction in second-order perturbation theory and to the wave function correction in first-order perturbation theory, the levels of the two configurations will be mixed together more strongly as the interconfiguration matrix elements are larger and the closer in energy to one another these two levels are. This allows us to introduce a new quantity for estimating the degree of mixing – the configuration interaction strength [120]:

$$T(K, K') = [\bar{E}(K) - \bar{E}(K')]^{-2} \sum_{\gamma\gamma' JM} \langle K\gamma J \,|\, H \,|\, K'\gamma' J\rangle^2, \tag{2.22}$$

where $\bar{E}(K)$ and $\bar{E}(K')$ are the average energies of the configurations. Summation over interconfiguration matrix elements can be done in general form [119] and the dependence $T(K, K')$ on the number of electrons in the shells can be given as binomial coefficients.

If $T(K, K')$ is divided by $\beta(K, K')$ – the number of states in the K configuration that are mixed with states of the K' configuration – a quantity that is approximately equal to the mean-squared value of the weight of the K configuration in the level function for the K' configuration is obtained:

$$c^2(K, K') = \frac{T(K, K')}{\beta(K, K')}. \tag{2.23}$$

Equation (2.22) can be made more accurate if, instead of $[\bar{E}(K) - \bar{E}(K')]^2$, the square of the energy difference between the interacting levels, averaged with due consideration for the interconfiguration matrix elements, is used:

$$\frac{\sum_{\gamma, \gamma' J} [J]\left(\langle K\gamma J|H|K\gamma J\rangle - \langle K'\gamma' J|H|K'\gamma' J\rangle\right)^2 \langle K\gamma J|H|K'\gamma' J\rangle^2}{\sum_{\gamma, \gamma' J} [J]\langle K\gamma J|H|K'\gamma' J\rangle^2}. \tag{2.24}$$

The main types of strongly interacting configurations in atoms are examined in Sec. 3.4. Taking several such configurations into account by the method of configuration interaction often makes it possible to make the energy values and other quantities more precise. However, when the basis is expanded, the configuration interaction method converges rather slowly – several hundreds or thousands of configurations must be calculated in order to calculate the correlation energy for the ground state to within a few percent [121]. For light atoms there are calculations in a basis that contains more than six thousand configurations.

The many-configuration method [122, 45, 123] is more effective and converges much more rapidly, in which not only the c_γ coefficients but the wave functions Ψ_γ are optimized in the expansion (namely this method in a narrow sense bears its name)

$$\Phi = \sum_\gamma c_\gamma \Psi_\gamma \tag{2.25}$$

Substituting Eq. (2.25) into the energy equation

$$E = \langle\Phi|H|\Phi\rangle = \sum_\gamma c_\gamma^2 \langle\Psi_\gamma|H|\Psi_\gamma\rangle + \sum_{\gamma\neq\gamma'} c_\gamma c_{\gamma'} \langle\Psi_\gamma|H|\Psi_{\gamma'}\rangle \tag{2.26}$$

and adding Lagrangian multipliers that guarantee the orthonormality of one-electron radial functions gives us a many-configuration functional. When this functional is varied relative to the $P_{nl}(r)$ the Hartree–Fock–Jucys equations [122, 45] are obtained. These have the same general form as the Hartree–Fock equations, but contain additional interconfiguration terms that are derived from the second sum in Eq. (2.26) [depending on the type of interconfiguration integral and the number of $P_{nl}(r)$ functions it contains, these terms are added to the exchange function, the potential function, or the $\varepsilon_{nl,nl}$]. The Hartree–Fock equations are solved self-consistently with the calculation of c_γ coefficients by diagonalizing the many-configuration energy matrix.

In practice, the many-configuration method can be combined with the configuration interaction method: from the first the mixing of the lower-lying, strongly interacting configurations, e.g., configurations belonging to the same complex, can be determined more precisely and the configuration interaction method is then used to calculate a number of the less important terms in the expansion given by Eq. (2.25) [45]. This corresponds to dividing the wave function into two parts

$$\Phi = |0\rangle + |1\rangle, \quad \langle 0|1\rangle = 0. \tag{2.27}$$

the first of which is calculated via the many-configuration (MC) method, and the second is calculated by the configuration interaction (CI) method.

The main part of the correlations between electrons has a two-electron nature [123, 95]; therefore, $|1\rangle$ can be written approximately as a sum over the electron pairs contained in $|0\rangle$:

$$|1\rangle = \sum_{\beta} |1_\beta\rangle. \tag{2.28}$$

Here, $|1_\beta\rangle$ is a pair correlation function

$$\begin{aligned} 1_\beta\rangle &\equiv |1_{n_1 l_1 n_2 l_2 L_1 S_1 \gamma_2 L_2 S_2 LS}\rangle = \\ &= \sum_{n_3 l_3 n_4 l_4} c_{n_3 l_3 n_4 l_4 L_1 S_1} |n_3 l_3 n_4 l_4 L_1 S_1 \gamma_2 L_2 S_2 LS\rangle, \end{aligned} \tag{2.29}$$

where $n_1 l_1 n_2 l_2$ is an electron pair in the configuration being examined that have been recoupled into common angular momenta $L_1 S_1$; the quantum numbers $\gamma_2 L_2 S_2$ denote the subsystem formed by the atom's remaining electrons.

The admixed pair $n_3l_3n_4l_4$ is distinguished from the $n_1l_1n_2l_2$ pair by at least one quantum number. The coefficients c are determined by diagonalizing the energy matrix with regard for all configurations defining $|\ 0\rangle$ and $|1\rangle$. When this is done the correction ε_β to the energy E^{MC}

$$E_\beta = E^{MK} + \varepsilon_\beta, \quad E^{MK} = \langle 0 | H | 0 \rangle \tag{2.30}$$

is called the pair correlation energy.

The main advantage of the method of pair correlations is the additivity of the quantities ε_β, their weak dependence on the quantum numbers of "passive" electrons and the configuration for which they are defined, as well as their monotonic dependence on Z. This allows us to use the formula

$$E = E^{MK} + \sum_\beta \varepsilon_\beta . \tag{2.31}$$

to calculate the energy. The operations of summation and angular momentum coupling are commutative among themselves on the right-hand side of Eq. (2.29); therefore, the summation can be carried out over the two-electron wave functions.

The dual sum over the principal quantum numbers of the electrons in the correction pairs

$$\sum_{nn'} c_{nn'} | nl\, n'\, l'\, LS \rangle \tag{2.32}$$

is reduced to the single sums [45]:

$$\sum_n \sum_{n'(\geq n)} c_{nn'} | nl\, n'\, lLS \rangle = \sum_n c'_n | nl^2 LS \rangle_{nt}, \quad L+S \ - \tag{2.33}$$

$$\sum_n \sum_{n'(\geq n)} c_{nn'} | nl\, n'\, lLS \rangle = \sum_{\substack{n \\ (\Delta n = 2)}} c'_n | nl\,(n+1)\, lLS \rangle_{nt}, \quad L+S \ - \tag{2.34}$$

$$\sum_{n(\geq n_1)} \sum_{n'(\geq n_1)} c_{nn'} | nl\, n'\, l'\, LS \rangle = \sum_{n(\geq n_1)} c'_n | nln''\, l'\, LS \rangle_{nt}, \ l \neq l', \ n'' = n + (n_2 - n_1). \tag{2.35}$$

when the so-called natural orbitals [97] – linear combinations of radial functions that diagonalize the coefficients $c_{nn'}$ – are used. The symbol *nt* on the wave function indicates that natural radial orbitals were used in its construction.

A transition to natural orbitals lets us exclude many terms from the expansion of the pair correlation function. This offers us no advantage in the configuration interaction method, since all wave functions must be known in order to find the natural orbitals themselves. However, in the many-configuration method, the natural orbitals are obtained automatically as the best one-electron wave functions [45]; therefore, only those configurations that are maintained in the sums when the transition to natural orbitals is made need be considered in the expansion of the many-configuration function.

A similar expansion of the correlation energy into the contributions from the electron pairs is also used in the relativistic many-configuration method [73].

2.3. Stationary Perturbation Theory

Developing the techniques of perturbation theory and expanding the possibilities for implementing them on high-speed computers promotes more widespread use of perturbation theory for describing not only the simplest atomic systems, but atoms with open shells as well [47, 124–127]. Initially the greatest successes were achieved with the time-dependent perturbation theory and until now it is widely used even for calculating energy levels and other stationary quantities. However, a simpler stationary perturbation theory (the basis of which are the concepts contained in the works [47, 128, 129]), which we will acquaint the reader with in this section, has been rapidly developed.

Perturbation theory emerges from the possibilities of breaking up a system Hamiltonian into a simpler, but at the same time fairly realistic, model Hamiltonian H_0 and perturbation V:

$$H = H_0 + V. \tag{2.36}$$

Let the eigenfunctions and the eigenvalues of the Hamiltonian H_0 be known:

$$H_0 \Psi_\gamma^{(0)} = E_\gamma^{(0)} \Psi_\gamma^{(0)}, \tag{2.37}$$

and solutions of Schrödinger's equation for the total Hamiltonian

$$H \Psi_\gamma = E_\gamma \Psi_\gamma \tag{2.38}$$

be sought. We will divide the space of the functions $\Psi_\gamma^{(0)}$ into a model space P that contains the wave functions for the K configuration being examined (this is a single function when the configuration has closed shells) and an orthogonal space Q that contains all the remaining

$\Psi_\gamma^{(0)}$ functions. If there are open shells in K, then it is worthwhile to also include in P those functions of other configurations that strongly interact with K, including the functions of those configurations that can be obtained by redistributing the electrons in the open shells.

We will introduce the projection operators for the model and orthogonal spaces:

$$P=\sum_{\gamma\in P} |\Psi_\gamma^{(0)}\rangle\langle\Psi_\gamma^{(0)}|,\quad \gamma=1,\ \ldots,\ d, \tag{2.39}$$

$$Q=\sum_{\beta\notin P} |\Psi_\beta^{(0)}\rangle\langle\Psi_\beta^{(0)}|. \tag{2.40}$$

Because the basis of the wave functions is complete,

$$P+Q=1 \tag{2.41}$$

which allows us to divide Ψ_γ into a model function $\Psi_\gamma^{(0)}$ and a correction Φ_γ:

$$\Psi_\gamma=(P+Q)\Psi_\gamma=\Psi_\gamma^{(0)}+\Phi_\gamma,\quad \gamma=1,\ \ldots,\ d, \tag{2.42}$$

$$\Psi_\gamma^{(0)}=P\Psi_\gamma,\quad \Phi_\gamma=Q\Psi_\gamma. \tag{2.43}$$

Ordinarily, the so-called intermediate normalization is used: $\Psi_\gamma^{(0)}$ is considered to have been normalized and $\langle\Psi_\gamma^{(0)} \mid \Phi_\gamma\rangle = 0$; then, Ψ_γ is only approximately normalized.

Further investigation is done using the Brillouin–Wigner or Rayleigh–Schrödinger approaches. We initially obtain a series of Brillouin–Wigner perturbations for the nondegenerate case. We use Eq. (2.42) ($\Psi_\gamma^{(0)} \equiv \Psi_0$, $\Psi_\gamma \equiv \Phi$, $E_\gamma^{(0)} \equiv E_0$, $E_\gamma \equiv E$) to transform Eq. (2.38) to the form

$$(E-H_0)\Phi=V\Psi-(E-E_0)\Psi_0 \tag{2.44}$$

and operate from the left on both sides of Eq. (2.44) with Q, which commutes with H_0, to obtain

$$(E-H_0)\Phi=QV\Psi. \tag{2.45}$$

We define the resolvent operator T_E:

$$T_E(E-H_0)=Q,\quad T_E=\frac{Q}{E-H_0}. \tag{2.46}$$

If $E_0 = E$, we must add a small imaginary quantity $i\eta$ to the energy denominator and T_E can be thought of as the limit as $\eta \to 0$. Using the identity

$$T_E = \sum_{\beta} T_E | \Psi_\beta^{(0)} \rangle \langle \Psi_\beta^{(0)} | , \tag{2.47}$$

we can find the spectral expansion of the operator T_E:

$$T_E = \sum_{\beta \notin P} \frac{| \Psi_\beta^{(0)} \rangle \langle \Psi_\beta^{(0)} |}{E - E_\beta^{(0)}} . \tag{2.48}$$

Using the expression that follows from Eq. (2.45) for Φ and substituting this into Eq. (2.42) gives us an equation for Ψ:

$$\Psi = \Psi_0 + T_E V \Psi . \tag{2.49}$$

We introduce a wave operator Ω that will transform the model function Ψ_0 to an exact function Ψ (this operator acts only in the model space):

$$\Psi = \Omega_E \Psi_0 . \tag{2.50}$$

The equation for Ω_E follows from Eq. (2.49) when Eq. (2.50) is used:

$$\Omega_E = 1 + T_E V \Omega_E . \tag{2.51}$$

Solving this by an iterative method

$$\Omega_E = 1 + T_E V + T_E V T_E V + \ldots = \sum_{n=0}^{\infty} \left(\frac{Q}{E - H_0} V \right)^n \tag{2.52}$$

and substituting into Eq. (2.50) yields a Brillouin–Wigner series for the wave function:

$$\Psi = \sum_{n=0}^{\infty} \left(\frac{Q}{E - H_0} V \right)^n \Psi_0 . \tag{2.53}$$

If we operate on Eq. (2.44) from the left with P and then multiply by $\langle \Psi_0 |$, we obtain an equation for the energy,

$$E = E_0 + \langle \Psi_0 | V\Omega_E | \Psi_0 \rangle, \tag{2.54}$$

which is transformed into a series for the energy when the expansion given by Eq. (2.52) is used. The series in compact operator form given by Eqs. (2.52) and (2.53) can easily be transformed to common matrix form if the operators are put in the form of Eq. (2.48). The projection operator Q eliminates some matrix elements from every term in the sum. For example, the familiar energy series

$$E = E_0 + \langle \Psi_0 | V | \Psi_0 \rangle + \sum_{\beta \notin P} \frac{|\langle \Psi_0 | V | \Psi_\beta^{(0)} \rangle|^2}{E - E_\beta^{(0)}} + \ldots \tag{2.55}$$

is obtained.

In the case of degeneracy in the level E_0 being examined, which is removed in first-order perturbation theory, the basic formulas remain valid if the zeroth-order Hamiltonian H_0 is replaced by $H_0 + P_0VP_0$, the perturbation V is replaced by $V - P_0VP_0$, and Ψ_0 is replaced by Ψ_0', where Ψ_0' is the proper wave function for the level being examined and is slightly distorted when a perturbation acts (it is an eigenfunction of the $V\Omega_E$ operator in a given approximation), and P_0 is a projection operator into the subspace of eigenfunctions that correspond to the level E_0.

Every term in a Brillouin–Wigner series contains a dependence on the unknown exact energy E; therefore, the calculations require a self-consistency procedure and each level is solved individually. The Rayleigh–Schrödinger series is more convenient for multilevel configurations.

Let the model space contain more than one function. Similar to what we did for the Brillouin–Wigner series we introduce the projection operators P [Eq. (2.39)] and Q [Eq. (2.40)], and a wave operator Ω:

$$\Psi_\gamma = \Omega \Psi_\gamma^{(0)}, \quad \gamma = 1, 2, \ldots, d. \tag{2.56}$$

Using these operators transforms Schrödinger's equation, Eq. (2.38), to the equivalent, generalized Bloch operator equation [47]:

$$[\Omega, H_0]P = V\Omega P - \Omega P V \Omega P. \tag{2.57}$$

The square brackets signify the commutator of Ω with H_0. The operator P on the right-hand side shows that the equation is satisfied for functions in the model space. Starting from $\Omega^{(0)} = 1$ and using an iterative method to solve this equation gives us a series for Ω:

$$\Omega = 1 + \Omega^{(1)} + \Omega^{(2)} + \ldots \tag{2.58}$$

the nth term of the series is found from an equation that follows from Eq. (2.57):

$$[\Omega^{(n)}, H_0]P = QV\Omega^{(n-1)}P - \sum_{m=1}^{n-1} \Omega^{(n-m)}PV\Omega^{(m-1)}P. \tag{2.59}$$

Just as with a Brillouin–Wigner series, the energy of an atom is found from Eq. (2.54), but with a wave operator that is no longer dependent on E. This equation can easily be transformed by introducing an effective Hamiltonian that will yield the exact energy when it operates on the model function:

$$E_\Upsilon = \langle \Psi'^{(0)}_\Upsilon \mid H^{ef} \mid \Psi'^{(0)}_\Upsilon \rangle, \tag{2.60}$$

$$H^{ef} = PH\Omega P = PH_0P + PV\Omega P. \tag{2.61}$$

If there are nondiagonal elements in the matrix for the H^{ef}, defined for wave functions in the model space, appropriate model functions that satisfy Eqs. (2.56) and (2.60) are found by diagonalizing this matrix.

The use of graphical techniques makes calculating the terms of the wave operator or energy expansion much easier. We will first introduce this technique for representing the operators. When we do this, writing the operators in second quantization form, Eqs. (1.86) and (1.87), proves to be very convenient.

We operate on a determinant function having a one-electron state ξ (other states are not shown) with a V-type one-electron operator, Eq. (1.86):

$$V \mid \xi \rangle = \sum_{\zeta\vartheta} a^\dagger_\zeta a_\vartheta v_{\zeta\vartheta} \mid \xi \rangle = \sum_\zeta v_{\zeta\xi} \mid \zeta \rangle. \tag{2.62}$$

The operator a_ξ annihilates an electron in the ξ state and the $a_\zeta{}^\dagger$ operator creates an electron in the ζ state. As a result of an interaction that can be described by the operator V, the electron makes a transition from the ξ state into the ζ state. The quantity $v_{\zeta\xi} = \langle \zeta \mid v \mid \xi \rangle$ describes the probability of this transition, which is shown in graphical form in Fig. 2.1a.

If an interaction is described by a two-electron operator, Eq. (1.87), this interaction can change the states of two electrons (Fig. 2.1b).

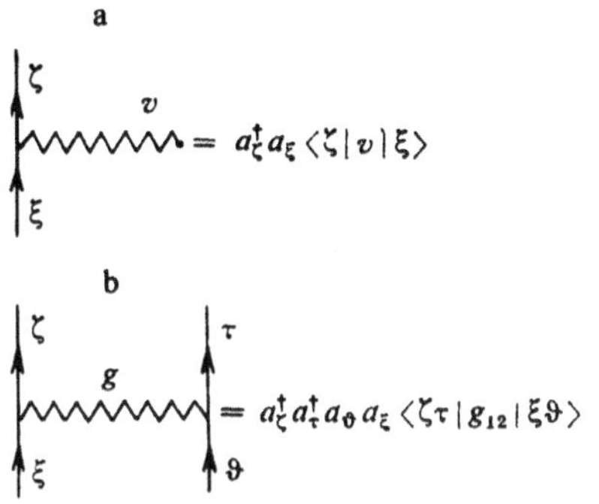

Fig. 2.1. Diagrams of one-electron (a) and two-electron (b) operators.

The action of an operator is interpreted in the language of excitations if the operator satisfies certain conditions. In order to formulate these conditions, we must introduce the concepts of core, virtual, and open states. Core states exist in all determinants of the model space, open states exist in only some determinants, and virtual states do not exist in this space. We will use the letters *a, b, c, d, e,*. . . to designate core states and the letters ξ, ζ, ϑ, ε, $\varkappa$, ν. . . to denote virtual and open states, leaving *i, j, k, l*. . . for nonspecific states.

We will use determinant functions in second quantization representation (the transition to functions of coupled angular momenta is made only at the final step of calculating the contributions for the graphs [47]). If the electron creation operator in a core state $a_{\text{core}}{}^\dagger$ or annihilation operator in a virtual state a_{virt} act on a determinant function from the model space, then, by definition

$$a^\dagger_{\text{core}}\, P = a_{\text{virt}}\, P = 0. \tag{2.63}$$

We can easily see that, for example, the operator

$$a_b^\dagger a_i = -a_i a_b^\dagger + \delta_{bi} \tag{2.64}$$

does not correspond to excitation, because the first term on the right-hand side goes to zero according to Eq. (2.63) and the second term, acting on the determinant, does not change it if $i = b$ or is zero if $i \neq b$. An operator that acts on a wave function $\Psi_\gamma{}^{(0)}$ creates one- or many-electron excitations (or yields zero) only when it is given in normal form: the $a_{\text{core}}{}^\dagger$ and a_{virt} operators stands on the right, respectively, of a_{core} and $a_{\text{virt}}{}^\dagger$. As for the a_{open} operators that act upon electrons in open shells, there is a possibility for selection; we will agree that they must be on the right of $a_{\text{open}}{}^\dagger$. The normal form of an operator is denoted by using braces around it, e.g.,

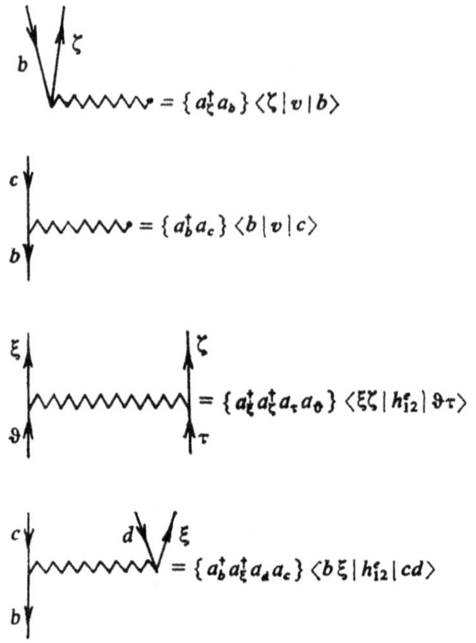

Fig. 2.2. Graphical representation of operators written in normal form.

$$\{ a_\xi^\dagger a_\zeta \} = a_\xi^\dagger a_\zeta, \quad \{ a_b^\dagger a_c \} = -a_c a_b^\dagger = -\{ a_c a_b^\dagger \}. \tag{2.65}$$

Transforming the operator into normal form and using Eq. (2.63) allows us to expand this operator into the scalar, the one-, and the two-electron, etc. parts.

We are now in a position to formulate the rules for representing the operators graphically. Creation and annihilation operators in a virtual state are shown by an arrow directed upward: for an a_ζ operator, upward to the vertex (the interaction point), for an $a_\zeta^\dagger$ operator, upward from the vertex. The creation or annihilation of an electron in a core state can respectively be thought of as the annihilation or creation of a vacancy, which we will denote by an oppositely directed arrow – downward: a_b downward to the vertex, and $a_b^\dagger$, upward from the vertex. We will agree to treat open states as particle states and designate them by arrows directed upward (if we need to distinguish them, we can use lines with two arrows).

Notice that these same diagrams are frequently used to represent only the amplitudes of the processes (the terms on the far right in Figs. 1 and 2); the solid line directed upward is an electron and the line directed downward is a vacancy in a state of an atomic system.

If the Hartree–Fock potential for a Coulomb interaction between electrons $\Sigma u(r_i)$ is included into the Hamiltonian for the nonrelativistic model Hamiltonian, then H can be separated into H_0 and V according to Eqs. (1.159) and (1.160). Putting the operators that make up V into second quantization form and reducing them to normal form, we can separate V into scalar, one-, and two-electron operators:

$$V = V_0 + V_1 + V_2. \tag{2.66}$$

Here,

$$V_0 = \sum_{b}^{\text{core}} \langle b | -u | b \rangle + \sum_{bc}^{\text{core}}{}' [\langle bc | h_{12}^e | bc \rangle - \langle cb | h_{12}^e | bc \rangle], \tag{2.67}$$

$$V_1 = \sum_{ij} \{a_i^\dagger a_j\} \langle i | v | j \rangle, \tag{2.68}$$

$$V_2 = \sum_{ijkl}{}' \{a_i^\dagger a_j^\dagger a_l a_k\} \langle ij | h_{12}^e | kl \rangle, \tag{2.69}$$

where

$$\langle i | v | j \rangle = \langle i | -u | j \rangle + \sum_{b}^{\text{core}} [\langle ib | h_{12}^e | jb \rangle - \langle bi | h_{12}^e | jb \rangle]. \tag{2.70}$$

The prime over the summation sign over the one-electron states means that only matrix elements that are not identically equal are considered (the $\langle ij \mid h_{12}{}^e \mid kl\rangle$ are distinguished from the $\langle ij \mid h_{12}{}^e \mid lk\rangle$, but are equal to the $\langle ji \mid h_{12}{}^e \mid lk\rangle$).

If $u(r)$ is a Hartree–Fock potential, both terms in Eq. (2.70) are compensated and the matrix element for the V_1 operator goes to zero, which corresponds to Brillouin's theorem for determinant functions.

Figure 3 shows topologically nonequivalent diagrams that represent V_0, V_1, and V_2 (generally speaking, diagrams that can be transformed into one another by a reflection in a vertical plane or a distortion which keeps all vertices on the same horizontal level are topologically equivalent and make the same contributions). The diagrams of the scalar operator V_0 do not contain outer (free) lines.

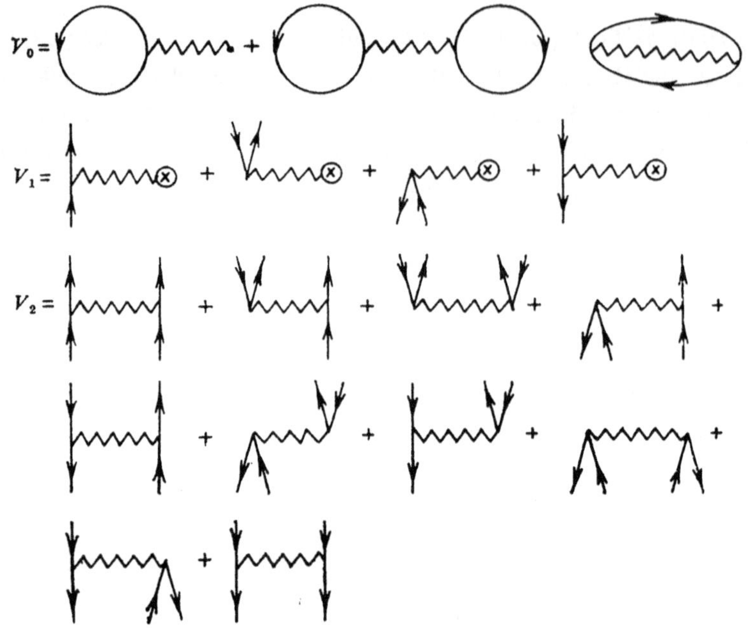

Fig. 2.3. Topologically nonequivalent diagrams of the scalar (V_0), the one-electron (V_1), and the two-electron (V_2) parts of the perturbation operator V, Eq. (2.66). The zigzag lines with a cross correspond to an interaction that can be described by the matrix element, Eq. (2.70).

The principal task of the perturbation technique is to find a wave operator Ω. In a second quantization representation, it too can be expanded into scalar ($\Omega_0 = 1$), one-electron (Ω_1), two-electron (Ω_2), etc. parts [47]:

$$\Omega = 1 + \Omega_1 + \Omega_2 + \ \ldots, \tag{2.71}$$

$$\Omega_1 = \sum_{ij} \{ a_i^\dagger a_j \} x_j^i, \tag{2.72}$$

$$\Omega_2 = \sum_{ijkl}{}' \{ a_i^\dagger a_j^\dagger a_l a_k \} x_{kl}^{ij}. \tag{2.73}$$

The coefficients x are found from the generalized Bloch equation, Eq. (2.57). The commutator on the left-hand side of Eq. (2.57) is reduced to normal form by using

$$\langle \Psi_\beta^{(0)} | [\Omega, \ H_0] | \Psi_\gamma^{(0)} \rangle = (E_\gamma^{(0)} - E_\beta^{(0)}) \langle \Psi_\beta^{(0)} | \Omega | \Psi_\gamma^{(0)} \rangle, \tag{2.74}$$

which is valid for any wave functions of the model Hamiltonian. Substituting Eqs. (2.71)–(2.73) into the right-hand side of Eq. (2.74) and considering the fact that the differences in the eigenvalues of the Hamiltonian H_0, Eq. (1.159), are equal to the differences of the one-electron energies ($E_\gamma^{(0)} - E_\beta^{(0)} = \varepsilon_j - \varepsilon_i$), we obtain

$$[\Omega, H_0] = \sum_{ij} \{a_i^\dagger a_j\}(\varepsilon_j - \varepsilon_i)\, x_j^i + \\ + \sum_{ijkl}{}' \{a_i^\dagger a_j^\dagger a_l a_k\}(\varepsilon_k + \varepsilon_l - \varepsilon_i - \varepsilon_j)\, x_{kl}^{ij} + \dots \tag{2.75}$$

The product of the operators on the right-hand side of Bloch's equation is graphically reduced to normal form [47]. Diagrams corresponding to the product of UT operators are obtained by mapping the diagram of U onto the diagram of T and combining the 0, 1, 2, 3, . . . outer lines found at the bottom of the diagram of U by all possible means with identically directed outer lines in the upper part of the diagram of T (Fig. 2.4).

As an example of using the graphical technique, we will derive an expression for the first-order wave operator $\Omega^{(1)}$. According to Eq. (2.59), this operator is defined by

$$[\Omega^{(1)}, H_0]\, P = QVP. \tag{2.76}$$

The operator V, Eq. (2.66), consists of scalar, one- and two-electron terms. The scalar term makes no contribution to Eq. (2.76), since there are projection operators into different spaces on the right- and left-hand sides of V. Taking this fact, as well as Eqs. (2.74) and (2.76) into account, we see that $\Omega^{(1)}$ also has one- and two-electron components. For closed shells, these components in normal form may contain only electron annihilation operators in the core state, and only electron creation operators in the virtual state:

$$\Omega^{(1)} = \Omega_1^{(1)} + \Omega_2^{(1)} = \sum_{b\xi} a_\xi^\dagger a_b\, x_b^{\xi\,(1)} + \sum_{bc\zeta\xi}{}' a_\xi^\dagger a_\zeta^\dagger a_c a_b\, x_{bc}^{\xi\zeta\,(1)}. \tag{2.77}$$

Figure 2.5 shows both terms of the $\Omega^{(1)}$ operator.

Using Eqs. (2.75) and (2.77) to express the commutator contained in Eq. (2.76) and substituting the expansion for V given by Eqs. (2.66)–(2.69) into the left-hand side of this equation we can transform this equation to the form (the operators Q and P merely give concrete form to the indices of the $a_i^\dagger$ and a_j in V):

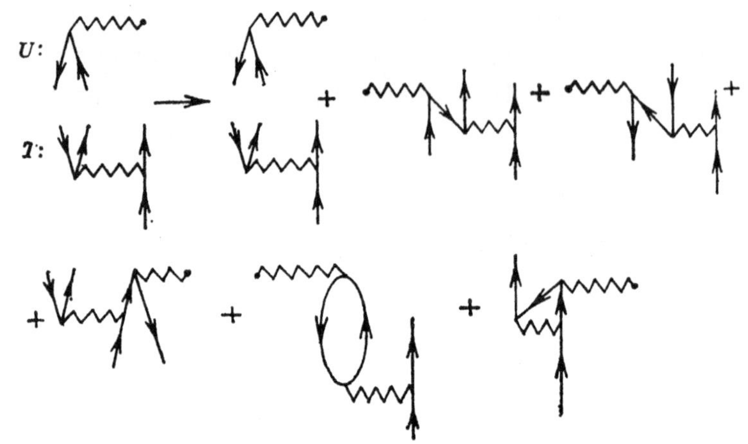

Fig. 2.4. Graphical derivation of the diagrams of an operator which is the product of the operators U and T.

$$\sum_{b\xi} a_\xi^\dagger a_b(\varepsilon_b-\varepsilon_\xi)x_b^{\xi\,(1)}+\sum_{bc\xi\zeta}{}' a_\xi^\dagger a_\zeta^\dagger a_c a_b(\varepsilon_b+\varepsilon_c-\varepsilon_\xi-\varepsilon_\zeta)x_{bc}^{\xi\zeta\,(1)}=$$

$$=\sum_{b\xi} a_\xi^\dagger a_b\langle\xi|v|b\rangle+\sum_{bc\xi\zeta}{}' a_\xi^\dagger a_\zeta^\dagger a_c a_b\langle\xi\zeta|h_{12}^e|bc\rangle. \quad (2.78)$$

Equating terms that contain one- and two-electron operators, we find the unknown coefficients x in the expansion for the operator $\Omega^{(1)}$ (2.77):

$$\Omega^{(1)}=\sum_{b\xi} a_\xi^\dagger a_b\frac{\langle\xi|v|b\rangle}{\varepsilon_b-\varepsilon_\xi}+\sum_{bc\xi\zeta}{}' a_\xi^\dagger a_\zeta^\dagger a_c a_b\frac{\langle\xi\zeta|h_{12}^e|bc\rangle}{\varepsilon_b+\varepsilon_c-\varepsilon_\xi-\varepsilon_\zeta}. \quad (2.79)$$

In keeping with Fig. 2.5, an auxiliary energy denominator is assigned to the interaction line on the diagrams that represent the wave operator (for a more complex diagram containing several interaction lines, the multiplier in the denominator)

$$D=\Sigma\varepsilon_\downarrow-\Sigma\varepsilon_\uparrow. \quad (2.80)$$

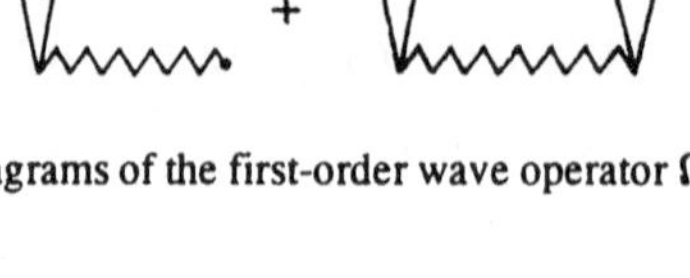

Fig. 2.5. Diagrams of the first-order wave operator $\Omega^{(1)}$ for closed shells.

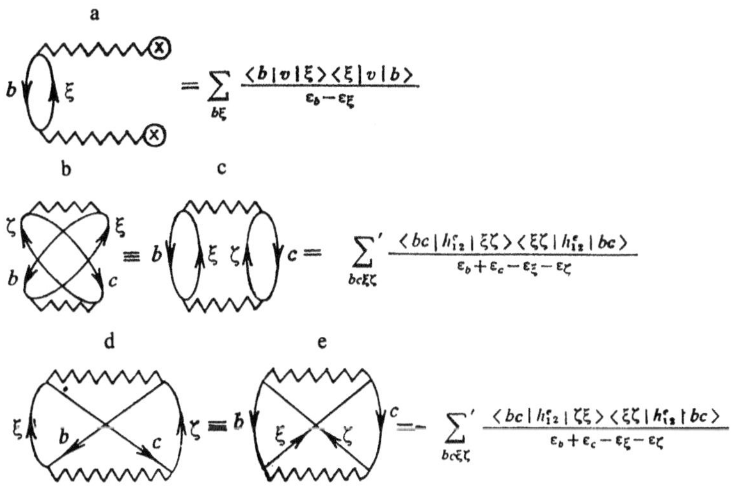

Fig. 2.6. Second-order energy $E^{(2)}$ diagrams and their contributions for an atom with closed shells; the zigzag line with the cross has the same meaning as in Fig. 2.3.

Here $\varepsilon_{\downarrow}$ and $\varepsilon_{\uparrow}$ are one-electron energies that correspond to the downward or upward lines that are intersected by the horizontal line above the respective interaction line.

The operators V and $\Omega^{(1)}$ allow us to construct a first- and second-order effective operator, Eq. (2.61), and thereby find the energy corrections $E^{(1)}$ and $E^{(2)}$:

$$E_{\gamma}^{(1)} = \langle \Psi_{\gamma}^{(0)} \mid H^{ef\,(1)} \mid \Psi_{\gamma}^{(0)} \rangle = \langle \Psi_{\gamma}^{(0)} \mid V \mid \Psi_{\gamma}^{(0)} \rangle = \langle \Psi_{\gamma}^{(0)} \mid V_0 \mid \Psi_{\gamma}^{(0)} \rangle = V_0 . \quad (2.81)$$

Because the operators V_1 and V_2 are written in normal form, their matrix elements go to zero according to Eq. (2.63) and $E_{\gamma}^{(1)}$ is equal to the constant V_0. Thus, the diagrams of V_0 (Fig. 2.3) are also diagrams for the energy $E^{(1)}$.

When the series for Ω and the energy $E = \Sigma E^{(k)}$ are substituted into Eqs. (2.60) and (2.61) and terms of the same order are set equal, we obtain the expression for $E_{\gamma}^{(k)}$ [47]:

$$E_{\gamma}^{(k)} = \langle \Psi_{\gamma}^{(0)} \mid (V_1 + V_2)\,\Omega^{(k-1)} \mid \Psi_{\gamma}^{(0)} \rangle, \quad k \geqslant 2. \quad (2.82)$$

When $E^{(2)}$ is calculated for an atom with closed shells, the diagrams for V_1 and V_2 (Fig. 2.3) are combined with the diagrams of $\Omega^{(1)}$ (Fig. 2.5), and only those having no outer lines remain. The product of $\Omega_1^{(1)}$ and V_1 yields only one diagram (Fig. 2.6a), but the product of $\Omega_2^{(1)}$ and V_2 yields four diagrams (Fig. 2.6b–e). Algebraic expressions corresponding to the different diagrams are obtained by substituting the expression for

V_1, Eq. (2.68), V_2, Eq. (2.69), and $\Omega^{(1)}$, Eq. (2.79), into Eq. (2.82) and transforming them via the rules of anticommutation for the operators $a^\dagger$ and a. The diagrams of Fig. 2.6b and 6d are topologically equivalent to the diagrams of Fig. 2.6c and 6e, respectively; therefore, we can omit the diagrams of Fig. 2.6c and 6e. Notice for energy diagrams that the quantity D, Eq. (2.80), in the denominator is associated with every interaction line but an upper, or, in other words, with every intermediate virtual state in the system.

Whereas the definition of the energy denominator differs for diagrams of different quantities, the other rules for evaluating them, discussed in this book, are general in nature and can be used for every diagram.

The matrix element of the operator $h_{12}{}^e$ (for nonantisymmetrized wave functions) corresponds to each zigzag interaction line and it is written from right to left according to how the the lines are arranged from top to bottom in the diagram. The states of a matrix element are written for the diagram from left to right:

$$\langle \text{outgoing lines} \mid h^e{}_{12} \mid \text{incoming lines} \rangle$$

The contributions of a diagram are summed over the quantum numbers of all inner lines (summation over virtual states also includes an integration over the continuous spectrum). The matrix elements $\langle ij \mid h_{12} \mid kl \rangle$ and $\langle ji \mid h_{12} \mid lk \rangle$ that are equal to one another are considered in the sums only once (denoted by a prime on the summation sign). The contribution from a diagram is multiplied by the phase factor $(-1)^{h+l}$, where h is the number of inner lines directed downward and l is the number of closed partial-hole loops in the diagram.

In some cases the energy denominator in the expression for the diagram can be zero. By way of example, we will consider diagram c in Fig. 2.6. If the ξ and ζ states correspond to exciting an electron into inner open or unfilled shells as well as into the continuum, then $\varepsilon_b + \varepsilon_c$ are always less than $\varepsilon_\xi + \varepsilon_\zeta$. However, when there is an inner vacancy present in the configuration one electron may be excited downward and the other electron excited into a continuous spectrum state then an anomaly appears for some value of its energy – the energy denominator D becomes zero. This corresponds to the possibility of an actual Auger transition. Then a small imaginary quantity $i\eta$ ($\eta > 0$) must be added to D and the contribution from the diagram is defined as

$$E^{(2)}(b, c, \xi) = \lim_{\eta \to 0} \sum_{\zeta} \frac{|\langle \xi\zeta | h^e_{12} | bc \rangle|^2}{D + i\eta} . \tag{2.83}$$

The limit is found from Dirac's formula [54]:

$$\lim_{\eta \to 0} (D + i\eta)^{-1} = P\left(\frac{1}{D}\right) - i\pi\delta(D), \tag{2.84}$$

where P means that the principal value of the integral is used when integrating over the continuum:

$$P\int_0^\infty \frac{f(\varepsilon)}{\varepsilon' - \varepsilon} d\varepsilon = \lim_{\Delta \to 0} \left[\int_0^{\varepsilon' - \Delta} + \int_{\varepsilon' + \Delta}^\infty \right] \frac{f(\varepsilon)}{\varepsilon' - \varepsilon} d\varepsilon, \tag{2.85}$$

and $\delta(D)$ is Dirac's delta function.

Applying Eq. (2.84) to (2.83) yields

$$E^2(b, c, \xi) = P\sum_{\zeta} \frac{|\langle \xi\zeta | h_{12}^e | bc \rangle|^2}{D} - i\pi \sum_{lm\mu} |\langle \xi\, \varepsilon lm\mu | h_{12}^e | bc \rangle|^2, \tag{2.86}$$

$$\varepsilon = \varepsilon_b + \varepsilon_c - \varepsilon_\xi > 0, \quad lm\mu \in \zeta.$$

The contribution from the exchange diagram (Fig. 2.6e) can be found in a similar manner. Their sum is

$$E^{(2)}(b, c, \xi) = \mathscr{E}^{(2)}(b, c, \xi) - \frac{i\Gamma^{(1)}(bc \to \xi\varepsilon)}{2}, \tag{2.87}$$

where $\mathscr{E}^{(2)}$ is a real second-order perturbation theory correction to the energy level; $\Gamma^{(1)}$ is the partial Auger width of the level and corresponds to the Auger transition $bc \to \xi\varepsilon(\xi^{-1} \to b^{-1}c^{-1}\varepsilon)$; dividing Γ by $\hbar$ gives the probability of this transition in first-order perturbation theory. The imaginary components of the contributions from higher-order energy diagrams give correlational corrections to the probabilities of Auger transitions [128]. Thus, we see that calculating the energy diagrams yields not only the energy levels, but the probabilities of Auger transitions.

Stationary perturbation theory can also be used to calculate the probabilities that an atom will absorb or emit a photon. Because the operator for interaction between an atom and an electromagnetic field contains a small parameter α, the process amplitude can only be found in a first-order approximation when calculated by methods from perturbation theory, and only the wave functions for the initial and final states of the atom can be refined. The process amplitude for an exact wave function is given by the matrix element of the effective operator for the model functions via the wave operator, Eq. (2.56):

$$M_{21} = \langle \Psi_2 | O^{(t)} | \Psi_1 \rangle = \langle \Psi_2'^{(0)} | O^{ef}\, \Psi_1'^{(0)} \rangle, \tag{2.88}$$

$$O^{ef} = \Omega_2^{\dagger} O^{(t)} \Omega_1 . \tag{2.89}$$

Here, $O^{(t)}$ is the transition operator, and Ω_1 and Ω_2 are the wave operators for the initial and final states, respectively. When intermediate normalization is used the functions Ψ_i that are obtained are only approximately normalized, which results in the wave and effective operators being only approximately Hermitian ($\Omega^{\dagger} \approx \Omega$). This can be made more accurate by using special normalized diagrams [128]. Diagrams for the effective operator and the transition amplitude are obtained from the diagrams of Ω and $O^{(t)}$ (the latter is a straight line having an interaction point) in a similar manner for the effective operator. The energy denominator D is defined by the rule, Eq. (2.80), only for the zigzag interaction lines below the straight line. If an interaction line in the diagram is higher, the energy of the absorbed photon, $\hbar\omega$, must be added to Eq. (2.80). Amplitude diagrams for a one-electron transition must contain two outer lines that describe how an electron disappears in the initial state and appears in the final state. By way of example, a photoabsorption amplitude diagram that takes the first-order correction to the wave function of the initial state into account is shown in Fig. 2.7. Amplitude diagrams for radiative transitions will be examined in more detail in Chapters 5, 6, 7, and 10.

Note the opportunity for a descriptive interpretation of the diagrams: they can be thought of as the distribution of excitations in an atom caused by interactions between electrons from the initial state through the intermediate virtual states to the final state (from right to left in the mathematical expressions and from bottom to top in the diagrams, in which the vacancy lines run in opposite directions to the electron lines). For example, the diagram of Fig. 2.7 is explained thus: a Coulomb interaction between electrons in an atom creates two virtual "electron–vacancy" pairs: vacancies in the b and c states, and electrons in the ζ and $\varkappa$ states. One electron absorbs a photon $\hbar\omega$ and fills a vacancy, and the other electron remains excited; thus, in the final state there is a vacancy b^{-1} and an excited electron in a discrete or continuous spectrum ζ state.

A similar interpretation emerges from nonstationary perturbation theory, but, because the same diagrams are used to map this and the other series, it is often carried over into stationary theory, not only for transition amplitude diagrams, but for energy diagrams (here the electrons return to their initial states after several virtual excitations).

A Brillouin–Wigner series is also explained by the same diagrams, and their evaluation differs only in how the energy denominator is defined (Sec. 5.3). An important

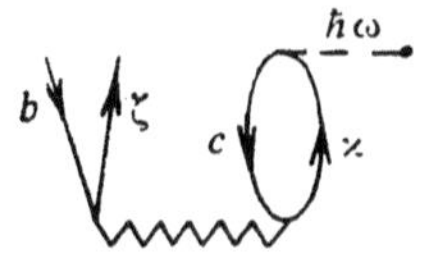

$$= \sum_{bc\xi\zeta}{}' \frac{\langle c|o|\varkappa\rangle\langle\zeta\varkappa|h_{12}^{e}|bc\rangle}{\varepsilon_b+\varepsilon_c-\varepsilon_\varkappa-\varepsilon_\zeta}$$

Fig. 2.7. Diagram showing the effective operator or photoabsorption amplitude in lowest-order perturbation theory, and the contribution of the diagram.

property of the Rayleigh – Schrödinger series must be noted – diagrams having unlinked parts without outer core or virtual lines can be omitted because the contributions from these diagrams compensate one another. This is not true for a Brillouin – Wigner series.

Feynman diagrams in the basis of determinant functions are topologically equivalent to Jucys – Levinson – Vanagas diagrams that describe the addition of the angular momenta. This allows us to formulate the rules for how a transition is made from quantities in a determinant basis to the corresponding quantities that have been defined in a coupled-angular momenta function basis by multiplying the contributions of the diagrams by the respective spin-angular momentum coefficients [47].

2.4. Interaction between a Discrete State and a Continuum

The expansion of the wave function in the basis of one-configuration functions, Eq. (2.8), contains not only a summation over the discrete spectrum, but an integration over the continuous spectrum or continuum. The effect of the continuum is frequently neglected for a ground configuration and weakly excited configurations. This effect does, however, become significant for the autoionizing levels that are found in the continuum with respect to another ionization limit and which correspond to the atom's quasistationary states. For example, the $1s2s^22p^6\ {}^2S$ level belongs to the discrete spectrum of this configuration, but is found in the $1s^22s^02p^6\varepsilon l\ {}^2S$, the $1s^22s2p^5\varepsilon l\ {}^2S$, and the $1s^22s^22p^4\varepsilon l\ {}^2S$ continua.

Autoionizing levels corresponding to configurations with two excited outer electrons are well known, e.g., the He$2s2p\ {}^1P$ that are found in the $1s\varepsilon p\ {}^1P$ continuum. The levels of a configuration having an inner vacancy and some levels obtained when a single outer electron is excited and the many-electron quantum numbers of an open shell are changed at the same time [e.g., the $p^3({}^2D)ns\ {}^3D$ levels may be found in the continuum for $p^3({}^4S)\varepsilon d\ {}^3D$] are also autoionizing levels.

Mixing of discrete and continuous states having the same energy leads to the possibility that the excited state will decay in a radiationless process. Such a process is called autoionization, or – when a state of configuration having an inner vacancy decays – an Auger transition.

The interaction between discrete and continuous spectrum states is determined by the magnitude of the interconfiguration matrix element of the system Hamiltonian. Consequently, autoionization is only possible when parity and total angular momentum are conserved. Because discrete and continuous spectrum configurations differ only by the states of two electrons or an electron and a core, only the two-electron part of the Hamiltonian contributes to the interconfiguration matrix element, mainly, the electrostatic interaction operator between two electrons.

We will first examine an isolated discrete state that interacts with a single continuum [130].

In a zeroth one-configuration approximation, let φ be the wave function for the discrete state of an atom, and $\psi_{\mathscr{E}}$ be the wave function for the continuous spectrum having energy $\mathscr{E}$ (one or more electrons are replaced by free electrons). In an ion + free electron continuum $\mathscr{E} = E_0 + \varepsilon$, where E_0 is the ion's energy, and ε is the energy of the free electron. We denote the energy matrix elements by

$$\langle\varphi|H|\varphi\rangle = E_{\varphi}, \quad \langle\psi_{\mathscr{E}'}|H|\psi_{\mathscr{E}''}\rangle = \mathscr{E}'\delta(\mathscr{E}' - \mathscr{E}''), \tag{2.90}$$

$$\langle\psi_{\mathscr{E}}|H|\varphi\rangle = V_{\mathscr{E}}. \tag{2.91}$$

To simplify future expressions, we will say that the part of the energy matrix between the continuous spectrum states is first diagonalized (techniques for diagonalizing it are discussed, for example, in [131]).

The wave function of an autoionization state must be the superposition of functions from the discrete and continuous spectra:

$$\Psi_{\mathscr{E}} = a\varphi + \int b_{\mathscr{E}'}\psi_{\mathscr{E}'}\,d\mathscr{E}', \tag{2.92}$$

where a and $b_{\mathscr{E}'}$ are the coefficients of expansion (they also depend on $\mathscr{E}$, but for the sake of brevity we will omit their designation).

We require that $\Psi_{\mathscr{E}}$ be an eigenfunction of H:

$$H\Psi_{\mathscr{E}} = \mathscr{E}\Psi_{\mathscr{E}}. \tag{2.93}$$

Substituting Eq. (2.92), multiplying by $\psi_{\mathscr{E}''}$ and then φ, and integrating over all points (φ and $\psi_{\mathscr{E}}$ are assumed to be orthogonal to one another) gives us the following system of equations for finding the coefficients a and $b_{\mathscr{E}'}$:

$$\begin{cases} V_{\mathscr{E}'}a+b_{\mathscr{E}'}\mathscr{E}'=\mathscr{E}b_{\mathscr{E}'}, & (2.94) \\ E_{\varphi}a+\int V^{*}_{\mathscr{E}'}b_{\mathscr{E}'}d\mathscr{E}'=\mathscr{E}a. & (2.95) \end{cases}$$

When $\mathscr{E}\neq\mathscr{E}'$, $b_{\mathscr{E}'}$ is given by the first equation

$$b_{\mathscr{E}'}=\frac{V_{\mathscr{E}'}a}{\mathscr{E}-\mathscr{E}'}. \tag{2.96}$$

With due regard for the singularity, the coefficient $b_{\mathscr{E}'}$ is sought as

$$b_{\mathscr{E}'}=\left[P\,\frac{1}{\mathscr{E}-\mathscr{E}'}+\pi\,\varkappa(\mathscr{E})\,\delta(\mathscr{E}-\mathscr{E}')\right]V_{\mathscr{E}'}a, \tag{2.97}$$

where P means that the principal value of the integral, Eq. (2.85), is taken when integrating this term over $\mathscr{E}'$. The unknown real function $\varkappa(\mathscr{E})$ is found from Eq. (2.95):

$$\varkappa(\mathscr{E})=\frac{\mathscr{E}-E_{\varphi}-F(\mathscr{E})}{\pi\,|V_{\mathscr{E}}|^{2}}, \tag{2.98}$$

and the coefficient a is found from the normalization condition for the function $\Psi_{\mathscr{E}}$[130]:

$$|a_{\mathscr{E}}|^{2}=\frac{|V_{\mathscr{E}}|^{2}}{[\mathscr{E}-E_{\varphi}-F(\mathscr{E})]^{2}+\pi^{2}\,|V_{\mathscr{E}}|^{4}}. \tag{2.99}$$

The notation

$$F(\mathscr{E})=P\int\frac{|V_{\mathscr{E}'}|^{2}}{\mathscr{E}-\mathscr{E}'}\,d\mathscr{E}'. \tag{2.100}$$

is used in Eqs. (2.98) and (2.99).

Choosing a positive a and substituting a and $b_{\mathscr{E}'}$ into Eq. (2.92) yields an expression for the wave function of an autoionizing state:

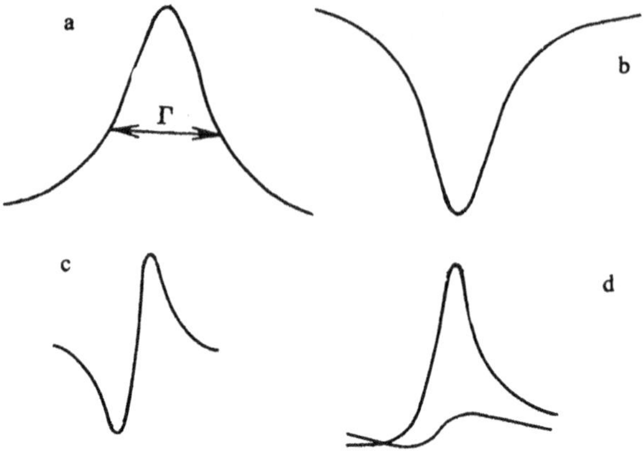

Fig. 2.8. Autoionization resonances of different shapes are possible in an atom's ionization cross section near an autoionizing level.

$$\Psi_{\mathscr{E}} = a\left\{ \varphi + \frac{1}{V^*_{\mathscr{E}}} [\mathscr{E} - E_\varphi - F(\mathscr{E})]\psi_{\mathscr{E}} \right\} + aP \int \frac{V_{\mathscr{E}'}\psi_{\mathscr{E}'}}{\mathscr{E} - \mathscr{E}'} d\mathscr{E}'. \quad (2.101)$$

The coefficient a in Eq. (2.92) in the discrete state function is equal to unity in a zeroth approximation and, when an interaction with a continuum is included, is "diluted" and becomes a function of $\mathscr{E}$. The probability of finding a level $|a_{\mathscr{E}}|^2$ reaches a maximum at an energy of $\mathscr{E} = E_r$:

$$E_r = E_\varphi + F(E_r), \quad (2.102)$$

called the resonance energy. Thus, interaction with a continuum not only expands the level, but shifts it by an amount F. If E_r is distant from an ionization edge, $V_{\mathscr{E}}$ and F are weakly dependent upon $\mathscr{E}$, and the contributions to F from $\mathscr{E}' > E_r$ and $\mathscr{E}' < E_r$, which have opposite signs, almost completely compensate one another.

If $V_{\mathscr{E}}$ and F in the vicinity of a resonance can be considered constants, then $|a_{\mathscr{E}}|^2$, Eq. (2.99), is mapped by a symmetric Lorentz function (Fig. 2.8a) whose width at the half height is

$$\Gamma = 2\pi V_{\mathscr{E}}^2 = 2\pi |\langle \psi_{\mathscr{E}} | H | \varphi \rangle|^2. \quad (2.103)$$

The equation for the probability of radiationless decay of an autoionization state (per unit time) follows from the relationships between the level width and the total probability of decay of an excited state (9.6):

$$W = \frac{2\pi}{\hbar} \,|\langle \psi_{\mathscr{E}} | H | \varphi \rangle|^2. \tag{2.104}$$

It was assumed when the initial equations, Eqs. (2.94) and (2.95), were derived that the functions φ and $\psi_{\mathscr{E}}$ were orthogonal to one another. If this is not the case, a more general expression for W follows:

$$W = \frac{2\pi}{\hbar} \,|\langle \psi_{\mathscr{E}} | H - \mathscr{E} | \varphi \rangle|^2. \tag{2.105}$$

The assumption that the quantities $F(\mathscr{E})$ and $V_{\mathscr{E}}$ are constant in the vicinity of resonance becomes invalid when there is a strong interaction with a continuum – primarily for $F(\mathscr{E})$. Equation (2.99) can then be transformed by expanding the function

$$f(\mathscr{E}) = \mathscr{E} - E_{\varphi} - F(\mathscr{E}) \tag{2.106}$$

in a Taylor series in the neighborhood of the resonance energy E_r and keeping the first nondisappearing term in the series:

$$f(\mathscr{E}) \approx \frac{\partial f(\mathscr{E})}{\partial \mathscr{E}} \bigg|_{\mathscr{E}=E_r} (\mathscr{E} - E_r) = [z(E_r)]^{-1} (\mathscr{E} - E_r). \tag{2.107}$$

Here z is the so-called spectroscopic factor

$$z \equiv z(E_r) = \left[1 - \frac{\partial}{\partial \mathscr{E}} F(\mathscr{E}) \bigg|_{\mathscr{E}=E_r} \right]^{-1}. \tag{2.108}$$

Substituting Eq. (2.107) into (2.99) yields

$$|a_{\mathscr{E}}|^2 = \frac{z^2 |V_{\mathscr{E}}|^2}{(\mathscr{E} - E_r)^2 + (\pi z |V_{\mathscr{E}}|^2)^2} \tag{2.109}$$

and, as a result,

$$W = \frac{2\pi}{\hbar} \, z \,|\langle \psi_{\mathscr{E}} | H | \varphi \rangle|^2. \tag{2.110}$$

We will examine the excitation of an atom from a discrete spectrum state γ into an autoionizing state and the interference effects that arise when this happens. Using the expression for $\psi_{\mathscr{E}}$, Eq. (2.101), we can write the amplitude of a transition, which can be described by the operator O, in the form

$$\langle \Psi_{\mathscr{E}} | O | \Psi_\gamma \rangle = a^* \langle \Phi_{\mathscr{E}} O | \Psi_\gamma \rangle + a^* \pi V^*_{\mathscr{E}} \varkappa \langle \psi_{\mathscr{E}} | O | \Psi_\gamma \rangle =$$
$$= \frac{1}{\pi V_{\mathscr{E}} (1+\varkappa^2)^{1/2}} \langle \Phi_{\mathscr{E}} O \Psi_\gamma \rangle + \frac{\varkappa}{(1+\varkappa^2)^{1/2}} \langle \psi_{\mathscr{E}} | O | \Psi_\gamma \rangle, \qquad (2.111)$$

where the reduced energy $\varkappa$ is determined according to Eq. (2.98), and $\Phi_{\mathscr{E}}$ is a modified φ function (with due regard for the mixing of continuum wave functions):

$$\Phi_{\mathscr{E}} = \varphi + P \int \frac{V_{\mathscr{E}'} \psi_{\mathscr{E}'}}{\mathscr{E} - \mathscr{E}'} d\mathscr{E}'. \qquad (2.112)$$

If excitation is possible into the φ state only and the matrix elements of $\langle \psi \ | O | \ \Psi_\gamma \rangle$ are zero, an excitation line is a transition line between two discrete spectrum levels that has been shifted and expanded by an interaction with a continuum (Fig. 2.8a).

If the matrix element $\langle \varphi O \ | \ \Psi_\gamma \rangle$ is zero the cross section for excitation into a continuum (the ionization cross section) has a gap at an energy equal to the energy of the autoionizing level (Fig. 2.8b).

In general, the square of the transition amplitude contains an interference term in addition to the terms corresponding to an excitation into the φ state and a continuum. The quantity $\varkappa(\mathscr{E})$ becomes zero when $\mathscr{E} = E_r$ and changes sign when $\mathscr{E}$ goes beyond this value. Because the second term in Eq. (2.111) contains $\varkappa$ to the first degree and the first term contains its square, the interference from one side is destructive and constructive from the other side (Fig. 2.8c, d).

We will introduce the Fano parameter

$$q = \frac{\langle \Phi_{\mathscr{E}} | O | \Psi_\gamma \rangle}{\pi V^*_{\mathscr{E}} \langle \psi_{\mathscr{E}} | O | \Psi_\gamma \rangle}. \qquad (2.113)$$

The quantity $\pi q^2/2$ is approximately equal to the ratio of the excitation cross section to the ionization cross section in the Γ wide band of an unperturbed continuum.

According to Eqs. (2.111) and (2.113) the cross section near an autoionizing level can be factorized:

$$|\langle \Psi_{\mathscr{E}} | O | \Psi_\gamma \rangle|^2 = |\langle \psi_{\mathscr{E}} | O | \Psi_\gamma \rangle|^2 \frac{(q+\varkappa)^2}{1+\varkappa^2}. \qquad (2.114)$$

The first term on the right-hand side is a background cross section, which is weakly dependent upon $\mathscr{E}$, and the second term describes the resonance change in the cross section near the autoionizing level. The quantities q and $\varkappa$ can quite often be considered independent of $\mathscr{E}$.

If the interaction between a discrete state and a continuum is weak, then when $V_{\mathscr{E}}$ is small the q parameter is fairly large, and the resonance is described by a symmetric Lorentz curve. When q is on the order of a few units, a vivid asymmetry is characteristic of resonance [130].

Equation (2.114) is widely used for not only theoretical, but also semiempirical, descriptions of isolated autoionization resonances in the cross sections of photoionization or inelastic scattering of electrons by atoms. This formula was generalized for resonances in the angular and energy distributions of electrons emitted by atoms in [132, 133].

The case of a single, isolated discrete state interacting with several continua $\beta_{\mathscr{E}}$ comes down to the situation examined earlier – one discrete state + one continuum [131, 130]. By using a unitary transformation of the $\psi_{\beta\mathscr{E}}$ functions, we can introduce modified continua, only one of which interacts with the discrete state being examined. The wave function for this interacting continuum is given in terms of $\psi_{\beta\mathscr{E}}$ as follows [131]:

$$\psi'_{1\mathscr{E}} = \frac{\sum\limits_{\beta} V_{\beta\mathscr{E}} \psi_{\beta\mathscr{E}}}{\left[\sum\limits_{\beta} |V_{\beta\mathscr{E}}|^2\right]^{1/2}}, \quad V_{\beta\mathscr{E}} = \langle \psi_{\beta\mathscr{E}} | H | \varphi \rangle . \tag{2.115}$$

The use of $\psi_{1\mathscr{E}}'$ instead of $\psi_{\mathscr{E}}$ in Eqs. (2.91) and (2.92) and subsequent expressions extends these equations to the case of many continua. In particular, the total probability of a radiationless decay of a discrete state is found to be equal to the sum of the transition probabilities into the different continua:

$$W = \frac{2\pi}{\hbar} |\langle \psi'_{1\mathscr{E}} | H | \varphi \rangle|^2 = \frac{2\pi}{\hbar} \sum_{\beta} |\langle \psi_{\beta\mathscr{E}} | H | \varphi \rangle|^2 . \tag{2.116}$$

Because a resonance change in the cross section near an autoionizing level occurs in only one modified channel when an atom is ionized and transitions into other noninteracting continuums are unaffected by a discrete state, the spectrum will be the superposition of the resonance function onto a monotonically varying background.

In order to describe overlapping resonances, we must use a more general theory for many discrete levels interacting with one [130] or several [130, 134, 135] continua. This is a rather complex mathematical problem that has thus far been solved in practice for only a few of the simplest cases. The levels in atoms having inner vacancies of the same symmetry are usually distant from one another; in addition, the number of levels that are important to the excitation process is reduced because of the existence of an effective potential barrier (Sec. 3.3). For this reason, a model of an isolated resonance that interacts with many continua is used as a rule to interpret x-ray absorption spectra and Auger spectra.

The description of the autoionizing levels can be generalized by considering not only the radiationless, but also the radiation, channels for their decay [135–137]. For this atom must be considered together with the electromagnetic field and the Hamiltonian as well as the wave functions for the entire system being used. The radiation width is then added to the autoionization width and their sum is the total natural width.

Chapter 3

THE STRUCTURE OF ATOMIC ELECTRON SHELLS. CONFIGURATIONS WITH VACANCIES

Configurations with vacancies as well as normal atomic configurations take part in producing x-ray and electron spectra. In this chapter we will devote our attention to configurations with internal vacancies because the normal configurations have been more thoroughly investigated and are simpler.

Configurations having vacancies in the inner shells exhibit several peculiarities: their states are short-lived autoionizing states, relativistic effects are significant for them, and their energy spectrum has a specific (for the case of a single vacancy – quasiparticle) nature. We have examined a few of these topics in previous chapters (a theoretical description of autoionizing levels, a consideration of relativistic effects). The short lifetime of highly excited atomic states and its consequences (the large width of the spectral lines, the connection between the processes of excitation and decay of the states) will be discussed in Chapters 5, 7, 8, and 9. We will examine the symmetry that exists between vacancies and electrons, the quasiparticle properties of a vacancy, and how vacancies affect an atom's characteristics in Sec. 3.1. The basic types of coupling (Sec. 3.2), the correlations between electrons in different configurations (Sec. 3.4), and the collapse of an excited electron and the anomalies that this causes in atomic properties (Sec. 3.3) will be examined.

3.1. The Symmetry between Vacancies and Electrons. A Vacancy as a Quasiparticle

In conjunction with the notations for shells and subshells that have been used in the preceding chapters, the Bohr—Coster symbols are often used. These are the letters *K, L,*

$M, N, O, P, Q, \ldots$ which designate the layers having principal quantum numbers $n = 1, 2, 3, 4, 5, 6, \ldots$, and the subshells are numbered in order of increasing l, and for a given l — the quantum number j:

$1s_{1/2}$	$2s_{1/2}$	$2p_{1/2}$	$2p_{3/2}$	$3s_{1/2}$	$3p_{1/2}$	$3p_{3/2}$	$3d_{3/2}$	$3d_{5/2}$
K	L_1	L_2	L_3	M_1	M_2	M_3	M_4	M_5

$4s_{1/2}$	$4p_{1/2}$	$4p_{3/2}$	$4d_{3/2}$	$4d_{5/2}$
N_1	N_2	N_3	N_4	N_5

The same subscripts are used to denote the states of an atom having vacancies in a given subshell. For excited configurations with open shells, it is easier to indicate the distribution of the electrons in the shells (usually only in the open shells) or only the vacancies in them ($nl^{-N} \equiv nl^{4l+2-N}$).

Electrons can be replaced by vacancies in not only the designation of a configurations, but it has a more profound significance. Second quantization representation is especially convenient for examining the symmetry that exists between electrons and vacancies. For an atom (or core) with closed shells, an electron annihilation operator is at the same time a vacancy creation operator, and an electron creation operator is a vacancy annihilation operator. This feature allows us to make a complete transition from describing electrons to describing vacancies [47, 49, 53]. An atom's ground state is then the vacancy's vacuum state, etc.

This symmetry is destroyed by an interaction between electrons; when this interaction is taken into account, a vacancy must be thought of as a quasiparticle that moves not in the nuclear field, but in the atom's effective field. However, in the one-configuration approximation of a spherically symmetric field, the symmetry between electrons and vacancies is only destroyed in radial space; symmetry is maintained in spin and orbital spaces in which the electrons are described by the same wave functions used when no interaction takes place. This allows us to derive the relations between the spin-angular parts of the matrix elements for conjugated configurations (having the same number of electrons and vacancies in the same shells, e.g., l^N and l^{4l+2-N}).

We will introduce a conjugation operator C, under the influence of which an operator $a^{(qls)}$ changes the projection of the quasispin rank [40]:

$$Ca^{(qls)}_{pm\mu}C^{-1} = (-1)^{q-p}a^{(qls)}_{-pm\mu}. \tag{3.1}$$

In keeping with Eq. (1.92), C transforms an electron's creation operator to its annihilation operator, and vice-versa:

$$Ca^{\dagger}C^{-1} = \tilde{a}, \quad C\tilde{a}C^{-1} = -a^{\dagger}. \tag{3.2}$$

The rules for transforming the $V^{(k_1k_2k_3)}$ operators, Eq. (1.93), Q_z, Eq. (1.90), and its eigenfunction $|\gamma Q M_Q\rangle$ when acted upon by the operator C follow from Eq. (3.2) [40]:

$$CV^{(k_1k_2k_3)}_{q_1q_2q_3}C^{-1}=(-1)^{k_1-q_1}V^{(k_1k_2k_3)}_{-q_1q_2q_3}, \tag{3.3}$$

$$CQ_zC^{-1}=-Q_z, \tag{3.4}$$

$$C|\gamma QM_Q\rangle=(-1)^{Q-M_Q}|\gamma Q-M_Q\rangle. \tag{3.5}$$

With due regard for Eq. (1.91), Eq. (3.5) becomes

$$C|l^N\gamma\rangle=(-1)^{2l+1-\frac{N+v}{2}}|l^{4l+2-N}\gamma\rangle. \tag{3.6}$$

Thus, every state γ of a partially filled shell l^N corresponds to the same state of an almost filled shell; in other words, conjugated configurations have the same terms (Pauli's principle for vacancies).

The relationships between coefficients of fractional parentage and between matrix elements for partially filled and almost filled shells presented in Chapter 1 without proof follow from Eqs. (3.2)–(3.6).

In keeping with Eq. (1.137), the diagonal matrix elements of the spin–orbit interaction operator H^{so} for the l^N and l^{4l+2-N} configurations have opposite signs. If the nondiagonal matrix elements are small, the levels of the term are inverted on going from a partially filled to an almost filled shell.

The spin-angular part of the matrix element for the electrostatic interaction operator differ in magnitude for the l^N and l^{4l+2-N} configurations and is independent of the term. However, if the average energy $\bar{E}^e$ is excluded from H^e the reduced coefficient $f_k'(l^N)$ is found to be the same for conjugated configurations. The eigenvalues of the purely scalar H^k and H^p operators and the relativistic corrections H^m and H^d are proportional to the number of electrons in a shell and are not symmetric with respect to the substitution $N\rightarrow 4l+2-N$.

In a Hartree–Fock–Pauli interaction the energy of a shell having a single vacancy is, when basic interactions are considered,

$$E(nl^{4l+1}j)=(4l+1)[I(nl)+R(nl)]-\frac{2l}{2l+1}\sum_{k>0}\langle l\|C^{(k)}\|l\rangle^2F^k(nl,\ nl)+$$
$$+2l(4l+1)F^0(nl,\ nl)-\frac{1}{2}[j(j+1)-l(l+1)-s(s+1)]\zeta_{nl}, \tag{3.7}$$

where $R(nl)$ denotes a one-electron matrix element for the relativistic correction operators H^m and H^d, Eq. (1.214).

On the other hand, in the same approximation, the energy of a single electron in a nuclear field of charge Z is given by

$$E(nlj)=I(nl)+R(nl)+\frac{1}{2}[j(j+1)-l(l+1)-s(s+1)]\zeta_{nl}. \tag{3.8}$$

When algebraic expressions for the integrals with respect to hydrogenic wave functions and the virial theorem are used [54], Eq. (3.8) reduces [39] to the standard formula

$$E(nlj)=-Rhc\left[\frac{Z^2}{n^2}+\frac{\alpha^2Z^4}{n^4}\left(\frac{n}{j+1/2}-\frac{3}{4}\right)\right]. \tag{3.9}$$

Here, R is the Rydberg constant for an infinite mass (we will ignore electron energy as a function of nuclear mass here and in what follows).

In order to transform Eq. (3.7) to the same form we must subtract from $E(l^{4l+1})$ the energy of a closed shell that had been calculated with the same radial wave functions:

$$\begin{aligned}E(nlj^{-1})&=E(nl^{4l+1}j)-E(nl^{4l+2})=\\&=-[I(nl)+R(nl)]-\frac{1}{2}[j(j+1)-l(l+1)-s(s+1)]\zeta_{nl}\\&-(4l+1)F^0(nl,nl)+[l]^{-1}\sum_{k>0}\langle l\|C^{(k)}\|l\rangle^2F^k(nl,nl).\end{aligned} \tag{3.10}$$

We see that Eq. (3.10) has more terms than Eq. (3.8). These terms are the result of an electrostatic interaction within a shell [if there are other closed shells K_0, then when we subtract the energy $E(K_0l^{4l+1})$ from $E(K_0l^{4l+2})$ the other terms that correspond to the interaction between an l-electron and a closed shell appear]. The wave function $P_{nl}(r)$ is given approximately by a one-electron function having an effective nuclear charge $Z_{nl}^* = Z - \sigma_{nl}$, where σ_{nl} is the screening constant, defined by Hartree [37]:

$$\sigma_{nl}=Z-\frac{3n^2-l(l+1)}{2\bar{r}_{nl}}; \quad \bar{r}_{nl}=\langle nl|r|nl\rangle. \tag{3.11}$$

By writing the integrals in Eq. (3.10) for hydrogenic functions in powers of $(Z - \sigma)$, we can write Eq. (3.7) as

$$E(nlj^{-1}) = Rhc\left[\frac{(Z-\sigma'_{nl})^2}{n^2} + \frac{\alpha^2(Z-\sigma_{nlj})^4}{n^4}\left(\frac{n}{j+1/2} - \frac{3}{4}\right)\right]. \quad (3.12)$$

The quantum number j is added in the second term, which describes relativistic corrections to σ_{nl}. The quantity σ_{nlj} is sometimes called the inner screening constant – it is basically a function of the charge in those shells which are deeper than the nl shells that screen the nucleus. Additional Coulomb terms that are contained in Eq. (3.10) are included in $\sigma_{nl'}$. These terms correspond to an interaction with all closed shells and have a different dependence on Z than potential and kinetic energy; therefore, $\sigma_{nl'}$ is already a function of the atomic number and is determined by those atomic shells that are inside and outside of the nl shell. This is the total screening constant [1, 55].

The formula for the energy of an inner vacancy, Eq. (3.12), remains valid even when there are open shells in an atom if these shells interact weakly with the vacancy. The energy of a vacancy, Eq. (3.12), has the opposite sign of the electron energy, Eq. (3.9); therefore, the levels in an atom with vacancy correspond to the inverted spectrum of a one-electron atom: the deeper the shell containing the vacancy, the greater the positive energy of the level, and the ground level corresponds to a vacancy in the outer valence shell. Keeping only the first basic term in Eq. (3.12), we obtain Moseley's law for the level of a configuration having a vacancy.

Graphical perturbation theory is a common technique for describing a vacancy as a quasiparticle and, when this is done, a Brillouin–Wigner series is more convenient [138, 139]. A vacancy in the frozen field of other electrons is an idealized, or frozen, vacancy and is shown by the solid line directed downward in Fig. 3.1. The vacancy being produced attracts the surrounding electron charge. This relaxation in electron shells can be thought of as the scattering of a vacancy by electrons, which causes actual and virtual excitations. The latter are shown as graph fragments (containing no outer lines), which can be inserted into a vacancy line, cutting it in the middle. An interaction with electrons such as this changes a vacancy's energy: $-\varepsilon_a$ into $-\varepsilon_a - \Sigma_a$, where $-\Sigma_a$ is a correction called the self-energy of a vacancy. The contribution of an individual diagram to the self-energy is described by part of a diagram obtained by cutting off the vacant line from above and below. The contribution is determined according to the rules that were laid down in Section 2.3, except that the quantity E must be added to the energy denominator D (2.80) (the quantity E is the unknown energy of the state having the vacancy and is equal to the binding

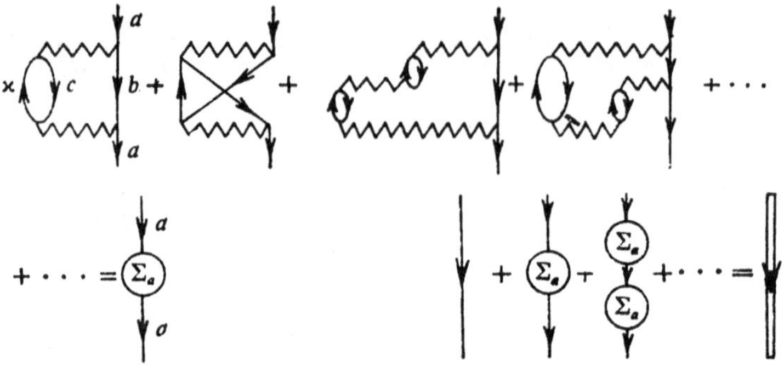

Fig. 3.1. Diagrams showing a vacancy as a quasiparticle taking into account its scattering by the surrounding electrons.

energy of the electron that was removed, but has opposite sign). For example, the contribution from the first diagram of the series shown in Fig. 3.1 is

$$-\sum_{\varkappa bc}{}' \frac{|\langle cb\,|\,h^{e}_{12}\,|\,\varkappa a\rangle|^2}{\varepsilon_c+\varepsilon_b-\varepsilon_\varkappa+E+i\eta}. \tag{3.13}$$

An infinite number of diagrams describe a vacancy-quasiparticle, shown in Fig. 3.1 by the downward directed heavy line. The self-energy of a quasiparticle is found by adding the contributions from the diagrams for this series, and contains both real and imaginary parts. The imaginary parts come from diagrams corresponding to excitations in which energy is conserved (the energy denominator then has a singularity, and a small imaginary quantity, $i\eta$, must be added to it). Excitations that obey the law of conservation of energy thus contribute to the real and imaginary parts of the self-energy, and excitations that are forbidden energetically contribute only to the real part.

To describe the shape of an autoionizing level, it is convenient to introduce a vacancy spectral function [138] [compare this with Eq. (2.99)]:

$$A_a(E)=\frac{1}{\pi}\,\mathrm{Im}\,\frac{1}{E-\varepsilon_a-\Sigma_a(E)}=\frac{1}{\pi}\,\frac{\mathrm{Im}\,\Sigma_a(E)}{[E-\varepsilon_a-\mathrm{Re}\,\Sigma_a(E)]^2+[\mathrm{Im}\,\Sigma_a(E)]^2}. \tag{3.14}$$

For a noninteracting vacancy, the spectral function is transformed to a δ-function:

$$A_a(E)\sim\delta(E-\varepsilon_a). \tag{3.15}$$

If the self-energy of a vacancy in the vicinity of a level is considered constant, Eq. (3.14) is a Lorentz function having an Auger width of

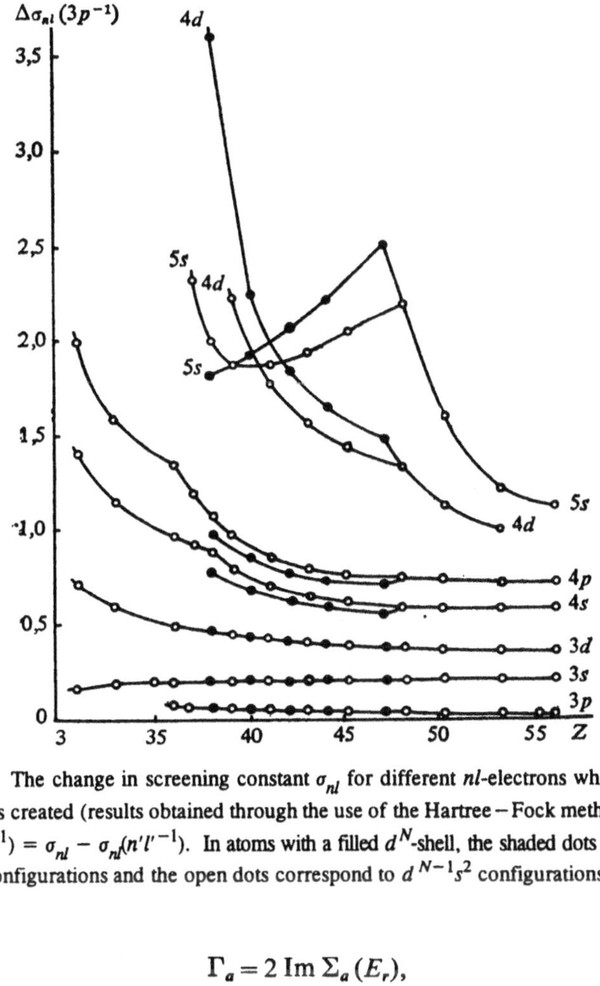

Fig. 3.2. The change in screening constant σ_{nl} for different nl-electrons when a $3p^{-1}$ vacancy is created (results obtained through the use of the Hartree–Fock method [140]); $\Delta\sigma_{nl}(n'l'^{-1}) = \sigma_{nl} - \sigma_{nl}(n'l'^{-1})$. In atoms with a filled d^N-shell, the shaded dots correspond to $d^N s$ configurations and the open dots correspond to $d^{N-1}s^2$ configurations.

$$\Gamma_a = 2\,\mathrm{Im}\,\Sigma_a(E_r), \tag{3.16}$$

where E_r is the resonance energy.

When $E - \varepsilon_a - \mathrm{Re}\,\Sigma_a$ as a function of energy is taken into account in the approximation of the first term in a Taylor series, as in Eq. (2.109), a more general expression that contains the spectroscopic factor z is obtained:

$$\Gamma_a = 2z_a(E_r)\,\mathrm{Im}\,\Sigma_a(E_r), \tag{3.17}$$

$$z_a(E_r) = \left(1 - \frac{\partial}{\partial E}\,\mathrm{Re}\,\Sigma_a(E)\right)^{-1}_{E=E_r}. \tag{3.18}$$

The relaxation of an atom when a vacancy is created can be divided into monopole relaxation, which conserves the orbital quantum number l of the vacancy when it is scattered by neighboring electrons, and multipole relaxation, in which l is not conserved. The fundamental part of monopolar relaxation – relaxation within a given ionic configuration – is accounted for by the Hartree–Fock approximation. In this approximation the self-energy of a vacancy is equal to the difference between the Koopmans and the Hartree–Fock binding energy and coincides with the relaxation energy which is frequently used as a measure of an atomic relaxation when a vacancy is created in a one-configuration approximation. Starting with the relaxation energy being defined as the difference of the Hartree–Fock total ion energies calculated with frozen wave functions for the atom $[E'(a^{-1})]$ and the functions for this ion, we find

$$\begin{aligned} E^{rx}(a^{-1}) &= E'(a^{-1}) - E(a^{-1}) = \\ &= \left(E'(a^{-1}) - E\right) - \left(E(a^{-1}) - E\right) = |\varepsilon_a| - I_a = \Sigma_a^{HF} > 0. \end{aligned} \tag{3.19}$$

Relaxation energy ranges from a few tenths of an electronvolt to roughly 100 eV [9], is higher when the vacancy is deeper, and increases almost linearly as the atomic number increases.

The creation of a vacancy affects the outer, weakly bound shells more strongly than it does the shells neighboring the vacancy [140]. The screening constant for a valence electron may change by as much as two or three when an inner vacancy is created (Fig. 3.2): the effect of a vacancy is to increase the nuclear charge in a neutral atom by a few units. This is because the asymptote of the self-consistent field potential changes when an electron is removed and the relative change in the potential increases as r increases.

The approximation [141]

$$\Delta\varepsilon_a(b^{-1}) \approx \Delta\varepsilon_b(a^{-1}), \tag{3.20}$$

is obtained through the use of the Koopmans theorem, according to which the change in the energy of an electron a when a vacancy b^{-1} is created is approximately (usually to within a few percent) equal to the change in energy of an electron b when a vacancy a^{-1} is created.

Consideration of not only monopolar or radial, but angular relaxations of electron shells corresponds to describing a state with a vacancy in a multiconfiguration approximation.

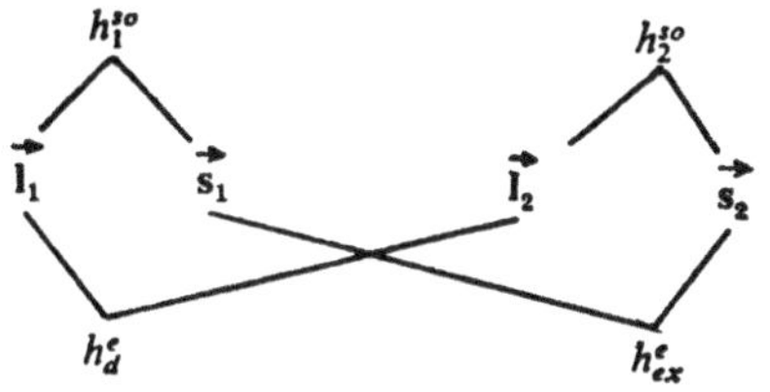

Fig. 3.3. Coupling between the orbital and spin angular momenta of two electrons caused by direct (h_d^e) and exchange (h_{ex}^e) electrostatic interaction, and spin-orbit interaction (h^{so}) between them.

When there is strong mixing of the configurations, the quasiparticle image is destroyed and even loses validity. For example, the $4p_{1/2}{}^{-1}$ level in xenon is due to a very strong interaction with $4d^{-2}\varepsilon l$ ($l = 1, 3$) continua and corresponding Rydberg series and dilutes into a wide band in the photoelectric spectrum [142, 139]. The concept of a vacancy as a quasiparticle is destroyed even for internal vacancies when mixing with many continua occurs [143]. Strong mixing of a wave function for a state with a vacancy with the functions of a continuous spectrum corresponds to a large complex part in the self-energy of the vacancy and, starting from the properties of a spectral function, we can formulate the following condition for using the quasiparticle representation [144]:

$$\mathrm{Re}\,\Sigma_a(E)| \gg |\,\mathrm{Im}\,\Sigma_a(E)|. \tag{3.21}$$

A less strict condition was proposed in [145].

3.2. Types of Coupling and the Energy Spectrum

Different types of electron angular momentum coupling – coupling schemes – are possible in an atom. Realizing one type of coupling or another depends on the relative amount of interactions that couple the electron angular momenta: angular momenta are primarily coupled by a stronger interaction [38, 46].

We will first examine the coupling of the angular momenta of two electrons in the nonrelativistic theory. Direct electrostatic repulsion between electrons h_d^e couples only their orbital angular momenta and does not affect spin angular momenta (Fig. 3.3). On the other hand, an exchange electrostatic interaction h_{ex}^e also couples the spin angular momenta. Spin–orbit interaction combines the orbital and spin angular momenta of each electron into a total one-electron angular momentum.

If the electrostatic interaction

$$h_d^e,\ h_{ex}^e \gg h_1^{so},\ h_2^{so}, \tag{3.22}$$

between electrons is dominant, an *LS* coupling is achieved:

$$\mathbf{l}_1+\mathbf{l}_2=\mathbf{L},\ \mathbf{s}_1+\mathbf{s}_2=\mathbf{S},\ \mathbf{L}+\mathbf{S}=\mathbf{J}. \tag{3.23}$$

If the inverse inequality

$$h_1^{so},\ h_2^{so} \gg h_d^e,\ h_{ex}^e, \tag{3.24}$$

is valid, a *jj* coupling is achieved:

$$\mathbf{l}_1+\mathbf{s}_1=\mathbf{j}_1,\ \mathbf{l}_2+\mathbf{s}_2=\mathbf{j}_2,\ \mathbf{j}_1+\mathbf{j}_2=\mathbf{J}. \tag{3.25}$$

These two schemes of coupling are called symmetric (both electrons participate in a symmetric, equivalent manner). Two asymmetric, or pair, types of couplings are possible. Under the condition

$$h_d^e \gg h_1^{so} \gg h_{ex}^e,\ h_2^{so} \tag{3.26}$$

an *LK* coupling

$$\mathbf{l}_1+\mathbf{l}_2=\mathbf{L},\ \mathbf{L}+\mathbf{s}_1=\mathbf{K},\ \mathbf{K}+\mathbf{s}_2=\mathbf{J}, \tag{3.27}$$

is achieved, and when

$$h_1^{so} \gg h_d^e \gg h_{ex}^e,\ h_2^{so} \tag{3.28}$$

a *jK* coupling

$$\mathbf{l}_1+\mathbf{s}_1=\mathbf{j}_1,\ \mathbf{j}_1+\mathbf{l}_2=\mathbf{K},\ \mathbf{K}+\mathbf{s}_2=\mathbf{J} \tag{3.29}$$

is achieved.

The values of the basic radial integrals $F^2(n_1l_1,n_2l_2), G^{|l_1-l_2|}(n_1l_1,n_2l_2), F^2(nl,nl)$, and ζ_{nl} [38] may serve as an approximate measure of the interactions in the inequalities given earlier.

In the relativistic theory the one-electron wave function is a function of the quantum number j; therefore, a unique type of coupling – a jj coupling – is possible.

The angular momenta of equivalent electrons are coupled only in an equivalent manner. Consequently, only an LS or jj coupling is possible within a shell or subshell. In light atoms and in the valence and subvalence shells of heavy atoms at low values of effective nuclear charge, the electrostatic interaction between electrons dominates and only an LS coupling occurs. As the effective nuclear charge increases, the spin–orbit interaction increases to the fourth power, whereas the electrostatic interaction increases only linearly; therefore, the coupling approximates a jj coupling. When this happens the magnitude of the orbital quantum number l is important: the electrons having larger l are closer to one another and repel each other more strongly, while, conversely, the spin–orbit interaction constant decreases with increasing l due to the increased distance from the nucleus caused by the centrifugal effect. Thus, a jj coupling is achieved only within a valence p-shell in heavy atoms and, in medium and heavy atoms, within inner shells with vacancies.

All four schemes of couplings are possible between shells. An inner vacancy is often coupled to an outer open shell via a jj or jK coupling. The almost purely nonhomogeneous jK and LK couplings are sometimes achieved between a core and an excited electron, especially if it has a large orbital angular momentum [38, 46]. In an isoelectron series, or when the degree of ionization in an atom increases, the coupling approximates a jj coupling.

The transition is made from wave functions for one type of coupling to wave functions for another type of coupling by using transformation matrices (Appendix 3); for example, in the nonrelativistic theory, from an LS to a JJ coupling between two shells or between two nonequivalent electrons via the matrix

$$\langle L_1 L_2 L,\ S_1 S_2 S\ J | L_1 S_1 J_1,\ L_2 S_2 J_2 J\rangle = \\ =[L,\ S,\ J_1,\ J_2]^{1/2} \begin{Bmatrix} L_1 & S_1 & J_1 \\ L_2 & S_2 & J_2 \\ L & S & J \end{Bmatrix}. \tag{3.30}$$

Because the transformation matrix is unitary, the inverse transformation is made via the same matrix.

An atomic coupling that is pure in the practical sense is comparatively rare, e.g., for the ground term of a ground configuration. The wave functions of a pure coupling are used only as an initial basis for forming the energy matrix. Linear combinations –

wave functions of an intermediate coupling – are obtained by diagonalizing this matrix and are used to find the transition probabilities and other atomic quantities. In an intermediate coupling the calculated results do not depend on the choice of initial pure coupling, although using a more realistic basis lets us use the quantum numbers of a pure coupling to designate the levels (by the quantum numbers of the largest weight).

Average coupling purity – the mean squared value of the weighted coefficient in the expansion of the wave function for an intermediate coupling [46] – is used to estimate the purity of a coupling.

Using the quantum numbers of a pure coupling to classify the levels is accompanied by some difficulties. Because most quantum numbers correspond to operators that do not commute with the Hamiltonian, the energy matrix for configurations having open shells will contain many nondiagonal elements, and terms are strongly mixed in all coupling types. The maximum weight in the expansions of wave functions for several levels having the same J may correspond to one and the same term, whereas some of the terms are not distinguished by weighting coefficient in even a single intermediate coupling. If a level in an isoelectronic series or an atomic sequence is examined, then it can change designation: in one atom it may correspond to a single, almost pure term, and in a different atom, to another term of the same coupling type. We will illustrate this with an example for two levels [46]. When the energy matrix

$$\begin{Vmatrix} E_1 & E_{12} \\ E_{12} & E_2 \end{Vmatrix}$$

is diagonalized (1 corresponds to the γ_1 quantum numbers and 2 corresponds to the γ_2), the following eigenvalues are obtained:

$$\mathcal{E}_{1,2} = \frac{E_1+E_2}{2} \pm E_{12}\left[\left(\frac{E_1-E_2}{2E_{12}}\right)^2+1\right]^{1/2}. \tag{3.31}$$

Let a nondiagonal matrix element be $E_{12} = 0$, and E_1 and E_2 be functions of a parameter χ. When χ is varied between zero and one, the levels E_1 and E_2 intersect and change places (the dashed straight lines in Fig. 3.4). When this happens, both states remain pure states γ_1 and γ_2. If a small perturbation to E_{12} is added, the change in the energies $\mathcal{E}_1$ and $\mathcal{E}_2$ is, according to Eq. (3.31), mapped by the solid lines in the figure. The levels do not intersect. However, because E_{12} is so small, the curves pass rather close to the straight lines, and the states correspond to the almost pure states γ_1 and γ_2 except in the area where they approach one another. Thus, as χ varies, the pure state γ_1 changes to the pure state γ_2.

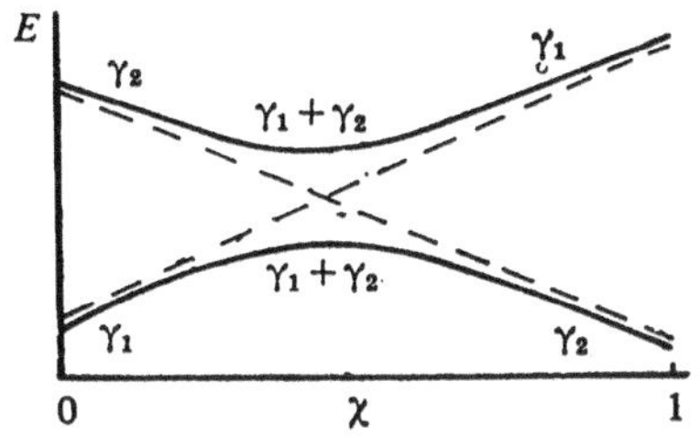

Fig. 3.4. The energy change of two interacting (solid lines) and two noninteracting (dashed lines) levels as a function of the energy parameter χ.

A more exact, but not very economical, way to designate levels in complex configurations is to indicate several mixed terms.

The results from calculating transition probabilities in an intermediate coupling may contain significant errors for some weak lines. This happens when contributions from the different terms of an expansion over the functions of a pure coupling have opposite sign and almost compensate one another. The compensation factor [46]

$$\tau = \left[\frac{\left| \sum_{\gamma} \sum_{\gamma'} c_{\gamma}^{\Gamma} \langle \gamma | O | \gamma' \rangle c_{\gamma'}^{\Gamma'} \right|}{\sum_{\gamma} \sum_{\gamma'} \left| c_{\gamma}^{\Gamma} \langle \gamma | O | \gamma' \rangle c_{\gamma'}^{\Gamma'} \right|} \right]^2, \tag{3.32}$$

can be used to test the reliability of the results, where c_{γ}^{Γ} is a weight coefficient in the expansion of a wave function for the Γ state over the functions of a pure coupling $|\ \gamma\rangle$, and O is the transition operator. The strength of the radiative transition line $S\ (\Gamma,\ \Gamma')$ may contain a significant error if $\tau < (0.05-0.1)$ [46].

Detailed calculations need not be performed for an intermediate or pure coupling in order to make estimates or a preliminary interpretation of a complex energy spectrum – its statistical moments, average energy, variance, etc. all yield useful general information about the spectrum. Formulas [Eqs. (1.144) and (1.199)] have already been presented for the average configuration energy in the nonrelativistic theory and the relativistic theory, and the individual terms of the variance corresponding to electrostatic and spin–orbit interactions within a shell and electrostatic interaction between shells are given by [146]:

$$\sigma_e^2(l^N) = \frac{N(N-1)(4l+2-N)(4l+1-N)}{(4l+2)(4l+1)4l(4l-1)} \sum_{k>0} \sum_{k'>0} \left[\frac{2\delta(k, k')}{2k+1} - \frac{1}{(2l+1)(4l+1)} - \right.$$

$$\left. - \begin{Bmatrix} l & l & k \\ l & l & k' \end{Bmatrix} \right] \langle l \| C^{(k)} \| l \rangle^2 \langle l \| C^{(k')} \| l \rangle^2 F^k(l, l) F^{k'}(l, l), \tag{3.33}$$

$$\sigma_{so}^2(l^N) = \frac{l(l+1)}{4(4l+1)} N(4l+2-N)\zeta_l^2, \tag{3.34}$$

$$\sigma^2(l_1^{N_1} l_2^{N_2}) = \frac{N_1(4l_1+2-N_1) N_2(4l_2+2-N_2)}{(4l_1+2)(4l_1+1)(4l_2+2)(4l_2+1)} \times$$

$$\times \left\{ \sum_{k>0} \sum_{k'>0} \frac{4\delta(k, k')}{2k+1} \langle l_1 \| C^{(k)} \| l_1 \rangle^2 \langle l_2 \| C^{(k')} \| l_2 \rangle^2 F^k(l_1, l_2) F^{k'}(l_1, l_2) + \right.$$

$$+ \sum_k \sum_{k'} \left[\frac{4\delta(k, k')}{2k+1} - \frac{1}{(2l_1+1)(2l_2+1)} \right] \langle l_1 \| C^{(k)} \| l_2 \rangle^2 \langle l_1 \| C^{(k')} \| l_2 \rangle^2 \times$$

$$\times G^k(l_1, l_2) G^{k'}(l_1, l_2) - 4 \sum_{k>0} \sum_{k'} \begin{Bmatrix} l_1 & l_1 & k \\ l_2 & l_2 & k' \end{Bmatrix} \times$$

$$\left. \times \langle l_1 \| C^{(k)} \| l_1 \rangle \langle l_2 \| C^{(k)} \| l_2 \rangle \langle l_1 \| C^{(k')} \| l_2 \rangle^2 F^k(l_1, l_2) G^{k'}(l_1, l_2) \right\}. \tag{3.35}$$

If a configuration contains several open shells, the variance is equal to the sum of the terms in Eq. (3.35) over all pairs of shells and to the sum of Eqs. (3.33) and (3.34) over all shells. For the inner shells, the nonrelativistic integrals must be replaced by corresponding superpositions of relativistic integrals, Eqs. (1.208)–(1.211).

The contributions from different interactions to the variance give a more accurate estimate of interaction strength than do the magnitudes of the integrals when a coupling scheme is chosen.

The variance of the energy spectrum describes its width ΔE, which can be estimated from the formula

$$\Delta E = 2[2 \ln 2\sigma^2]^{1/2} \approx 2{,}3548 \sqrt{\sigma^2} \tag{3.36}$$

as the width of the normal distribution of the energy levels.

3.3. The Collapse of an Excited Electron

When the atomic number, the degree of ionization, or the many-electron state of an

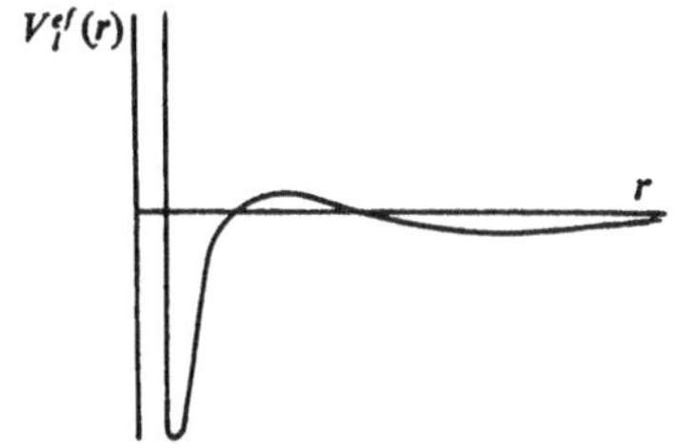

Fig. 3.5. The effective potential energy $V_l^{ef}(r)$ in the form of two wells separated by a barrier.

atom is changed in configurations having an excited electron, a drastic change – possibly several orders of magnitude – in the characteristics of this electron is possible. This phenomenon has been named wave function collapse, or, more briefly, electron collapse. This phenomenon is quite pronounced in configurations having a vacancy in the subvalent shell and may be the reason for various anomalies in x-ray and electron spectra [29, 147, 148].

The Hartree–Fock equation that determines the radial wave function $P_{nl}(r)$ is the equation for one-dimensional motion having an effective potential energy of

$$V_{nl}^{ef}(r) = V_{nl}(r) + \frac{l(l+1)}{2r^2}. \tag{3.37*}$$

Here, $V_{nl}(r)$ is the potential energy of an electron in the field due to the nucleus and the other electrons and includes its exchange component [we will say that it is approximated by a local potential function of the Slater's potential type or, for a nodeless $P_{nl}(r)$ function, is obtained by dividing the exchange component of the Hartree–Fock equation by this function]. The second term in Eq. (3.37) is the positive centrifugal energy which increases rapidly with increasing l and becomes greater than $V(r)$ for $l \geq 2$ (having negative sign) not only for small r, but in many atoms and at distances on the order of 0.5 to 5 a.u., where attraction to the nucleus is reduced by the screening produced by inner electrons. Because of this, the function for the effective potential energy looks like two potential wells separated by a positive potential barrier (Fig. 3.5).

The barrier for the upper ^1P term is significantly exceeded because of the large positive coefficient on the exchange integral for an electrostatic interaction in configurations having $nl_0^{4l_0} + nl$ vacancy (see Sec. 5.2). This coefficient is a small negative quantity for other terms and $V_{nl}^{ef}(r)$ is closely approximated by the potential function of the HF-av. equation.

For an excited electron the $V_{nl}^{ef}(r)$ function is weakly dependent on n and can be considered the same for an entire Rydberg series.

A consequence of the small area of the internal well is that it will contain only a few levels, and for some atoms no levels at all. In the latter case the main maxima for all $P_{nl}(r)$ having a given l are located far from the nucleus, near an outer one-electron well.

As the nuclear charge increases, an inner well becomes deeper and wider. When an energy level appears in this well and becomes lower than the ground energy level in an outer well, the radial wave function for the first state in the given series collapses — its main maximum is replaced into the inner well region. When this happens, any other function $P_{n'1}(r)$ for the given series becomes similar to the previous $P_{n'-1l}{}^{(r)}$ in the region of the outer well. Further deepening of an inner well leads to the collapse of the wave function for the second state, etc.

In neutral atoms having an excited valence nl-electron, collapse takes place at an atomic number that is one less than for an element in which an nl-shell appears in a ground configuration [149]. If an electron has been excited from an inner shell, the effective nuclear charge for this electron is higher than for a normal core; therefore, electron collapse tends to occur at a lower atomic number [this tendency can be negated by the aforementioned strong dependence of $V_l^{ef}(r)$ on the term].

The higher and wider the potential barrier, the more pronounced the collapse; therefore, the wave function and other characteristics of an f-electron change more drastically and sharply than those of a collapsing d-electron [149, 147]. The barrier decreases rapidly as the degree of ionization increases – it is more characteristic of neutral atoms, although in ions having a degree of ionization on the order of eight a nonmonotonic change in the oscillator strengths may take place in an isoelectron series.

The collapse of an l-electron in $l_0^{4l_0+1}l$ configurations depends on the many-electron quantum numbers: for the lower group of terms an electron may be collapsed, and for the upper ^1P term in the same atom it may be located in an outer well [150]. Using solutions of HF-av. equations can produce large errors, especially for the ^1P term. On the other hand, a comparison of the results of calculating the photoabsorption spectra for $4d^{10} \rightarrow 4d^9 4f$ with experimental spectra [147, 151] shows that the HF-t. approximation correctly describes the collapse of a single excited electron, even though correlation and relativistic effects may also be significant near the critical value of the effective charge. If an outer shell has several electrons we can study their localization only by removing the constraint that they be described by the same radial wave function. The results from calculating radial wave functions by the Hartree–Fock spin-polarization technique confirms that even for $N = 8$ there is one noncollapsed f-electron in the d^9f^N configurations of neutral, rare-earth atoms [100].

When a barrier is negative or has a small positive value, gradual collapse and partial penetration into an inner well is possible for excited nl-electrons with different n [29, 151]. In the "transition" range of Z values that corresponds to the omission of a quasi-independent level in an inner well in the range of energies for an external well, there is a successive resonance penetration of Rydberg series wave functions into the atomic core [152].

The existence of a positive potential barrier strongly affects the motion of a slow free electron in the atomic field, especially if there is a quasidiscrete level having positive energy in an inner well. When the energy of the free electron approaches the energy of such a level, $P_{\varepsilon l}(r)$ penetrates the atom's inner region in a resonance fashion [26, 147, 153]. This phenomenon is sometimes called the collapse of a continuous spectrum wave function.

The effects of electron collapse and a potential barrier on x-ray and electron spectra will be examined in subsequent chapters.

3.4. Strongly Mixed Configurations

We will briefly examine the conditions under which the one-configuration approximation is inadequate for describing electron shell structure — strong mixing of the configurations occurs, especially when an internal vacancy is present.

It is well known that configurations of the discrete spectrum are mixed more strongly when the magnitude of the interconfiguration matrix elements is greater and when the distance between levels is smaller. The interaction between a discrete state and the continuum in which it is located is, according to Eq. (2.99), determined by the matrix element $V_{\mathscr{E}}$ and by the distance of the level from the continuum edge.

Values of the interconfiguration integrals depend on the degree to which the wave functions contained in them overlap: the magnitude of the integrals increases if most or all of the wave functions have the same principal quantum numbers.

The interaction between configurations corresponding to the same principal quantum numbers (belonging to a given complex of configurations) plays a major role in the subvalent shells when the effective nuclear charge is fairly high. The higher the value of Z^*, the more hydrogenlike are the states, and if their energy is mainly determined by electrostatic interaction, the configurations in the complex become quasidegenerate. The R^k integrals, just as the G^k and F^k integrals, are linear functions of Z^*, and the distance between the energy levels of two configurations with vacancies behave differently if they belong to the same complex or to different complexes. This dependence can be estimated, starting from Eq. (3.12) and leaving the basic quadratic term in it. In the first case we have

$$\Delta E = E_1 - E_2 \sim \frac{(Z-\sigma_1')^2}{n^2} - \frac{(Z-\sigma_2')^2}{n^2} = \frac{2(\sigma_2'-\sigma_1')}{n^2}\left[Z - \frac{\sigma_1'+\sigma_2'}{2}\right] \sim Z^*. \quad (3.38)$$

On the other hand, when the principal quantum numbers are different, the term with the larger n may be neglected and a square law dependence on Z^* is obtained. Thus, $(R^k)^2/(E_1 - E_2)^2$ for configurations in the same complex is almost independent of Z^* and decreases as $(Z^*)^{-2}$ for an admixed configuration that does not belong to the complex. Mixing of configurations in a complex is important in multiply charged ions without outer electrons and in atoms having vacancies in their intermediate shells. For example, the mixing of the final configurations $2p^{-q-1} + 2s^{-2}2p^{-q+1}$ during the radiation decay of $1s^{-1}2p^{-q}$ ions obtained in ion–atom collisions leads to the appearance of intense satellites of two-electron radiative rearrangement [154].

If the configurations differ by a single electron the terms in the interconfiguration matrix element for Brillouin configurations are partially or totally compensated, and when $l' \neq l$ levels having the same parity are often distant from one another. Among correlations of this type, the $(s + d)^N$ mixing in transition element atoms having an open d-shell must be noted [155, 156, 120]:

$$(s+d)^N \equiv nd^N + nd^{N-1}(n+1)s + nd^{N-2}(n+1)s^2. \tag{3.39}$$

In normal configurations this mixing is stronger for the end of the groups, but is essentially the same for the beginning of the group when there is an inner vacancy, especially the elements with a collapsing d-electron that precede it [29, 120].

Two-electron and one-electron excitations from a single shell into another and two-electron excitations into the nearest unfilled shells are usually important in configurations having open outer shells [45, 123, 157].

A configuration having an inner $n_1 l_1 j_1{}^{-1}$ vacancy interacts strongly with configurations of the continuous spectrum (in which it is found) and a discrete spectrum of the type [158–161, 113]

$$n_1 l_2^{-1} n_1 l_3^{-1} \binom{\varepsilon}{n} l, \quad n_1 l_2^{-2} \binom{\varepsilon}{n} l, \tag{3.40}$$

and to lesser degree with the configurations

$$n_1 l_2^{-1} n_2 l_3^{-1} \binom{\varepsilon}{n} l, \quad n_2 > n_1. \tag{3.41}$$

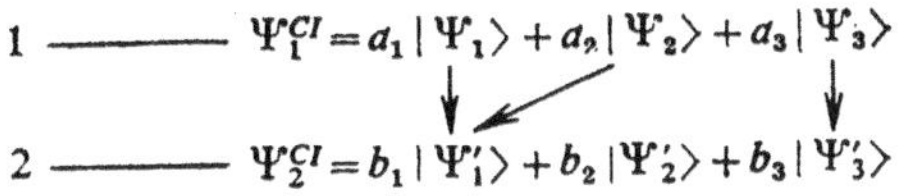

Fig. 3.6. Diagram illustrating the appearance of additional possibilities for a transition between the E_1 and E_2 levels when they are described by the wave functions Ψ^{CI} as an expansion of one-configuration functions Ψ_i; a_i and b_i are the coefficients of the expansion.

With respect to the $n_1 l_1 j_1^{-1}$, the continua given by Eq. (3.40) are called super Coster–Kronig continua, and those given by Eq. (3.41) are called Coster–Kronig continua (the nomenclatures arise from the corresponding radiationless transitions from the $n_1 l_1 j_1^{-1}$ into these configurations, scc Scc. 8.1).

The magnitude of the square of the interconfiguration matrix element when $n_2, n_3 \neq n_1$ is

$$V_\varepsilon^2 \equiv \langle n_1 l_1^{-1} | H^e | n_2 l_2^{-1} n_3 l_3^{-1} \varepsilon l \rangle^2 \approx 10^{-3} - 10^{-4} \text{ a.u.} \qquad (3.42)^*$$

When $n_1 = n_2$ and $n_3 \neq n_1$ this quantity increases by roughly an order of magnitude, and when $n_2 = n_3 = n_1$ it increases by still another order of magnitude [159]. On the other hand, the $n_1 l_1 j_1^{-1}$ level is far from the double ionization threshold $n_2 l_2^{-1} n_3 l_3^{-1}$ $(n_2, n_3 > n_1)$; therefore, V_ε is weakly dependent upon ε and the contributions to the level shift, Eq. (2.100), when $E_r = E(n_1 l_1 j_1^{-1})$, are

$$\Delta E(n_1 l_1 j_1^{-1}) \equiv F(n_1 l_1 j_1^{-1}) = P \int \frac{|V_{\varepsilon'}|^2 \, d\varepsilon'}{E(n_1 l_1 j_1^{-1}) - E(n_2 l_2^{-1} n_3 l_3^{-1}) - \varepsilon'} \qquad (3.43)$$

from $\varepsilon' < E(n_1 l_1 j_1^{-1}) - E(n_2 l_2^{-1} n_3 l_3^{-1})$ and $\varepsilon' < E(n_1 l_1 j_1^{-1}) - E(n_2 l_2^{-1} n_3 l_3^{-1})$ almost compensate one another. If $n_2 = n_1 \neq n_3$ and especially when all three principal quantum numbers are equal, $E(n_1 l_1 j_1^{-1})$ approaches a double ionization threshold and this compensation does not occur.

Super-Coster–Kronig or Coster–Kronig autoionization continua exist for a configuration with a vacancy $n_1 l_1 j_1^{-1}$ if the layer with n_1 contains other shells or subshells having an orbital momentum $l > l_1$ and their double ionization thresholds lie below the ionization threshold of $n_1 l_1 j_1$.

The interaction between the discrete spectrum configurations $n_1 l_1 j_1^{-1}$ and $n_2 l_2^{-1} n_3 l_3^{-1} nl$ also increases substantially if $n_1 = n_2$ and especially if $n_2 = n_3 = n_1$. From among this type of correlations we must single out [158–162]

$$ns^{-1} + np^{-2}n'd + np^{-2}n''s + np^{-1}(n+1)d^{-1}n'''f,$$
$$np^{-1} + nd^{-2}n'f, \tag{3.44}$$
$$nd^{-1} + nf^{-2}n'g.$$

They all correspond to a so-called symmetric change of symmetry [158] – the l for one electron decreases by unity and for another electron increases by unity.

Correlations between electrons that have been pointed out above appear in not only the energy level spectra, but in x-ray, electron, and photoabsorption spectra, which leads to redistribution of intensities and the appearance of strong satellite lines and their groups. It must be noted that the choice of configurations when transition probabilities are calculated is governed not only by the extent to which they are mixed, but by the magnitude of the matrix elements for the transition operator between the main and the admixed configurations — the opportunity for taking additional process channels (Fig. 3.6) into account. For example, the second term in the expansion of Ψ_1^{CI} can, even when $a_2 < b_2$, be more important for the transition probability than the second term in the expansion of Ψ_2^{CI}. Whereas the configuration interaction method almost always refines the one-configuration results when energy levels are calculated, for the transition magnitudes a random choice of several admixed configurations may worsen the agreement with experimental data.

Chapter 4

THE CREATION OF EXCITED STATES

If the formation of an excited state and the relaxation of the electron shells takes place in a time period that is much shorter than the lifetime of the excited state, and if we can ignore interference effects between the various processes that connect the same initial and final states, the creation and decay of an excited state having a vacancy can be examined as individual parts of a two-stage process. This approximation has been widely adopted in x-ray and electron spectroscopy and will be used in this chapter. Some possibilities for refining the two-step model – describing excitation and decay as a unified process – will be examined in Secs. 5.3, 7.2, and 8.3.

Excited states that decay originate x-ray and Auger spectra are obtained when atoms interact with x rays, electrons, and ions and as a result of nuclear transformations. Inelastic collisions between atoms and various particles is a distinct, broad class of microphysics and numerous monographs and review articles have been devoted to this subject (see [163]). In this chapter we will briefly describe only the most widely used theoretical methods for interpreting x-ray and electron spectra.

The basic methods of obtaining x-ray and electron spectra are to impact an atom with a beam of x rays or electrons. Electron excitation is subject to less rigid selection rules whereas during excitation by x rays, especially soft x rays, the levels associated with the initial levels by dipole selection rules are preferentially populated. A focused, high-intensity electron beam produces electron excitation more easily; however, electrons are less selective than photons in how they act upon an atom, and cause additional phenomena, including background bremsstrahlung. A theoretical description of how electrons interact with atoms is also more complex.

General expressions for the populations of the levels when the system is in dynamic equilibrium are presented and discussed in Sec. 4.1. The excitation and ionization of an atom by x rays is examined in Chapter 5. Section 4.2 is devoted to the filling of excited levels during inelastic collisions between atoms and electrons and the widely used first Born approximation is given fundamental attention. Some of the techniques from the rapidly developing (but as yet comparatively undeveloped in a theoretical sense) branch of collision physics, ion–atom collisions, are discussed in Sec. 4.3.

In addition to single ionization and one-electron excitation of an atom, many-body processes also play a significant role in producing the various spectra. Because describing these processes by means of the correlation methods is rather complicated, the probabilities of these processes in many-electron atoms can be estimated by the method of a sudden perturbation (Sec. 4.4). This method can be applied to any short-time action on electron shells, especially during a nuclear transformation.

4.1. Populations of the Levels when the System is in Dynamic Equilibrium

The population density of atoms – the number of atoms in a given state per unit volume – is an important characteristic of an atomic excited state. The relative population densities (e.g., with respect to the atomic population density in the ground state) are normally used. If the total population density is normalized to unity, the probabilities of finding atoms in the different states are obtained.

As a rule, x-ray or Auger spectra are recorded in the stationary mode. The system is in dynamic equilibrium: the number of atoms making a transition into the $i(\Delta\mathcal{N}_i^{in})$ state per unit time is equal to the number of atoms leaving the $i(\Delta\mathcal{N}_i^{out})$ state in the same time period:

$$\Delta\mathcal{N}_i^{in} = \Delta\mathcal{N}_i^{out}. \tag{4.1}$$

This equation is the basis for a general expression for the population density of a state [39]. The number of atoms making a transition into the state i is

$$\Delta\mathcal{N}_i^{in} = \sum_t \sum_k \mathcal{N}_k W^{(t)}(k \to i), \tag{4.2}$$

where $\mathcal{N}_k$ is the population density of the state k and $W^{(t)}$ is the probability of a transition from the state k into the state i by means of the process t. The summation is carried out over all processes and initial states.

On the other hand, the number of atoms leaving a given state is

$$\Delta\mathcal{N}_i^{out} = \mathcal{N}_i \sum_t \sum_k W^{(t)}(i \to k) = \mathcal{N}_i W_i. \tag{4.3}$$

The double summation in Eq. (4.3) gives the total probability of decay from an excited state per unit time or the rate of decay W_i (in atomic units, it is equal to the total width of the level Γ_i).

A comparison of Eqs. (4.2) and (4.3) yields

$$\mathcal{N}_i = \frac{\sum_t \sum_k \mathcal{N}_k W^{(t)}(k \to i)}{W_i} = \sum_t \sum_k \mathcal{N}_k K^{(t)}(k \to i). \tag{4.4}$$

The quantity $K^{(t)}(k \to i)$ is called the branching coefficient.

If there is no alignment of the atoms (all states in the same level have the same population density) or we are only interested in population densities that have been averaged over the projections of the angular momenta then, using the relations

$$\mathcal{N}(\gamma JM) = \frac{\mathcal{N}(\gamma J)}{2J+1}, \quad W(\gamma JM) = W(\gamma J),$$
$$\frac{1}{2J'+1} \sum_{MM'} W^{(t)}(\gamma' J' M' \to \gamma JM) = W^{(t)}(\gamma' J' \to \gamma J), \tag{4.5}$$

and summing Eq. (4.4) over M, we obtain an expression for the population density of a level:

$$\mathcal{N}(\gamma J) = \frac{\sum_{\gamma' J'} \mathcal{N}(\gamma' J') \sum_t W^{(t)}(\gamma' J' \to \gamma J)}{W(\gamma J)} =$$
$$= \sum_{\gamma' J'} \mathcal{N}(\gamma' J') \sum_t K^{(t)}(\gamma' J' \to \gamma J). \tag{4.6}$$

If the atoms are ionized or excited by particles having an energy distribution function $F(\varepsilon)$ then, with due regard for Eq. (1.253), the probability of the process is given by the cross section $\langle v\sigma^{(t)} \rangle$ of a corresponding process t that has been averaged over the particle distribution:

$$W^{(t)} = N_p \langle v \sigma^{(t)} \rangle, \quad \langle v \sigma^{(t)} \rangle = \int v \sigma^{(t)} (\varepsilon) F(\varepsilon) d\varepsilon, \tag{4.7}$$

where N_p is the density of the incident particles.

When atoms are excited by a fairly monochromatic beam whose energy is far from the process threshold, the cross section is weakly dependent on the energy ε and we can take the term $\sigma^{(t)}$ outside the integral in Eq. (4.7); i.e., we can say that $W^{(t)} \sim \sigma^{(t)}$.

X-ray or Auger spectra for free atoms are usually generated in a gas stream by a beam of several particles in which excited atoms and ions are only a small fraction of all the atoms, and we can say that the particles excite atoms from only the levels of a ground term (as a rule, energetically remote from other levels) even at fairly low temperatures (which can be estimated from the relation 1 eV/k = 11604.5 K, where k is Boltzmann's constant and K is Kelvin) only from the lowest level of an atom's ground configuration. The population density of the initial levels may be considered dependent on only the temperature in the active zone, and for only one initial level the relative population density may be taken as unity. The population densities of the excited levels are governed by the processes by which atoms are excited and ionized by particles and by the radiation and radiationless transitions that occur during spontaneous deexcitation of the atoms. In this model, the population densities of the excited levels may be calculated consequently, starting from the upper levels and examining all levels from which transitions can directly or indirectly affect the spectrum being studied. To perform these calculations the cross sections and probabilities of the different processes must be known, including the many radiation and Auger transitions that determine the natural line widths. In order to reduce the number of calculations, it is worthwhile to first find the total population densities of the configurations and identify the major processes and detailed calculations to be done for the main part of the cascade of processes.

In some cases we can consider all states in the same configuration to be equally populated and the population densities of the levels to be proportional to the statistical weights. This approximation is used, e.g., when single ionization by a beam of particles having an energy that is far from the threshold is the main process that populates the states of an atom having an inner vacancy and the outer shells of the atom have been closed or interact weakly with the vacancy. Because the ionization cross section far from the threshold is proportional to the number of electrons in the $2j + 1$ subshell and the width of the levels $\Gamma(nlj^{-1})$ is weakly dependent on j (Sec. 9.2), we find from Eq. (4.6) that

$$\mathcal{N}(nlj^{-1}) \sim 2j + 1. \tag{4.8}$$

The population densities of all levels in one configuration can be considered in an approximation to be proportional to their statistical weights if filling occurs via the different processes and their channels (because of the complex cascade of processes, especially during ion – atom collisions).

When the system is in thermodynamic equilibrium (which, for example, ordinarily occurs only locally in a dense plasma) the relative population densities depend on only the absolute temperature:

$$\mathcal{N}(K\gamma J)=\mathcal{N}[J]\exp\left(-\frac{E(K\gamma J)}{kT}\right)\left[\sum_{K'\gamma' J'}[J']\exp\left(-\frac{E(K'\gamma' J')}{kT}\right)\right]^{-1}. \tag{4.9}$$

Here, all levels and configurations correspond to the same degree of ionization (the population densities as a function of the degree of ionization are given by Saha's equation); k is Boltzmann's constant; $E(K\gamma J)$ is the energy measured from the ground level energy, and $\mathcal{N}$ is the total population density in the levels for a given degree of ionization. If $E << kT$, the population densities of all states can be considered equal [146].

4.2. The Excitation and Ionization of Atoms by Electrons

Collisions between electrons and atoms are subdivided into slow, fast, and intermediate energies; the criterion is the ratio of an incident electron's velocity to the orbital velocity of the bound electron taking part in the collision [164-166].

In a slow collision the incident electron and atom target form a compound system, all electrons of which must be described in the same way; therefore, the interaction between a target and the electron being scattered cannot be considered as a perturbation. Then the connection between the different states of the compound system and the channels of its decay, which is considered in the method of strong coupling – a generalization of the many-configuration method for an atom + scattered electron system – is significant [66].

The system wave function is sought as

$$\Psi_j(x_1, \ldots, x_N, x_{N+1}) = A \sum_i \Phi_i(x_1, \ldots, x_N)\varphi_{ij}(x_{N+1}), \tag{4.10}$$

where A is the antisymmetrization operator; $x_k \equiv (\mathbf{r}_k, \sigma_k)$; Φ_i is the wave function of an N-electron atom target; φ_{ij} is the wave function of an electron being scattered from the system's initial state or input channel j into a final state or scattering channel i. A channel is described by the complete sets of atomic and electron quantum numbers. The summation in Eq. (4.10) is done over all possible scattering channels – open, in which φ_{ij} corresponds to a scattered electron's final state having positive energy, and closed, in which it has negative energy.

The wave function of an electron that has been scattered into the ith open channel is far from the atom the superposition of plane and spherical waves:

$$\varphi_{ij}(x) \underset{r\to\infty}{\sim} \delta(i, j)\delta(\mu_i, \sigma)\, e^{i\mathbf{k}_j\cdot\mathbf{r}} + \frac{e^{ik_i r}}{r} f_{ij}(\hat{\mathbf{r}}, \sigma). \tag{4.11}$$

Here f_{ij} is the scattering amplitude and it uniquely defines the differential scattering cross section

$$\frac{d\sigma_{ij}}{d\Omega} = |f_{ij}(\hat{\mathbf{r}}, \sigma)|^2. \tag{4.12}$$

In the spherical atomic field, the function φ_{ij} can be conveniently expanded over partial waves having a fixed orbital angular momentum. Hereafter, by φ_{ij} we will mean the single term of this expansion:

$$\varphi_{ij}(x) = Y_{l_i m_i}(\hat{\mathbf{r}})\, \delta(\mu_i, \sigma)\, \frac{F_{ij}(r)}{r}. \tag{4.13}$$

In order for the asymptote of Eq. (4.11) to be satisfied, the radial function $F_{ij}(r)$ must satisfy the condition that

$$F_{ij}(r) \underset{r\to\infty}{\sim} k_i^{-\frac{1}{2}} [\delta(i, j) \sin\eta_i + K_{ij} \cos\eta_i], \tag{4.14}$$

where K_{ij} is an element of the real-valued K matrix associated with the scattering matrix S by the relation

$$S = \frac{1 + iK}{1 - iK}. \tag{4.15}$$

The phase η_i of an ion's field, considering its long-range nature, is given by

$$\eta_i = k_i r - \frac{l_i \pi}{2} + \frac{z}{k_i} \ln(2k_i r) + \arg\Gamma\left(l_i + 1 - i\,\frac{z}{k_i}\right), \tag{4.16*}$$

where $z = Z - N$ is the ion's charge. For a neutral atom, the last two terms are zero.

The angular momenta of the atom target and the scattered electron in the system wave function can conveniently be coupled into a total angular momentum. Making the transition from a coupling scheme $\gamma_i L_i S_i M_{L_i} M_{S_i} l_i m_i \mu_i$ to a scheme $\gamma_i L_i S_i l_i LSM_L M_S$, we have

$$\Psi_j(x_1, \ldots, x_N, x_{N+1}) = A \sum_i \Theta_i(x_1, \ldots, x_N, \hat{\mathbf{r}}_{N+1}\sigma_{N+1}) \frac{F_{ij}(r_{N+1})}{r_{N+1}}, \quad (4.17)$$

$$\Theta_i = \sum_{M_{L_i} M_{S_i} m_i \mu_i} \begin{bmatrix} L_i & l & L \\ M_{L_i} & m_i & M_L \end{bmatrix} \begin{bmatrix} S_i & \frac{1}{2} & S \\ M_{S_i} & \mu_i & M_S \end{bmatrix} \times$$
$$\times \Phi_i(x_1, \ldots, x_N)\, Y_{l_i m_i}(\hat{\mathbf{r}}_{N+1})\, \delta(\mu_i, \sigma_{N+1}). \quad (4.18)$$

In practice, the summation over i can be limited to a finite number of the most important channels. Known wave functions, e.g., Hartree–Fock, are usually used as the Φ_i. To find the unknown radial functions $F_{ij}(r)$, Kohn's variational principle [167] is used:

$$\delta \left\{ \langle \Psi_i | (H-E) | \Psi_j \rangle - \frac{1}{2} K_{ij} \right\} = 0, \quad (4.19)$$

where Ψ is a trial wave function of the type given by Eq. (4.17) and K_{ij} is an element of the K matrix. When Eq. (4.17) is substituted into (4.19) and the variations of the separate terms are expressed, equations for strong coupling are obtained that have the form [66]

$$\left[\frac{d^2}{dr^2} + 2V_i(r) - \frac{l_i(l_i+1)}{r^2} + 2\varepsilon_i \right] F_{ij}(r) = \sum_{i'} W_{ii'} F_{i'j}. \quad (4.20)^*$$

Here $W_{ii'}$ is an integral operator which, when it operates on the $F_{i'j}$, yields a nonlocalized potential function that takes the relationship between the scattering channels, as well as the exchange interaction between the atom and electron scattered in the ith channel, into account. The direct part of this interaction is described by the potential function $V_i(r)$. If we can ignore the relationship between the channels, Eq. (4.20) transforms to Eq. (1.165).

A modification of the method of strong coupling is the R-matrix method [167], in which solutions of the strong coupling equations are considered only inside the sphere where a strong exchange interaction between the atom and the electron being scattered takes place and are linked with the asymptotic wave functions at the boundary of the sphere.

In principle, the methods of strong coupling and an R-matrix can be used in the range of intermediate energies; however, as the energy decreases, the number of channels that must be taken into account when the wave function is computed increases, which makes the numerical solution of both methods difficult. On the other hand, perturbation theory, the most widely used method for theoretically describing collisions between atoms and fast electrons, often gives satisfactory results for this range, although without a reasonable theoretical basis [164–166].

If the energy of an incident electron is much higher than the binding energy of an excited electron or an electron removed from the atom, they may be examined differently. A fast electron interacts strongly with an atom for only a brief period of time; therefore, the probability of ionizing an electron shell is small and can be found by methods of perturbation theory. This approach to describing collisions between particles is called the Born approximation. If we confine ourselves to first-order perturbation theory, we have a first Born approximation.

The excitation of an atom is usually examined in the nonrelativistic theory because an electron's transition into an outer open shell has higher probability from a subvalent shell and only under an action by nonrelativistic electrons. We will also initially examine the ionization cross section in this approximation.

The interaction between an atom and an incident electron is described by the Hamiltonian

$$H' = \sum_{k=1}^{N} \frac{e^2}{r_{ek}} - \frac{Ze^2}{r_e}, \quad r_{ek} = |\mathbf{r}_e - \mathbf{r}_k|, \tag{4.21}$$

where $\mathbf{r}_e$ is the radius vector of a free electron and $\mathbf{r}_k$ is the radius vector of the atom's bound electron.

If the entire H' is thought of as a perturbation, then, in a zero approximation, we have an atom and a free electron that do not interact with one another. Because exchange effects are small when a fast collision occurs, the system wave function can be written in a zero approximation as the non-antisymmetrized product of the atomic wave function – the plane wave $\psi_{\mathbf{q}}$ and electronic wave functions φ_γ

$$\Psi_i^{(0)} = \Psi_\gamma \varphi_{\mathbf{q}}. \tag{4.22}$$

Let γ_1 and γ_2 be atomic states and $\mathbf{q}_1$ and $\mathbf{q}_2$ be the electronic wave vectors before and after a collision. According to Eq. (1.244), in first-order perturbation theory the probability of a $\gamma_1\mathbf{q}_1 \to \gamma_2\mathbf{q}_2$ transition is

$$dW(\gamma_1\mathbf{q}_1 \to \gamma_2\mathbf{q}_2) = \frac{2\pi}{\hbar} |\langle\gamma_2\mathbf{q}_2|H'|\gamma_1\mathbf{q}_1\rangle|^2 \delta(E_1 - E_2)\, d\mathbf{q}_2. \tag{4.23}$$

Here E_1 and E_2 are the energies of the system's initial and final states in a zero approximation:

$$E_1 = E(\gamma_1) + \frac{\hbar^2 q_1^2}{2m}, \quad E_2 = E(\gamma_2) + \frac{\hbar^2 q_2^2}{2m}. \tag{4.24}$$

In (4.23) the wave function $\varphi_{\mathbf{q}_1}(\mathbf{r})$ is normalized to unit flux (then the probability of excitation is equal to the excitation cross section), and $\varphi_{\mathbf{q}_2}(\mathbf{r}_e)$ to $\delta(\mathbf{q}_2 - \mathbf{q}'_2)$:

$$\varphi_{\mathbf{q}_1}(\mathbf{r}_e) = v_1^{-1/2} e^{i\mathbf{q}_1 \cdot \mathbf{r}_e}, \quad \varphi_{\mathbf{q}_2}(\mathbf{r}_e) = (2\pi)^{-3/2} e^{i\mathbf{q}_2 \cdot \mathbf{r}_e}. \tag{4.25}$$

We will say that the atom's wave functions $\Psi_{\gamma 1}$ and $\Psi_{\gamma 2}$ are orthogonal to one another themselves; then only the two-electron part of the H operator in Eq. (4.21) contributes to the probability of inelastic scattering.

After a collision an atom's state is determined by the total excitation cross section. Substituting Eqs. (4.21) and (4.25) into Eq. (4.23) and integrating over $d\mathbf{r}_e$ using the formula

$$\int e^{i\mathbf{q}\cdot\mathbf{r}} \frac{1}{|\mathbf{r} - \mathbf{r}_k|} d\mathbf{r} = \frac{4\pi}{q^2} e^{i\mathbf{q}\cdot\mathbf{r}_k}, \tag{4.26}$$

and over the wave vector $\mathbf{q}_2$ of the scattered electron, with due regard for the fact that when $\mathbf{q}_1$ is the direction of the z axis

$$q_1 q_2 \sin\theta_2 \, d\theta_2 = q \, dq, \tag{4.27}$$

($\mathbf{q} = \mathbf{q}_1 - \mathbf{q}_2$ is the transferred momentum), we finally obtain

$$\sigma_{\varepsilon_1}^{ex}(\gamma_1 J_1 M_1 \to \gamma_2 J_2 M_2) = $$
$$= 8\pi \left(\frac{me^2}{\hbar^2 q_1}\right)^2 \int_{q_1 - q_2}^{q_1 + q_2} |\langle \gamma_2 J_2 M_2 | \sum_k e^{i\mathbf{q}\cdot\mathbf{r}_k} | \gamma_1 J_1 M_1 \rangle|^2 \frac{dq}{q^3}, \tag{4.28}$$

where ε_1 is the kinetic energy of the incident electron.

If we examine population of levels and not the individual states, Eq. (4.28) must be summed over M_2 and averaged over M_1:

$$\sigma_{\varepsilon_1}^{ex}(\gamma_1 J_1 \to \gamma_2 J_2) = \frac{1}{2J_1 + 1} \sum_{M_1 M_2} \sigma_{\varepsilon_1}^{ex}(\gamma_1 J_1 M_1 \to \gamma_2 J_2 M_2). \tag{4.29}$$

It is convenient to introduce the dimensionless quantity

$$f_q(\gamma_1 J_1, \gamma_2 J_2) = \frac{2m}{\hbar^2} \frac{\Delta E(\gamma_2 J_2, \gamma_1 J_1)}{(2J_1+1)q^2} \times$$
$$\times \sum_{M_1 M_2} \left| \left\langle \gamma_2 J_2 M_2 \right| \sum_k e^{i\mathbf{q}\cdot\mathbf{r}_k} \left| \gamma_1 J_1 M_1 \right\rangle \right|^2, \tag{4.30}$$

the generalized oscillator strength, where ΔE is the excitation energy ($\Delta E > 0$). The operator on the right-hand sides of Eqs. (4.28) and (4.30) can be reduced to an irreducible tensor by expanding the exponent over the spherical harmonics:

$$e^{i\mathbf{q}\cdot\mathbf{r}_k} = \sum_{t=0}^{\infty} i^t [t] j_t(qr_k) \sum_{p=-t}^{t} (-1)^p C_{-p}^{(t)}(\hat{\mathbf{q}}) C_p^{(t)}(\hat{\mathbf{r}}_k). \tag{4.31}$$

Here, $j_t(qr)$ is a spherical Bessel function; $\hat{\mathbf{n}}$ is a unit vector in the $\mathbf{n}$ direction and is defined by the spherical angles $\theta_{\mathbf{n}}$ and $\varphi_{\mathbf{n}}$.

The matrix element for the operator of Eq. (4.31) is, according to Eq. (1.56), equal to

$$\langle \gamma_2 J_2 M_2 | e^{i\mathbf{q}\cdot\mathbf{r}} | \gamma_1 J_1 M_1 \rangle = \sum_t i^t [t] \sum_p (-1)^p C_{-p}^{(t)}(\hat{\mathbf{q}}) (-1)^{J_2 - M_2} \times$$
$$\times \begin{pmatrix} J_2 & t & J_1 \\ -M_2 & p & M_1 \end{pmatrix} \left\langle \gamma_2 J_2 \right\| \sum_k C^{(t)}(\hat{\mathbf{r}}_k) j_t(qr_k) \left\| \gamma_1 J_1 \right\rangle. \tag{4.32}$$

Substituting Eq. (4.32) into (4.28) and (4.29), summing over M_1 and M_2 by means of Eq. (1.13), and using the theorem for the addition of spherical functions, $\Sigma_p \mid C_p^{(t)} \mid^2 = 1$, we find

$$f_q(\gamma_1 J_1, \gamma_2 J_2) =$$
$$= \frac{2m\,\Delta E(\gamma_2 J_2, \gamma_1 J_1)}{\hbar^2 (2J_1+1) q^2} \sum_{t>0} [t] \left| \left\langle \gamma_2 J_2 \right\| \sum_k j_t(qr_k) C^{(t)}(\hat{\mathbf{r}}_k) \left\| \gamma_1 J_1 \right\rangle \right|^2. \tag{4.33}$$

The sum over t is restricted from above by the triangular condition $\{J_1 J_2 t\}$. Its individual terms have the same spin-angular coefficients as the submatrix element of the respective multipole transition operator given by Eq. (1.259).

As $q \to 0, j_t(qr)$ behaves like $(qr)^t$ and the main contribution to Eq. (4.33) for transitions that are allowed in a dipole approximation has only a term with $t = 1$. Then f_q turns into the usual oscillator strength for dipole transitions:

$$\lim_{q \to 0} f_q(\gamma_1 J_1, \gamma_2 J_2) = f(\gamma_1 J_1, \gamma_2 J_2). \tag{4.34}$$

The use of f_q is convenient because it is independent of the energy ε_1 of the incident electron. The excitation cross section as a function of ε_1 is found by integrating f_q over dq (the integral depends on ε_1 through the limits of integration):

$$\sigma^{ex}_{\varepsilon_1}(\gamma_1 J_1, \gamma_2 J_2) = \frac{8\pi a_0^2 (\mathrm{Ry})^2}{\varepsilon_1 \Delta E} \int\limits_{q_1 - q_2}^{q_1 + q_2} f_q(\gamma_1 J_1, \gamma_2 J_2) \frac{dq}{q}. \tag{4.35}$$

Here, Ry is the Rydberg, equal to 0.5 a.u. of energy, and a_0 is the Bohr radius.

If $\Delta E/\varepsilon_1 << 1$, for transitions allowed in a dipole approximation the integral is determined mainly by the term with $t = 1$ for small q. Substituting the asymptotic expression for $j_1(qr)$ into the integral, using approximate limits of integration [the lower, which follows from the relation

$$\Delta E = \frac{\hbar^2}{2m}(q_1^2 - q_2^2) \approx \frac{\hbar^2 q_1}{m}(q_1 - q_2), \tag{4.36}$$

and the upper q_0, defined, for example, from the requirement that the cross section for some energy coincide with the result calculated from Eq. (4.35)], we arrive at Bethe's formula:

$$\sigma^{ex}_{\varepsilon_1}(\gamma_1 \to \gamma_2) = \frac{4\pi a_0^2}{\varepsilon_1/\mathrm{Ry}} [A_{\gamma_1\gamma_2} \ln(\varepsilon_1/\mathrm{Ry}) - B_{\gamma_1\gamma_2}], \tag{4.37}$$

$$A_{\gamma_1\gamma_2} = \frac{\mathrm{Ry}}{\Delta E(\gamma_1, \gamma_2)} f(\gamma_1, \gamma_2), \quad B_{\gamma_1\gamma_2} = A_{\gamma_1\gamma_2} \ln\left[\frac{4(\mathrm{Ry})^2 q_0^2 a_0^2}{\Delta E(\gamma_1, \gamma_2)^2}\right]. \tag{4.38}$$

The A and B parameters are independent of the type (electron or proton) and energy of the incident particles and depend only on the properties of the target. The approximation given by Eq. (4.37) is the first term in an expansion of the Born cross section σ^{ex} in powers of ε_1^{-1}. The third term in the series in square brackets is equal to $C_{\gamma_1\gamma_2}\,\mathrm{Ry}/\varepsilon_1$ [165]. Equation (4.37) is frequently used as a semiempirical two- or three-parameter equation.

Taking the exchange between an incident electron and the atomic electrons (the Born–Oppenheimer approximation) into account makes the expression for the cross section more complex, but the refinement proves to be insignificant for transitions that are allowed in the dipole approximation. The exchange effect is more important for forbidden transitions [166].

The first Born approximation can be made more accurate by including a part of the interaction between the atom and incident electron, for example, the Hartree—Fock potential energy $u(r_e)$ in the zeroth-order Hamiltonian. Then the perturbation Hamiltonian H' in Eq. (4.23) is equal to

$$H' = \sum_{k=1}^{N} \frac{e^2}{r_{ek}} - u(r_e). \tag{4.39}$$

When an atom is excited or ionized, the matrix element of the one-electron operator $u(r_e)$ goes to zero and the cross section is defined by the electrostatic interaction operator between the atom and a free electron.

The one-channel wave function ψ^+ for a scattered electron that has been distorted by the atomic field must, at infinity, look like the superposition of plane and outgoing spherical waves, and the function ψ^- for a scattered electron (or an electron escaping from the atom) must be the superposition of plane and ingoing spherical waves [56]. The functions $\psi^\pm$ that describe an electron moving in the atom's spherically symmetric field and having a wave vector $\mathbf{q}$ and spin projection μ are expanded in partial waves:

$$\psi^{\mp}_{\mathbf{q}\mu}(\mathbf{r}, \sigma) = \frac{1}{q} \sum_{lm} i^l e^{\mp i\vartheta_l} Y^*_{lm}(\hat{\mathbf{q}}) Y_{lm}(\hat{\mathbf{r}}) \frac{P_{ql}(r)}{r} \chi_\mu(\sigma), \tag{4.40}$$

where Y_{lm} is the spherical harmonic, χ_μ is the spin function, and P_{ql} is the radial wave function; ϑ_l is the scattering phase; the $\psi^\pm$ are normalized to $\delta(\mathbf{q} - \mathbf{q}')$, and the $P_{ql}(r)$ are normalized to $\delta(q - q')$.

Substituting Eq. (4.40) into (4.23), integrating over the escape angles of the scattered electron, averaging over the angles of the incident electron, summing over the projections of the angular momentum of the final state, averaging over the projections of the angular momenta of the initial state, and recoupling the angular momenta of the atom and electron into a total angular momentum gives us the following expression for the total excitation cross section of an atom in an approximation of the distorted waves [66, 168]:

$$\sigma^{ex}_{\varepsilon_1}(\gamma_1 J_1, \gamma_2 J_2) = \frac{2\pi^3 e^4}{(2J_1+1) q_1^2} \times$$
$$\times \sum_{l_1 l_2 j_1 j_2 J} [J] \left| \left\langle \gamma_2 J_2 \varepsilon_2 l_2 j_2 J \right| \sum_k \frac{1}{r_{ek}} \left| \gamma_1 J_1 \varepsilon_1 l_1 j_1 J \right\rangle \right|^2, \quad \varepsilon_2 = \varepsilon_1 - \Delta E. \tag{4.41}$$

The wave functions $P_{\varepsilon l}(r)$ here have been normalized to $\delta(\varepsilon - \varepsilon')$. If the ε, q_1, e, and r_{ek} in Eq. (4.41) are defined in atomic units, the constant in the right-hand side of Eq. (4.41) is equal to $2\pi^3 a_0^2$.

There are two free electrons when ionization occurs; therefore, a theoretical description of ionization is more complicated than for excitation. When the same methods are used for the ionization cross section the results obtained are also in less satisfactory agreement with experimental values.

An expression for the ionization cross section in the first Born approximation of plane waves is found in a manner similar to that for the excitation cross section.

Instead of the generalized oscillator strength for transitions into the continuous spectrum, its density

$$\frac{df_q(\gamma_1 J_1, \gamma_2 J_2 \varepsilon l j J)}{d\varepsilon} = \frac{2m}{\hbar^2} \frac{\Delta E(\gamma_2 J_2 \varepsilon, \gamma_1 J_1)}{(2J_1+1) q^2} \times$$
$$\times \sum_{t>0} [t] \left| \left\langle \gamma_2 J_2 \varepsilon l j J \left\| \sum_k C^{(t)}(\hat{\mathbf{r}}_k) j_t(q r_k) \right\| \gamma_1 J_1 \right\rangle \right|^2, \tag{4.42}$$

$$\Delta E(\gamma_2 J_2 \varepsilon, \gamma_1 J_1) = [E^{ion}(\gamma_2 J_2) - E^{at}(\gamma_1 J_1)] + \varepsilon. \tag{4.43}$$

is used. The total ionization cross section is found as

$$\sigma_{\varepsilon_1}^{ion}(\gamma_1 J_1, \gamma_2 J_2) = \frac{8\pi a_0^2 (\mathrm{Ry})^2}{\varepsilon_1} \int_0^{\varepsilon_{max}} \frac{1}{\Delta E(\gamma_2 J_2 \varepsilon, \gamma_1 J_1)} \times \tag{4.44}$$
$$\times \int_{q_{min}}^{q_{max}} \sum_{ljJ} \frac{df_q(\gamma_1 J_1, \gamma_2 J_2 \varepsilon l j J)}{d\varepsilon} \frac{dq}{q} \, d\varepsilon,$$

where ε_1 is the energy of the incident electron and the other notations are

$$\varepsilon_{max} = \varepsilon_1 - [E^{at}(\gamma_1 J_1) - E^{ion}(\gamma_2 J_2)], \tag{4.45}$$

$$q_{min} = \frac{\sqrt{2m}}{\hbar} (\sqrt{\varepsilon_1} - \sqrt{\varepsilon_1 - \Delta E}), \quad q_{max} = \frac{\sqrt{2m}}{\hbar} (\sqrt{\varepsilon_1} + \sqrt{\varepsilon_1 - \Delta E}). \tag{4.46}$$

An expression for the total ionization cross section for a shell contains a triple integral in ε, q, and r. For the case of a closed shell, the general formula yields

$$\sigma_{\varepsilon_1}^{ion}(nlj^{-1}) = \frac{8\pi \mathrm{Ry}}{\varepsilon_1} \frac{2j+1}{2l+1} \sum_{l'} \sum_{t>0} [t]\langle l \| C^{(t)} \| l' \rangle^2 \times$$

$$\times \int_0^{\varepsilon_{max}} \int_{q_{min}}^{q_{max}} \frac{1}{q^3} \langle nl | j_t(qr) | \varepsilon l' \rangle^2 \, dq \, d\varepsilon. \tag{4.47}$$

Here $P_{\varepsilon l}(r)$ is a wave function normalized to $\delta(\varepsilon - \varepsilon')$ for a slow or a secondary electron. In contrast to the functions of incident or fast escaping electrons that are plane waves, $P_{\varepsilon l}(r)$ is usually calculated in the field of the target.

When the energies of the primary electrons are high (but nonrelativistic), σ^{ion} can also be approximated by Bethe's formula:

$$\sigma_{\varepsilon_1}^{ion} = \frac{4\pi a_0^2}{\varepsilon_1/\mathrm{Ry}} \left[A^{ion} \ln\left(\frac{\varepsilon_1}{\mathrm{Ry}}\right) + B^{ion} + \frac{C^{ion}\,\mathrm{Ry}}{\varepsilon_1} \right], \tag{4.48}$$

where

$$A^{ion} = \int_{\varepsilon_0}^{\infty} \frac{\mathrm{Ry}}{\varepsilon} \frac{df}{d\varepsilon} \, d\varepsilon, \tag{4.49}$$

and the constants B^{ion} and C^{ion} are independent of the incident particle's energy. Equation (4.42) can be made more accurate by adding a correction that is defined by the exchange interaction between the atom and scattered electron [165]:

$$\Delta\sigma_{\varepsilon_1}^{ion} = \frac{4\pi a_0^2}{(\varepsilon_1/\mathrm{Ry})^2} \sum_{nl} N_{nl} \left[1 + \ln\left(\frac{I_{nl}}{\varepsilon_1}\right) \right], \tag{4.50}$$

where I_{nl} is the binding energy and N_{nl} is the number of electrons in the shell or subshell being ionized.

The approximate semiempirical formulas of Lotz [169], Vainshtein et al. [170], the modified Mott formula [165], etc. are frequently used to calculate the total ionization cross section.

When the energy of the incident electron is $\varepsilon_1 > 10$ keV the electron mass as a function of its velocity begins to affect the ionization cross section, and when the inner shells are ionized the relativistic features of their structure become significant. Bethe's formula,

Eq. (4.48), refined with due regard for the relativistic expression for q and the contribution from the small component of a bound electron's relativistic wave function [171]

$$\sigma^{ion} = \frac{4\pi a_0^2 \alpha^2}{\beta^2} \left\{ A^{ion} \left[\ln \left(\frac{\beta^2}{1-\beta^2} \right) - \beta^2 \right] + B_1^{ion} \right\}, \tag{4.51}$$

$$\beta = \frac{v_1}{c}, \quad B_1^{ion} = A^{ion} \ln \left(\frac{2mc^2}{\text{Ry}} \right) + B^{ion}. \tag{4.52}$$

is applicable far away from the ionization threshold. Here v_1 is the incident particle's velocity, and A^{ion} and B^{ion} are nonrelativistic parameters that appear in Eq. (4.48). [Relativistic corrections to the A and B parameters such as these can also be included in the excitation cross section, Eq. (4.37).]

A more consistent accounting of relativistic effects is needed when a deep inner subshell is ionized by relativistic electrons. An expression for the ionization cross section of a subshell in a relativistic first Born approximation was given and the ionization cross sections for the K and L layers for atoms $18 \leq Z \leq 92$ were tabulated in [172] for a wide range of incident electron energies. Among the other relativistic methods for calculating σ^{ion}, we note the semiclassical method in which close and distant collisions are considered separately [173] and a method borrowed from nuclear physics that lets us give the cross section in terms of Lorentz-invariant structure functions in an approximation of a one-photon exchange [174].

If an atom has an outer open shell that interacts weakly with an inner vacancy, we can use an approximation of a ground term that assumes that the outer shell maintains its own term when the vacancy is created when ionization by photons or electrons occurs [113, 175]. Actually, in a pure coupling the ionization cross section is diagonal relative to the quantum numbers of the passive shell and if the interaction between this shell and the vacancy is weak, mixing of the terms may be neglected.

For an LS coupling the electrons in an outer shell approximately maintain only their spin angular momentum in a stronger interaction (because of the tendency for arrangement of terms having different multiplicity – from the bottom up in order of decreasing S [98] – and the fact that the transition operator is independent of the spin variables).

4.3. The Creation of States Having Inner Vacancies during Collisions between Atoms and Ions

The collision between an atom and an ion represents an exceptionally strong perturbation of the electron shells which results in the creation of vacancies in the different shells because of multiple ionization and excitation of the atom, and electron capture by the ion. A theoretical description of ion–atom collisions is still rather approximate and incomplete – equations obtained mainly from one-electron and semiclassical models that can be used in practice are known for only a few collision regimes. We can distinguish two areas in which theoretical methods can be used [176–178].

1. The Atomic Area. If a target's nuclear charge $Z_t e$ is much greater than an incident ion's nuclear charge $Z_p e$ and the relative velocity of the colliding particles v_p is not small in comparison with the average velocity v_e of an electron in an atom's inner shell, excitation or ionization of an atom can be thought of as a perturbation by the ion's Coulomb field. When this is done the ion is considered to be a structureless, and even a semiclassical, particle. The basic methods that can be used in the atomic area are the first Born approximation of plane waves, the semiclassical method of impact parameter, and the classical binary encounter approximation.

2. The Quasimolecular Area. If an ion's nuclear charge is comparable to that of the atom target and the collision is slow ($v_p << \bar{v}_e$), perturbation theory cannot be used and the creation of vacancies in the inner shells is examined in a quasimolecular model basis as an adiabatic change in the one-electron levels and states in the field of two nuclei.

Expressions for the excitation and ionization cross sections of an atom by an ion or proton are obtained in the first Born approximation from the corresponding formulas for electron–atom collisions [Eqs. (4.35, 4.37, 4.44, 4.47, 4.48)] by multiplying them by $(Z_p^* e)^2$ – the square of the ion's effective charge with due regard for screening of the nucleus, and using the quantity $\varepsilon_1 m/M$ instead of the incident particle's initial energy ε_1, where M is the particle mass and m is the electron mass. The difference in inelastic scattering cross sections of heavy charged particles in comparison with electrons is mainly determined by different limits on the integration with respect to transferred momentum [176]: when an atom collides with a heavy particle, $q_{\min}$ is fairly large and $q_{\max}$ can be considered infinite for all practical purposes. If the ionization cross sections for the different shells reach their maximal values at electron energies of from 0.1 to 100 keV during electron—atom collisions, these maxima will be reached during proton—atom collisions at energies of from 0.2 to 200 MeV, and at even higher energies when bombarded by ions [176].

The first Born approximation of plane waves for ion–atom collisions is made more accurate by taking relativistic effects for the atom's electron into account (an ion's motion is ordinarily considered to be nonrelativistic), increasing the binding energy of an electron in the ion's field, and using distorted, rather than plane, waves [176, 179].

When $Z_p << Z_t$ for a fairly large range of velocities, excitation of an atom by an ion's Coulomb field can be described by the quasiclassical method of impact parameter [176, 179, 180].

If we assume that an ion moves along a classical straight-line trajectory at a relative velocity v, then the atom's excitation cross section from a state β into a state γ is

$$\sigma_v(\beta \to \gamma) = 2\pi \int_0^\infty W_{b,v}(\beta \to \gamma)\, b\, db, \tag{4.53}$$

where $W_{b,v}$ is the probability of such a transition in a collision having a collision impact parameter b. The quantity $W_{b,v}$ is calculated by means of time-dependent perturbation theory by solving the system of equations [66]

$$i\hbar \frac{dc_\gamma(t)}{dt} = \sum_{\gamma'} \left\langle \gamma \left| -\sum_k \frac{e^2 Z_p^*}{|\mathbf{R}(t) - \mathbf{r}_k|} \right| \gamma' \right\rangle \exp\left[\frac{i}{\hbar}(E_\gamma - E_{\gamma'})t\right] c_{\gamma'}(t). \tag{4.54}$$

Here $c_\gamma(t)$ is a time-dependent coefficient in the expansion of the wave function for the atom's final state with respect to its unperturbed states, $\mathbf{R}(t)$ is the vector for the relative distance between the nuclei of the ion and the target, E_γ is the energy of an isolated atom in a state γ, and $Z_p^* e$ is the ion's effective charge.

The unknown probability of the $\beta \to \gamma$ transition is obtained in first-order perturbation theory by making the substitution $c_{\gamma'}^{(0)} = \delta(\gamma', \beta)$ on the right-hand side of Eq. (4.54) and integrating Eq. (4.54) with respect to time:

$$W_{b,v}(\beta \to \gamma) = \left| \frac{1}{i\hbar} \int_{-\infty}^{\infty} c_\gamma(t)\, dt \right|^2 = \tag{4.55}$$

$$= (Z_p^* e^2)^2 \left| \int_{-\infty}^{\infty} \left\langle \gamma \left| \sum_k |\mathbf{R}(t) - \mathbf{r}_k|^{-1} \right| \beta \right\rangle \exp\left[\frac{i}{\hbar}(E_\gamma - E_\beta)t\right] dt \right|^2 .$$

Using the Fourier transform

$$|\mathbf{R} - \mathbf{r}|^{-1} = \frac{1}{2\pi^2} \int e^{i\mathbf{q}\cdot(\mathbf{r}-\mathbf{R})} \frac{1}{q^2}\, d\mathbf{q}, \tag{4.56}$$

the probability can be factored into an integral over the incident particle's orbit and an atomic quantity, which are related by an integration over $d\mathbf{q}$ (this is the advantage of the method):

$$\sigma_v(\beta \to \gamma) = \frac{(Z_p^* e^2)^2}{2\pi^3} \int_0^\infty \left| \int_{-\infty}^{\infty} dt \exp\left[\frac{i}{\hbar}(E_\gamma - E_\beta)t\right] \times \right.$$

$$\left. \times \int \left\langle \gamma \left| \sum_k e^{i\mathbf{q}\cdot\mathbf{r}_k} \right| \beta \right\rangle e^{-i\mathbf{q}\cdot\mathbf{R}(t)} \frac{1}{q^2} d\mathbf{q} \right|^2 bdb. \tag{4.57}$$

With some simplifications (the ion's trajectory is a straight line and its velocity is constant) the integration in Eq. (4.57) can be done with respect to dt, db, and, in part, $d\mathbf{q}$. The excitation cross section then becomes similar to the cross section in the first Born approximation given by Eq. (4.35) [181].

We can also use a classical approach – the method of binary collisions – to examine an inelastic collision between an atom and an ion in the atomic area [182]. In this method an electron and ion are considered to be classical point charge particles. In using the laws of conservation of energy and momentum and the law of Coulomb interaction, the probability that an amount of energy ΔE is transferred to the electron is given as a function of the ion's initial velocity v_1, the velocity v_2 of the electron removed from the atom, the impact parameter, and the scattering angles. When integrating the probability with respect to the angles, the collision parameter, and, in the case of ionization, with respect to the transferred energy, an algebraic formula for $\sigma(v_1, v_2)$ is obtained. To find the total ionization cross section for an nl-shell, $\sigma(v_1, v_2)$ is averaged over a bound electron's velocity distribution and multiplied by the number of equivalent electrons:

$$\sigma_{v_1}(nl^{-1}) = N_{nl} \int_0^\infty \sigma(v_1, v_2)\, f_{nl}(v_2)\, dv_2. \tag{4.58}$$

Here, $f_{nl}(v_2)$ is a bound electron's velocity distribution function given as the square of the absolute value of the electron wave function in momentum space:

$$f_{nl}(v) = 4\pi p^2 \frac{m}{2l+1} \sum_{m_l=-l}^{l} |u_{nlm_l}(\mathbf{p})|^2, \tag{4.59}$$

where $u_{nlm_l}(\mathbf{p})$ is the Fourier transform of the one-electron wave function in coordinate space, Eq. (1.6) (see Sec. 10.2). An algebraic expression is obtained when one-electron wave functions are used for the $f_{nl}(v)$, e.g., for a $1s$-electron

$$f_{1s}(v)=\frac{32}{\pi}\frac{v_0^5 v^2}{(v^2+v_0^2)^4}, \quad v_0=\left[\frac{2I_{1s}}{m}\right]^{1/2}. \tag{4.60}$$

The following simple expression for the cross section of an nl^{-1} vacancy creation is valid in a hydrogenic approximation [177]:

$$\sigma_{v_1}(nl^{-1})=\frac{N_{nl}\pi(Z_p^* e^2)^2}{I_{nl}^2} G(V). \tag{4.61}$$

where V is the reduced velocity, equal to the ratio of v_1 to an electron's average orbital velocity $\bar{v}_e$. The universal function $G(V)$ has been tabulated, for example, in [177].

The basic method for calculating the population densities of the levels for a slow collision ($v_p << \bar{v}_e$) and a fairly strong perturbation ($Z_p \approx Z_t$) is the quasimolecular method, which treats a collision as the formation and decay of an unstable two-center molecule [183, 178]. When an atom and an ion approach one another slowly, their electrons adiabatically adapt to the two-center field and one-electron molecular orbitals are formed. X-ray spectra in the form of molecular bands being emitted during these types of collisions bear witness to this phenomenon [184]. The energies of the different molecular orbitals depend differently on the internuclear distance; therefore, the energy levels approach one another during a collision and the order in which they are arranged changes. When this happens the electrons in closed shells may make a transition to excited levels. The possibilities for transitions of these kinds are determined from correlation diagrams – diagrams of the changes in molecular levels as a function of the internuclear distance. The transition probabilities are given by the overlap integrals for the time-dependent wave function of a quasimolecular system with functions of its final separated state as $t \to \infty$. The quasimolecular method also makes it possible to describe a process that only occurs when atoms collide with ions or nuclei – the reversal of charge when the atom's electrons are captured by the incident particle [66]. This process is most likely when the charge $Z_p e$ is large, but its cross section falls off sharply as the velocity increases relative to the motion of the nuclei [66].

The probability that configurations with many vacancies will be formed during ion–atom collisions is high. The multiple ionization cross section for the outer and intermediate shells can be approximated by a single ionization cross section via the statistical quasiclassical method [185, 177]. The possibility of the statistical approach is based on the assumption that the characteristic relaxation time of electron shells is high in comparison with the duration of a collision; therefore, the interaction between the incident ion and an electron is only weakly dependent on the states of the other electrons in the target.

Let $w_b(n_1 l_1^{-1})$ be the probability that a vacancy $n_1 l_1^{-1}$ will be created in the one-electron state during an ion–atom collision having impact parameter b. The probability that this electron will not be removed is $1 - w_b(n_1 l_1^{-1})$. If we assume that $w_b(n_1 l_1^{-1})$ is independent of the number of electrons N_1 in the shell, the probability that m_1 vacancies will be created in the shell is

$$W_b(n_1 l_1^{-m_1}) = \binom{N_1}{m_1} w_b(n_1 l_1^{-1})^{m_1} [1 - w_b(n_1 l_1^{-1})]^{N_1 - m_1}. \tag{4.62}$$

The probability that vacancies will be created in two shells (or subshells) is given as the product of the probabilities of their independent ionization, Eq. (4.62):

$$W_b(n_1 l_1^{-m_1} n_2 l_2^{-m_2}) = W_b(n_1 l_1^{-m_1}) W_b(n_2 l_2^{-m_2}). \tag{4.63}$$

The total cross section for the process is obtained in the semiclassical approximation by substituting Eqs. (4.62) and (4.63) for a given impact parameter b into Eq. (4.53).

When atoms collide with multiply charged ions, especially when the relative velocities of the nuclei are low, we must take not only direct ionization into account, but the capture of the atom's electrons by the ion as well.

Because of the approximate nature of the methods for describing ion-atom collisions and the complexity of the x-ray and Auger spectra that are stimulated by an ion beam, their interpretation is often limited only by the group of lines. When this happens the population densities are either averaged over the many-electron quantum numbers, or it is assumed that all states in a given configuration are identically populated.

4.4. Using the Method of Sudden Perturbation to Calculate the Probabilities of Many-Electron Transitions

Not only can ion–atom collisions create several vacancies in an atom's electron shells, but – with lower probability – so can interactions between the atom and electrons and photons, many-electron radiative transitions and Auger transitions in the atoms, and nuclear transformations. A successive calculation of the probabilities of such many-electron processes requires that complex correlation methods be used (see Secs. 5.3, 7.2, and 8.3). For this reason, the results of such calculations are not abundant. A less accurate, but fairly simple and universal method of describing many-electron processes that uses a model of an atom's sudden perturbation is widely used for interpreting satellite lines in x-ray and electron spectra [186,189].

Assume that until time $t = 0$ an atom is in a stationary state n which can be described by the wave function of the Hamiltonian H_0

$$H_0 u_n = E_n u_n, \tag{4.64}$$

and that the atom has been suddenly perturbed during a small time interval $0 < t < \tau$ (for example, the nuclear charge has changed). The Hamiltonian for the new system is H_1 and corresponds to the eigenfunctions v_m:

$$H_1 v_m = E'_m v_m. \tag{4.65}$$

Because the perturbation was sudden, the state of the electron system has not changed and is described at time τ by the wave function

$$\Psi(\tau) = u_n \exp\left[-\frac{i}{\hbar} E_n \tau\right]. \tag{4.66}$$

Expanding this function over the total basis of wave functions for the Hamiltonian of the new system gives us

$$u_n \exp\left[-\frac{i}{\hbar} E_n \tau\right] = \sum_m c_m v_m \exp\left[-\frac{i}{\hbar} E'_m \tau\right]. \tag{4.67}$$

If the condition

$$\tau \ll \frac{\hbar}{E'_m - E_n}, \tag{4.68}$$

is satisfied, then

$$\exp\left[\frac{i}{\hbar}(E'_m - E_n)\tau\right] \approx 1. \tag{4.69}$$

Multiplying Eq. (4.67) by $v_{m'}{}^*$, integrating in coordinate space, and taking Eq. (4.69) into account, we find an expression for the probability of transition from a state n into a state m of the new Hamiltonian:

$$W(n \to m) = |c_m|^2 = |\langle v_m | u_n \rangle|^2 = |\langle u_n | v_m \rangle|^2. \tag{4.70}$$

We will now turn our attention to the fact that no assumption about the smallness of the perturbation was assumed when Eq. (4.70) was derived – only that it was of brief duration [Eq. (4.68)].

If n and m differ by the state of a single electron, the denominator on the right-hand side of Eq. (4.68) is approximately equal to the difference in the one-electron energies or, if we discard the smallest of them (the energy of an excited electron), it is equal to the binding energy I of an electron in the initial state. Thus, a perturbation can be said to be sudden if its duration is less than the period $\hbar/I$, which is on the order of $10^{-13}-10^{-17}$ sec.

In general, the shake model assumes that a system can be divided into two subsystems, the interaction between which is described by an average field; in one subsystem a quantum jump occurs, which acts like a sudden change in the second subsystem's field and leads to its excitation or ionization [186]. Thus, a many-particle process is divided into two steps and the total probability is the product

$$W = W_1 W_2, \tag{4.71}$$

where W_1 is the probability of changing the nuclear charge, producing a vacancy, or its transition from one shell into another when the other electrons are in a "frozen" state, and W_2 is the probability that the subsystem of passive electrons will be excited or ionized when shaken and is independent of the nature of the agitation.

Thus, a necessary condition for using the method of sudden perturbation is that not only must the perturbation be brief, but that the coupling between both subsystems be comparatively weak – their wave functions overlap only slightly. This comes down to the requirement that electrons taking part in a process that creates a perturbation, and electrons experiencing this perturbation, be in different shells and even different layers. This condition is hard to satisfy if the first process is an x-ray or Auger transition that affects several shells. For this reason the method of sudden perturbation is used less often and with less success to describe many-electron transitions in atoms than to describe multiple ionizations or the ionization and excitation of atoms when they interact with particles or x-ray beams.

Possible results of a sudden perturbation are the processes of shake-up – the excitation of a passive electron; shake-off – the supplemental ionization of the atom; and, much less probable, that several of the atom's electrons will be excited or removed. If an electron in an excited state is perturbed, it can also make a transition to a lower state (shake-down).

The probability of a primary process (W_1) that can be described by a one- (two-) electron operator is calculated in a one- (two-) electron approximation. If we are interested in only the probability that a supplemental vacancy will be created, and not in the atom's many-electron state, we can use the one-electron model to calculate W_2. Then the probability that a vacancy will be created is conveniently given as the probability that the electron will remain in its own shell or subshell. This latter quantity is equal to the square of the overlap integral of the electron's radial wave functions in the atom and ion having this vacancy (the overlap integral for an electron's spin—orbital wave functions is equal to the Kronecker δ-function). Raising this probability to a power equal to the number of electrons in the shell or subshell gives us the probability that all equivalent electrons will not change state during a sudden perturbation, and subtracting this probability from unity gives us the probability that at least one equivalent electron will be removed. In relativistic and nonrelativistic theories we have

$$W^{rel}(nlj^{-1})= \\ =1-\left\{\int [P^a_{nlj}(r)P^i_{nlj}(r)+Q^a_{n\bar{l}j}(r)Q^i_{n\bar{l}j}(r)]\,dr\right\}^{2N_{nlj}}-W^{rel}_F, \tag{4.72}$$

$$W(nl^{-1})=1-\left(\int P^a_{nl}(r)P^i_{nl}(r)\,dr\right)^{2N_{nl}}-W_F. \tag{4.73}$$

The indices a and i denote the wave functions for an atom and an ion (having a vacancy whose appearance caused the perturbation). The quantity W_F is a correction that takes the impossibility of shaking electrons up into occupied, one-electron states into account. For example, for Eq. (4.72),

$$W^{rel}_F=N_{nlj}\sum_{n'}\frac{N_{n'lj}}{2j+1}\int [P^a_{nlj}(r)P^i_{n'lj}(r)+Q^a_{n\bar{l}j}(r)Q^i_{n'\bar{l}j}(r)]\,dr. \tag{4.74}$$

Another approximate formula for the probability that a vacancy will be created is obtained from Eq. (4.62) for $m = 1$ when the probability's dependence on the b parameter is neglected and when $w(nl^{-1})$ is given in terms of the overlap of the one-electron wave functions $1 - \int P_{nl}^{a}(r)P_{nl}^{i}(r)dr$.

Because spin–orbital wave functions are orthogonal, transitions are only possible during a sudden perturbation in a one-configuration approximation when one-electron quantum numbers, except n, are preserved.

The results from calculating shake-off probabilities for the atoms of inert gases and other elements [189, 9] show that this process is more likely for outer shell electrons when a vacancy is created in an inner shell. The shake-off probability decreases as nuclear charge increases and the total probability of creating a vacancy in any shell is almost independent of Z. The shake-up of inner shell electrons has low probability because their wave functions weakly overlap the functions of the excited states and the probability of this process for outer electrons is comparable to the shake-off probability [189].

In some cases the electron shell shake-up and shake-off probabilities, which are functions of the many-electron quantum numbers, are also needed in calculating the populations of the levels. The corresponding expressions are obtained directly from Eq. (4.70) when the many-electron wave functions of an atom and an ion are only used for a sudden perturbation caused by a nuclear transformation, e.g., β-decay of the nucleus. If a perturbation results from the creation of a vacancy in an inner shell, then in the initial state the subsystem of passive electrons is not described by a proper wave function and a density matrix must be introduced.

We will assume that for $t < 0$ an N-electron atom is in the state γ that corresponds to the wave function $\Psi(\gamma)$ and that the state of the subsystem from the $(N-1)$-electron is described by a reduced density matrix of order $(N-1)$:

$$\Gamma_{\gamma}^{(N-1)}(x_1', \ldots, x_{N-1}', t \mid x_1, \ldots, x_{N-1}, t) = \\ = N! \int \Psi^*(\gamma \mid x_1', \ldots, x_{N-1}', x_N, t)\Psi(\gamma \mid x_1, \ldots, x_{N-1}, x_N, t)\, dx_N. \quad (4.75)$$

If a vacancy is created suddenly, the density matrix cannot be changed. Its expansion over the eigenfunctions Φ of the Hamiltonian for the $(N-1)$-electron system has the form

$$\Gamma_{\gamma}^{(N-1)}(x_1', \ldots, x_{N-1}', t \mid x_1, \ldots, x_{N-1}, t) = \\ = N! \sum_{\gamma'\gamma''} c_{\gamma\gamma'}^* c_{\gamma\gamma''} \Phi^*(\gamma' \mid x_1', \ldots, x_{N-1}', t)\Phi(\gamma'' \mid x_1, \ldots, x_{N-1}, t). \quad (4.76)$$

The summation in Eq. (4.76) contains an integration over the continuous spectrum. Comparing Eqs. (4.75) and (4.76) at $t = 0$, multiplying both sides of Eq. (4.76) by $\Phi(\gamma' \mid \gamma_1', \ldots, \gamma_{N-1}')\Phi^*(\gamma' \mid x_1, \ldots, x_{N-1})$, and integrating with respect to $dx_1, \ldots, dx_{N-1}$ and $dx_1', \ldots, dx_{N-1}'$ gives us the probability of a $\gamma \to \gamma'$ transition due to a sudden perturbation:

$$W(\gamma \to \gamma') = |c_{\gamma\gamma'}|^2 = \int dx_N \left| \int \Phi^*(\gamma' \mid x_1, \ldots, x_{N-1}) \times \right. \\ \left. \times \Psi(\gamma \mid x_1, \ldots, x_{N-1}, x_N)\, dx_1, \ldots, dx_{N-1} \right|^2. \quad (4.77)$$

Let an atom's configuration contain one open shell, and let an initial vacancy be formed in the closed shell $n_1 l_1^{4l_1+2}$. Equations are obtained for the probabilities of electron shake-up and shake-off from the $n_2 l_2^{N_2}$ shell from Eq. (4.77) by recoupling the angular momenta in the Ψ function and from the expression for the integral as a matrix element for an operator of zero rank in terms of the overlap integrals for one-electron wave functions. When this is done, the exchange interaction between subsystems is ignored along with the small terms that contain the overlap integrals $\langle nl \mid n'l \rangle$ $(n \neq n')$ in higher degrees:

$$W^{su}\left(K_0 n_1 l_1^{4l_1+2} n_2 l_2^{N_2} \gamma_2 L_2 S_2 J \to K_0 n_1 l_1^{4l_1+1} n_2 l_2^{N_2-1} \bar{\gamma}_2 \bar{L}_2 \bar{S}_2 (L_0 S_0) nlL'S'J'\right) =$$

$$= N_2 \delta(l_2, l) \Pi_1^2 [L_2, S_2, J'] \sum_j [j] \left[\sum_{\alpha_2'} c_{\gamma_2'}^{\gamma_2} (-1)^{\bar{L}_2'+\bar{S}_2'+L'+S'} \times \right.$$

$$\times [L_0', S_0', L', S']^{1/2} (l_2^{N_2} \gamma_2 L_2 S_2 \| l_2^{N_2-1} \bar{\gamma}_2' \bar{L}_2' \bar{S}_2' l_2) \times$$

$$\left. \times \begin{Bmatrix} L' & l_1 & L_2 \\ S' & 1/2 & S_2 \\ J' & j & J \end{Bmatrix} \begin{Bmatrix} l_2 & \bar{L}_2' & L_2 \\ l_1 & L' & L_0' \end{Bmatrix} \begin{Bmatrix} \frac{1}{2} & \bar{S}_2' & S_2 \\ \frac{1}{2} & S' & S_0' \end{Bmatrix} \right]^2, \tag{4.78}$$

$$W^{so}(K_0 n_1 l_1^{4l_1+2} n_2 l_2^{N_2} \gamma_2 L_2 S_2 J \to K_0 n_1 l_1^{4l_1+1} n_2 l_2^{N_2-1} \bar{\gamma}_2 \bar{L}_2 \bar{S}_2 L_0 S_0 J_0) =$$

$$= N_2 \Pi_2^2 [L_2, S_2, J_0] \sum_{jj'J'} [j, j', J'] \left[\sum_{\alpha_2'} c_{\alpha_2'}^{\alpha_2} [L_0', S_0']^{1/2} \times \right. \tag{4.79}$$

$$\left. \times (l_2^{N_2} \gamma_2 L_2 S_2 \| l_2^{N_2-1} \bar{\gamma}_2' \bar{L}_2' \bar{S}_2' l_2) \begin{Bmatrix} \bar{L}_2' & \bar{S}_2' & J' \\ l_2 & \frac{1}{2} & j' \\ L_2 & S_2 & J \end{Bmatrix} \begin{Bmatrix} L_0' & S_0' & J_0 \\ l_1 & \frac{1}{2} & j \\ \bar{L}_2' & \bar{S}_2' & J' \end{Bmatrix} \right]^2.$$

Here a_2 is the set of many-electron quantum numbers for the final state, with the exception of J_0; $c_{\alpha 2'}{}^{\alpha 2}$ is the expansion coefficient for the wave function of this state in an intermediate coupling in terms of a pure coupling (for the initial state the function for a pure coupling is used); Π_1 and Π_2 are the products of the radial overlap integrals:

$$\Pi_1 = \int_0^\infty P_{n_1 l_1}^a(r) P_{nl}^i(r)\, dr \prod_{k \in K} s(n_k l_k)^{N_k}, \tag{4.80}$$

$$\Pi_2 = [1 - s(n_2 l_2)^2] \prod_{k \in K} s(n_k l_k)^{2N_k}, \tag{4.81}$$

where $K = K_0 l_1{}^{4l_1+1} l_2{}^{N_2-1}$ and the indices a and i denote the wave functions for an atom and ion having a vacancy $n_1 l_1{}^{-1}$, and

$$s(nl) = \int_0^\infty P^a_{nl}(r)\, P^i_{nl}(r)\, dr. \tag{4.82}$$

The factor Π_2 is transformed so that it, just like Π_1, does not contain an integral over the continuous spectrum function; when this is done, the condition that Eq. (4.79), summed over the many-electron quantum numbers of the final state, should transform to an expression in a one-electron approximation similar to Eq. (4.62) when $m = 1$, is used.

If the $n_2 l_2^{N_2}$ shell is closed, the $n_1 l_1^{-1} n_2 l_2^{-1} \gamma J$ levels are populated proportionally to their statistical weights when the electrons are shaken off.

Chapter 5

X-RAY ABSORPTION SPECTRA

The reduction in the intensity of an x-ray beam as it passes through gases and vapors, especially those of the heavy elements, is caused by several processes: the ionization and excitation of the atoms, dispersion of the beams, and the formation of electron–positron pairs when the photon energy is high.

When the photon energy exceeds the sum of the rest masses of a positron and electron – roughly 1 MeV – and the corresponding radiation wavelength is $\lambda \lesssim 10^{-3}$ nm, pair creation becomes possible. This process relates more to nuclear physics than to atomic physics; hence, it will not be discussed here.

The attenuation of an x-ray beam from 1 MeV to approximately 100 keV is mainly due to the dispersion of the x rays by the atoms (Sec. 10.2). At lower energies corresponding to the binding energies of the electrons in an atom, x-ray absorption due to the photoeffect becomes the dominant process. This chapter will discuss this process and the photoexcitation of an atom, which is closely associated with it. Atomic and macroscopic quantities that describe x-ray absorption and general expressions for the photoionization and photoexcitation cross sections are introduced in Sec. 5.1. These are put into concrete form in Sec. 5.2 for atoms having open shells when one-configuration wave functions are used. Section 5.3 examines theoretical methods for calculating correlation effects in the absorption spectra.

5.1. The Photoabsorption Coefficient and Cross Section

We will examine the absorption of a monochromatic x-ray beam in a gas that consists of the atoms of only a single element. Let the direction of beam propagation be the x axis. The beam intensity $I_\omega(x)$ (the energy passing through a unit surface area positioned perpendicular to the direction of beam propagation per unit time) changes by an amount $dI_\omega(x)$ in the layer between x and $x + dx$, which is proportional to the intensity $I_\omega(x)$ and the layer thickness dx:

$$dI_\omega(x) = -\mu_\omega I_\omega(x)\,dx. \tag{5.1}$$

The negative sign indicates the reduction in intensity. The quantity μ_ω is the linear absorption coefficient for x rays having an energy of $\hbar\omega$ in a given material (in general, when the different beam attenuation processes are considered, this coefficient must be replaced by the sum of the absorption, beam dispersion, and pair creation coefficients).

Using the boundary condition that $I(0) = I_0$ and integrating Eq. (5.1) gives

$$I_\omega(x) = I_0 e^{-\mu_\omega x}. \tag{5.2}$$

According to Eq. (5.2), the absorption coefficient has a dimensionality of L^{-1} and is equal to $1/d$, where d is the distance at which the beam will be attenuated by a factor of e.

The mass absorption coefficient μ_ω/ρ is frequently used, where ρ is the density of the material. This coefficient describes the relative absorption of x rays per unit mass of material.

Along with the absorption coefficient, which is a macroscopic quantity, an atomic quantity – the effective absorption cross section, σ_ω, which describes how x rays of frequency ω are absorbed by a single atom – is introduced. According to the general definition, the photoabsorption cross section is equal to the ratio of the number of quanta, n_ω, absorbed by a single atom per unit time to the flux density of incident photons:

$$\sigma_\omega = \frac{n_\omega}{j_\omega}. \tag{5.3}$$

Multiplying the numerator and denominator by the quantum energy gives

$$\sigma_\omega = \frac{n_\omega \hbar\omega}{I_\omega}, \tag{5.4}$$

where $n_\omega \hbar\omega$ is the amount of energy absorbed per unit time by a single atom. If there are N_0 atoms per unit volume of material and each atom absorbs independently of the others, the intensity absorbed in the layer between x and $x + dx$ is

$$dI_\omega(x) = -n_\omega \hbar\omega N_0\,dx = -\sigma_\omega N_0 I_\omega(x)\,dx \tag{5.5}$$

By comparing Eqs. (5.5) and (5.1), we find the relationship between the photoabsorption cross section and the coefficient:

$$\sigma_\omega = \mu_\omega / N_0 . \tag{5.6}$$

On the other hand, $\rho = N_0 M$, where M is the mass of a single atom, and the mass absorption coefficient is given in atomic quantities:

$$\frac{\mu_\omega}{\rho} = \frac{\sigma_\omega}{M} . \tag{5.7}$$

The number of photons absorbed, n_ω, is equal to the number of ionized atoms and, because a single atom is being examined, it is the probability that the atom will be photoionized by absorbing a quantum at the respective frequency. Thus, Eq. (5.3) is equivalent to a relationship between the cross section and the probability of the process given by Eq. (1.253).

If there are n photons of a given type in a cube of volume V and the condition that an electromagnetic field be periodic at its boundaries is imposed, the photon flux density is

$$j = \frac{cn}{V} , \tag{5.8}$$

and the cross section is

$$\sigma = \frac{WV}{cn} . \tag{5.9}$$

We usually take $V = 1$ and $n = 1$ when examining photoabsorption; then

$$\sigma = W / c . \tag{5.10}$$

In first-order perturbation theory the probability that a photon will excite an atom into a discrete or continuous spectrum state is given by Eqs. (1.250) and (1.244).

If at the initial moment in time an atom is in a state γ, the probability of its one-photon ionization by producing an ion and of forming a photoelectron having a wave vector in the interval $\mathbf{q}$, $\mathbf{q} + d\mathbf{q}$ (we designate the quantum numbers for the ion + electron system as β) is, according to Eqs. (1.244) and (1.248),

$$dW(\gamma \to \beta) = \frac{2\pi}{\hbar} \langle \beta n_{\mathbf{k}\rho} - 1 \,|\, H' \,|\, \gamma n_{\mathbf{k}\rho} \rangle^2 \, \delta(E_\beta - E_\gamma - \hbar\omega) q^2 dq \, d\Omega , \tag{5.11}$$

where $n_{\mathbf{k}\rho}$ is the number of photons having wave vector $\mathbf{k}$ and unit polarization vector $\hat{\epsilon}_{\mathbf{k}\rho}$ in the system's initial state; $d\Omega$ is an element of the spatial photoelectron escape angle. The photoelectron's wave function has been normalized to $\delta(\mathbf{q} - \mathbf{q}')$.

In the relativistic theory the operator for an interaction between an atom and an electromagnetic field has the form given by Eq. (1.238). Only the main term, linear in $\mathbf{A}$, must be left in the nonrelativistic theory in H' [Eq. (1.240)], because the squared term in the first order describes only two-photon transitions which, because of a smallness parameter α that has been raised to the second degree in this operator, occurs with much lower probability and the absorption cross section may be neglected in them.

The wave function for the system atom + radiation is the product of the functions of the atom and the field; therefore, the matrix element for the H_1' is factored into its atomic and photon parts. Only one term in the expansion of the vector potential $\mathbf{A}$ given by Eq. (1.241), which contains the operator $a_{\mathbf{k}\rho}$, contributes to the photon matrix element. Taking Eq. (1.242) into account and setting $n_{\mathbf{k}\rho} = 1$ and $V = 1$, we find (for $t = 0$) that

$$\langle n_{\mathbf{k}\rho} - 1 \,|\, \mathbf{A}(\mathbf{r}) \,|\, n_{\mathbf{k}\rho} \rangle = \left[\frac{2\pi c^2 \hbar}{\omega} \right]^{1/2} \hat{\epsilon}_{\mathbf{k}\rho}\, e^{i\mathbf{k}\cdot\mathbf{r}}. \tag{5.12}$$

Because of the δ-function in Eq. (5.11), using

$$\delta \left[\frac{\hbar^2}{2m} (q^2 - q_0^2) \right] = \frac{m}{\hbar^2 q_0} \delta (q - q_0) \tag{5.13}$$

only the value of q that is determined by the law of conservation of energy gives the contribution to the integration over dq

$$q_0 = \left[\frac{2m}{\hbar^2} (E_\gamma - E_{\gamma'} + \hbar\omega) \right]^{1/2}, \tag{5.14}$$

where γ' is the ion's state.

Using the definition of H' in the relativistic and nonrelativistic theories, Eqs. (5.10), (5.12), and (5.13), we obtain the following expression for the differential photoionization cross section [190] in the nonrelativistic theory

$$\frac{d\sigma^{ion}(\gamma \to \beta)}{d\Omega} = \frac{4\pi^2 e^2 q}{m\omega c\hbar^2} \left| \langle \beta | \sum_j \mathbf{p}_j \cdot \hat{\epsilon}_{\mathbf{k}\rho} e^{i\mathbf{k}\cdot\mathbf{r}_j} | \gamma \rangle \right|^2 \tag{5.15}$$

and in the relativistic ($\hbar^2 q^2/2m << mc^2$) theory:

$$\frac{d\sigma^{ion}_{rel}(\gamma \to \beta)}{d\Omega} = \frac{4\pi^2 e^2 cmq}{\omega\hbar^2} \left| (\beta | \sum_j \boldsymbol{\alpha}_j \cdot \hat{\epsilon}_{\mathbf{k}\rho} e^{i\mathbf{k}\cdot\mathbf{r}_j} | \gamma) \right|^2. \tag{5.16}$$

The summation in Eqs. (5.15) and (5.16) is carried out over all of the atom's electrons, and $d\Omega$ is an element of the spatial angle of photoelectron escape.

Equation (1.275) can be used to transform the nonrelativistic operator in Eq. (5.15) from its p-form to the more widely used r-form:

$$\frac{d\sigma^{ion}(\gamma \to \beta)}{d\Omega} = \frac{4\pi^2 e^2 mqk}{\hbar^2} \left| \langle \beta | \sum_j \mathbf{r}_j \cdot \hat{\epsilon}_{\mathbf{k}\rho} e^{i\mathbf{k}\cdot\mathbf{r}_j} | \gamma \rangle \right|^2. \tag{5.17}$$

We will use the wave function ψ^-, Eq. (4.40), to describe a nonrelativistic photoelectron moving with a wave vector $\mathbf{q}$ and spin projection μ. If only the state of the ion is important, then, by substituting this equation for ψ^- into Eq. (5.17) and using the property of orthogonality in spherical harmonics, we can integrate the differential cross section with respect to the angles of the escaping photoelectron and obtain the total cross section:

$$\sigma^{ion}(\gamma \to \gamma' q\mu) = 4\pi^2 \frac{e^2 m}{\hbar^2} \frac{k}{q} \sum_{lm_l} \left| \langle \gamma' qlm_l \mu \sum_j \mathbf{r}_j \cdot \hat{\epsilon}_{\mathbf{k}\rho} e^{i\mathbf{k}\cdot\mathbf{r}_j} | \gamma \rangle \right|^2. \tag{5.18}$$

From our discussion in Sec. 1.8, in the nonrelativistic theory the exponential $e^{i\mathbf{k}\cdot\mathbf{r}_j}$ may be taken as unity when describing the absorption of soft x rays by an atom (for photon energies less than 1 keV the contribution from the other terms in its expansion is not more than 1%, and for energies up to 10 keV it is several percent [191]).

The dipole transition operator $\mathbf{r}_j \cdot \hat{\epsilon}_{\mathbf{k}\rho}$ is reduced to an irreducible operator by an equation for the scalar product in terms of the spherical components of the vectors:

$$\mathbf{r}_j \cdot \hat{\epsilon}_{\mathbf{k}\rho} = \sum_p (-1)^p C^{(1)}_{-p}(\hat{\epsilon})\, C^{(1)}_p(\hat{\mathbf{r}}_j)\, r_j. \tag{5.19}$$

If atoms are excited by an unpolarized beam and the population, rather than the states, of the levels is examined, then Eq. (5.18) must be summed over the projections of the angular momenta of the final state (ion and electron) and averaged over the projections of the angular momenta of the initial state, as well as over the two components of the polarization vector $\rho = 1, 2$. It is convenient when we do this to change the coupling scheme: we combine the electron's **l** and **s** angular momenta into a total angular momentum, which is then combined with the ion's total angular momentum. Finally, using the Wigner–Eckart theorem and the addition of spherical functions for $C_p^{(1)}(\hat{\epsilon})$

$$\sum_p (-1)^p C^{(t)}_{-p}(\theta_1, \varphi_1)\, C^{(t)}_p(\theta_2, \varphi_2) = P(\cos\theta_{12}) \tag{5.20}$$

[$P(\cos\theta_{12})$ is a Legendre polynomial which becomes unity when the angle θ_{12} between the directions $\theta_1\varphi_1$ and $\theta_2\varphi_2$ is zero] and the relation

$$\sum_{MM'} |\langle\gamma JM|C_p^{(t)}|\gamma' J' M'\rangle|^2 = [t]^{-1}|\langle\gamma J\|C^{(t)}\|\gamma' J'\rangle|^2, \tag{5.21}$$

we obtain

$$\sigma^{ion}(\gamma J \to \gamma' J' \varepsilon) = \frac{4\pi^2\alpha}{3e^2}\,\frac{\hbar\omega}{2J+1}\sum_{ljJ''}|\langle\gamma' J'\varepsilon ljJ''\|D^{(1)}\|\gamma J\rangle|^2. \tag{5.22}$$

Here, $P_{\varepsilon l}(r)$ has already been normalized to $\delta(\varepsilon - \varepsilon')$, a is the constant for a thin structure, and $D^{(1)} = -\Sigma er_j^{(1)}$ is the dipole transition operator. If it is then divided by a_0e, whereby it becomes dimensionless (atomic units are used in its matrix element), the coefficient in Eq. (5.22) then becomes $4\pi^2\alpha a_0{}^2/3 = 2.68909\cdot 10^{-22}\ \mathrm{m}^2$. The total cross section can also be given by the oscillator strength density, Eq. (1.290):

$$\sigma^{ion}(\gamma J - \gamma' J' \varepsilon) = \frac{2\pi^2\alpha\hbar^2}{m}\sum_{ljJ''}\frac{df(\gamma J, \gamma' J'\varepsilon ljJ'')}{d\varepsilon}. \tag{5.23}$$

The photoionization cross section in the relativistic theory is obtained in a similar manner, i.e., by starting from Eq. (4.16) when the four-component wave function ψ^- for the photoelectron is used [78] (along with this we will give an expression for the function ψ^+, which will be needed in Chapter 10):

$$\psi_{\mathbf{q}}^{\mp}(\mathbf{r}, \sigma) = \frac{1}{q}\sum_{jlm_j} i^l e^{\mp i\vartheta_{jl}}\left(\Omega_{jlm_j}^{*}(\hat{\mathbf{q}})\,\chi_\mu\right)\Phi_{qjlm_j}(\mathbf{r}). \tag{5.24}$$

Here, Φ is the one-electron relativistic wave function, Eq. (1.176): Ω is the generalized spherical function, Eq. (1.177), of the angles for the wave vector $\mathbf{q}$; χ_μ is a unit spinor, equal to $\begin{pmatrix}1\\0\end{pmatrix}$ when $\mu = 1/2$ and $\begin{pmatrix}0\\1\end{pmatrix}$ when $\mu = -1/2$, and ϑ_{jl} is the scattering phase.

The final expression for the total cross section has the form [192]

$$\sigma_{rel}^{ion}(\gamma \to \gamma' \varepsilon) = \frac{4\pi^2 e^2}{k}\sum_{jlm_j}\left|(\gamma' \varepsilon jlm_j|\,\hat{e}_{\mathbf{k}p}\cdot\sum_t \alpha_t e^{i\mathbf{k}\cdot\mathbf{r}_t}|\gamma)\right|^2, \tag{5.25}$$

where the γ are quantum numbers for an atom in the initial state, and γ' and εjlm_j are, respectively, the quantum numbers of the ion and the photoelectron in the system's final state; the photoelectron's radial wave function has been normalized to $\delta(\varepsilon - \varepsilon')$.

The probability that an atom can be photoexcited into a discrete level is comparable in magnitude to the probability of photoionization only when x rays act on the outer or intermediate shells whose wave functions more strongly overlap the wave functions of the open shells; therefore, photoexcitation is usually treated in the nonrelativistic theory. This process has a resonance nature – it can only occur when the energy of the x rays is equal to the excitation energy. The probability that an atom will undergo one-photon excitation from a state γ into a state γ' per unit time is given by Eq. (1.250). In a similar manner to that for photoionization we find that

$$\sigma^{ex}(\gamma \to \gamma') = 4\pi^2 \hbar \omega \alpha \left| \langle \gamma' | \hat{\mathbf{e}}_{\mathbf{k}p} \cdot \sum_j \mathbf{r}_j e^{i\mathbf{k}\cdot\mathbf{r}_j} | \gamma \rangle \right|^2. \tag{5.26}$$

by calculating the photon component of the matrix element of the H' operator for Eq. (1.242) and using Eq. (1.275) to convert the transition operator from its p-form into its r-form.

When unpolarized light is absorbed, the cross section for excitation from a level γJ into a level $\gamma' J'$ is given in the dipole approximation by the line strength S or the oscillator strength f, Eq. (1.289):

$$\sigma^{ex}(\gamma J \to \gamma' J') = \frac{4}{3} \pi^2 \frac{\alpha}{e^2} \hbar\omega \frac{1}{2J+1} S(\gamma J, \gamma' J') = 2\pi^2 \alpha \frac{\hbar^2}{m} f(\gamma J, \gamma' J'). \tag{5.27}$$

Notice that the total excitation cross sections are measured in units of energy per square meter.

When high-energy x-ray beams excite an atom, its final states are autoionizing states. From Sec. 2.4, the wave function for such a state is the superposition of discrete and continuous spectrum functions; e.g., for a final state that interacts with one continuum, it has the form given by Eq. (2.101). Substituting this equation into the matrix element for the transition operator, according to Eq. (2.111), we have

$$M^{ph} = a_{\mathscr{E}}^* M^{ex} + a_{\mathscr{E}}^* \pi V_{\mathscr{E}}^* \varkappa M^{ion}, \quad \mathscr{E} = E(nl^{-1}) + \varepsilon. \tag{5.28}$$

Here, M^{ph} and M^{ion} are the photoabsorption and photoionization amplitudes, respectively. The M^{ex} is only approximately equal to the photoexcitation amplitude, because it is calculated through the modified wave function for a discrete state, Eq. (2.112). The κ and $V_{\mathscr{E}}$ are given, respectively, by Eqs. (2.98) and (2.91). The $a_{\mathscr{E}}$ coefficient, Eq. (2.109), describes the "diluting" of a discrete level because of its interaction with a continuum.

By squaring the absolute value of Eq. (5.28), we find the following expression for the photoabsorption cross section in the vicinity of a level:

$$\sigma^{ph} = \sigma^{ion} \left| a^*_{\mathscr{E}} \frac{M^{ex}}{M^{ion}} + a^*_{\mathscr{E}} \pi V^*_{\mathscr{E}} \varkappa \right|^2, \tag{5.29}$$

where σ^{ion} is the photoionization cross section, which does not take into consideration the existence of an autoionizing level in the continuum. The second factor, which describes the resonance change in the cross section of this level, can be given by the quantities q, Eq. (2.113), and $\varkappa$, Eq. (2.98) which are weak functions of energy:

$$\sigma^{ph} = \sigma^{ion} \frac{(q+\varkappa)^2}{1+\varkappa^2}. \tag{5.30}$$

If continua that do not interact with the given state exist, we must add to Eq. (5.30) the photoionization cross section $\sigma_2{}^{ion}$ [26]

$$\sigma^{ph} = \sigma_1^{ion} \frac{(q+\varkappa)^2}{1+\varkappa^2} + \sigma_2^{ion}. \tag{5.31}$$

in these continua. This formula is also valid for a final discrete state that interacts with several continua.

In some cases, when an electron is excited and its principal quantum number does not change (see Sec. 5.2), the photoabsorption amplitude may be much greater than the photoionization amplitude. Then, by leaving only the first term of Eq. (5.29) in brackets, we obtain

$$\sigma^{ph} \approx | a_{\mathscr{E}} |^2 \sigma^{ex}. \tag{5.32}$$

The coefficient $| a_{\mathscr{E}} |^2$ given by Eq. (2.112) describes the shape of a photoexcitation line [it is the density of the final states $\rho(\mathscr{E})$ in the equation for the photoexcitation probability, Eq. (1.246)]. If the interconfiguration element $V_{\mathscr{E}}$, Eq. (2.91), in the resonance area is weakly dependent on energy, the line has Lorentzian form and is determined by the width of the autoionization level and its energy. Then the excitation spectrum can be approximated from Eq. (5.27): the total transition cross sections are first determined and they are then distributed with due regard for the natural width of the lines, as well as instrumental response (Sec. 9.1). The total photoionization cross sections or transition oscillator strengths may be compared with the heights of the experimental spectrum lines only when the line widths are approximately equal.

5.2. X-Ray Absorption by a Many-Electron Atom. The One-Configuration Approximation

The photoabsorption cross section as a function of many-electron quantum numbers is determined mainly by the submatrix element of the transition operator; expressions for this element in both the relativistic and nonrelativistic theories were presented in Sec. 1.8. The basic transitions made when an atom is excited by an x-ray beam are the $n_1l_1^{4l_1+2}n_2l_2^{N_2}\rightarrow n_1l_1^{4l_1+1}n_2l_2^{N_2+1}$ transitions. The probabilities of these transitions reach their greatest magnitude if both shells belong to the same layer and $l_2 = l_1 + 1$ [26]. In the final state, a coupling which is nearly LS is achieved between neighboring outer shells, and in the nonrelativistic theory the matrix element for the dipole transition operator is given by Eq. (1.266).

Equation (1.268) must be used when photoexcitation into a Rydberg series or ionization into the respective continuum occurs and Eq. (1.268) must be replaced by

$$n_1l_1^{4l_1+2}n_2l_2^{N_2}\rightarrow n_1l_1^{4l_1+1}n_2l_2^{N_2+1}. \tag{5.33}$$

if there is an LS coupling instead of a jJ coupling between a vacancy and an open shell.

The submatrix element for the general operator of an electric multipole transition is presented because the substitution $t = 1$ does not lead to a simplification of any kind.

When an outer open shell is photoionized, the matrix element becomes

$$\begin{aligned}&\langle n_1l_1^{4l_1+2}n_2l_2^{N_2}\gamma_2L_2S_2J_2\,\|\,O^{(t)}\,\|\,n_1l_1^{4l_1+1}n_2l_2^{N_2}\gamma_2'L_2'S_2'L'S'J'\varepsilon ljJ\rangle=\\&=\delta(\gamma_2L_2S_2,\ \gamma_2'L_2'S_2')(-1)^{L'+S_2+J+l+1}[L',\ S',\ J',\ j,\ J_2,\ J]^{1/2}\times\\&\times\sum_{L''}[L'']\begin{Bmatrix}L_2 & l_1 & L'\\ l & L'' & t\end{Bmatrix}\begin{Bmatrix}L_2 & J_2 & S_2\\ J & L'' & t\end{Bmatrix}\begin{Bmatrix}L' & S' & J'\\ l & \frac{1}{2} & j\\ L'' & S_2 & J\end{Bmatrix}s^{(t)}_{n_1l_1,\,\varepsilon l}.\end{aligned} \tag{5.34}$$

The summation over $3nj$ symbols in Eqs. (5.33) and (5.34) can be done; however, this leads to the appearance of the $12j$ symbol, which is more complex. When the matrix element is substituted into the photoionization cross section, summation over the quantum numbers j and J can also be done.

In the nonrelativistic dipole approximation the photoionization of a closed electron shell is given by the simple formula

$$\begin{aligned}&\langle n_1l_1^{N_1}\gamma_1L_1S_1J_1\,\|\,O^{(t)}\,\|\,n_1l_1^{N_1-1}\gamma_1'L_1'S_1'J_1'\varepsilon ljJ\rangle=(-1)^{L_1'+S_1+J+l_1}\times\\&\times[L_1,\ S_1,\ J_1,\ j,\ J_1',\ J]^{1/2}\sqrt{N_1}\,(l_1^{N_1}\gamma_1L_1S_1\,\|\,l_1^{N_1-1}\gamma_1'L_1'S_1'l_1)\,s^{(t)}_{n_1l_1,\,\varepsilon l}\times\end{aligned}$$

$$\times \sum_{L''} [L''] \begin{Bmatrix} L_1' & S_1' & J_1' \\ l & \frac{1}{2} & j \\ L'' & S_1 & J \end{Bmatrix} \begin{Bmatrix} L_1 & J_1 & S_1 \\ J & L'' & t \end{Bmatrix} \begin{Bmatrix} L_1 & t & L'' \\ l & L_1' & l_1 \end{Bmatrix}.$$

$$\sigma_{nlj}^{ion}(\omega) \equiv \sigma^{ion}(nl^{4l+2} \to nl^{4l+1} j \varepsilon) =$$
$$= \frac{4}{3}\pi^2 \alpha \frac{2j+1}{2l+1} \hbar\omega [l \langle nl | r | \varepsilon l-1 \rangle^2 + (l+1) \langle nl | r | \varepsilon l+1 \rangle^2], \tag{5.35}$$
$$\hbar\omega = I_{nlj} + \varepsilon,$$

where ε is the photoelectron energy and I_{nlj} is the binding energy of an nlj electron in the atom. If the energy and radial integrals in Eq. (5.35) are calculated in atomic units, and the cross section in megabarns, the coefficient is equal to $4\pi^2\alpha a_0^2/3 = 2.68909$ Mb.

By summing Eq. (5.35) over the j and ignoring the dependence of ε on j, we obtain an expression for the total photoionization cross section of a closed shell:

$$\sigma_{nl}^{ion}(\omega) \equiv \sigma^{ion}(nl^{4l+2} \to nl^{4l+1} \varepsilon) = \sum_j \sigma_{nlj}^{ion}(\omega) \approx$$
$$\approx \frac{8}{3}\pi^2 \alpha \hbar\omega [l \langle nl | r | \varepsilon l-1 \rangle^2 + (l+1) \langle nl | r | \varepsilon l+1 \rangle^2]. \tag{5.36}$$

In the relativistic theory the total photoionization cross section is given by the submatrix elements of the multipole transition operators ${}_eO^{(t)}$, Eq. (1.279), and ${}_mO^{(t)}$, Eq. (1.278). Using an expansion for the exponent, Eq. (1.256), making a number of transformations [67, 93], and summing over the projections of the angular momenta and the directions of polarization, $\rho = 1, 2$, we find the relationship

$$\sum_{MM'\rho} \left| (\gamma' J' M' | \hat{e}_{k\rho} \cdot \sum_j \alpha_j e^{-i\mathbf{k}\cdot\mathbf{r}_j} | \gamma J M) \right|^2 =$$
$$= c^{-2} \sum_t \left(\frac{\omega}{c}\right)^{2t} \frac{(2t+1)(t+1)}{[(2t+1)!!]^2 t} \left| (\gamma' J' \| ({}_eO^{(t)} + {}_mO^{(t)}) \| \gamma J) \right|^2. \tag{5.37}$$

In order to apply this relationship to Eq. (5.25), we must first recouple the angular momenta of the ion and the photoelectron in the final state into a total angular momentum, average the cross section over M and ρ, and sum over the M':

$$\sigma_{rel}^{ion} = 2\pi^2 \sum_t \left(\frac{\omega}{c}\right)^{2t-1} [J]^{-1} \frac{(2t+1)(t+1)}{[(2t+1)!!]^2 t} \times$$

$$\times \sum_{ljJ''} |(\gamma' J' \varepsilon ljJ'' \| ({}_eO^{(t)} + {}_mO^{(t)}) \| \gamma J)|^2. \tag{5.38}$$

The submatrix element for the multipole radiation operator is given by Eqs. (1.279)–(1.283). When a closed shell is photoionized, the cross section becomes

$$\sigma_{rel}^{ion}(n_1 l_1 j_1^{2j_1+1} \to n_1 l_1 j_1^{2j_1} \varepsilon) = 2\pi^2 \sum_t \left(\frac{\omega}{c}\right)^{2t-1} [t] \sum_{lj} [j_1, j] \times$$

$$\times \left\{ \frac{t+1}{t(2t+1)^2} \begin{pmatrix} j_1 & t & j \\ -\frac{1}{2} & 0 & \frac{1}{2} \end{pmatrix}^2 R_t^2(e) + \begin{pmatrix} j_1 & t & j \\ -\frac{1}{2} & 1 & -\frac{1}{2} \end{pmatrix}^2 R_t^2(m) \right\}, \tag{5.39}$$

where $R_t(e)$ and $R_t(m)$ are the radial integrals, Eqs. (1.282) and (1.283). Equation (5.39) is equivalent to the one given in [192]; however, it is more convenient because the coefficients in the integrals are independent of the orbital angular momenta of the electrons.

The relative probability of creating vacancies having $j = l + 1/2$ and $j = l - 1/2$ during photoionization is called the branching coefficient. If, in the nonrelativistic theory, we can ignore the differences in the electron binding energies and say that a pure jj coupling exists between a vacancy in a closed shell and a photoelectron, the branching coefficient is then equal to the ratio of the statistical weights of the final states of the vacancy, $(l + 1)/l$. The strong dependence of the branching coefficient on photon energy that occurs for some shells, especially at a threshold and near a Cooper minimum, is explained by correlation effects and at low energies comparable to the splitting of the levels for a state having an nlj^{-1} vacancy also by the difference in the energies of photoelectrons escaping subshells having $j = l \pm 1/2$ [190].

If the time that a photoelectron takes to escape from an atom, $t_e \approx a_0\sqrt{m/\varepsilon}$, is less than the atom's relaxation time when a vacancy is created, $\tau \approx \hbar/E^{rx}$, where E^{rx} is the relaxation energy, Eq. (3.19) (this occurs when the photoelectron energy ε is about 100 eV and higher), the atom's electron wave functions do not relax and the ion wave function must be constructed from the atom's frozen radial orbitals. The initial and final states are described by different wave functions for a slow photoelectron and during photoexcitation.

Because they are not orthonormal, a factor appears in the matrix element of a transition operator. This factor contains overlap integrals of functions of passive electrons in the initial and final states, as well as additional terms that correspond to the interchange of electrons having the same orbital quantum number in the wave function and containing integrals of the $\int P_{nl}^{at}(r)P_{nl}^{ion}(r)dr$ type. The calculations become much more difficult when these factors are taken into consideration. As was noted in Sec. 1.8, in a one-configuration approximation it is advantageous to do the calculations using different radial wave functions for the initial and final states in the radial integral and energy, but neglecting the partially compensating corrections for overlap and exchange.

If the potential used to calculate nonrelativistic wave functions is nonlocal (e.g., a Hartree–Fock potential), the radial integral for the radiative transition and the excitation or ionization cross section will assume different numerical values in terms of distance and velocity (in general, it depends on the gauge of the electromagnetic field [48]). The greater the nonlocal exchange part of the potential, the more substantial is the difference. This exchange part is determined by the degree to which the wave function of the electron being examined overlaps the function of the other (primarily open) shells. Thus, the results from calculating cross sections by the Hartree–Fock method in terms of distance and velocity may be quite different for transitions between neighboring outer shells, especially those having the same n, but these results are usually in good agreement for photoionization or excitation from an inner shell whose wave function weakly overlaps the function of an outer open shell. Because results from calculating the cross section in terms of distance and velocity near a threshold are often located in different areas from the experimental curve for a strong exchange interaction, in the final configuration it is advantageous to average geometrically these two forms [193].

Radial integrals corresponding to excitation into a Rydberg series and photoionization into the same channel in a continuous spectrum are related by the asymptotic condition established in [26, 46], starting from the monotonic nature of the change in oscillator strength density during a transition through an ionization edge [26]

$$\frac{df}{d\varepsilon}\bigg|_{\varepsilon=0} = \lim_{n\to\infty} f_n \left(\frac{dE_n}{dn}\right)^{-1}. \tag{5.40}$$

Here, $E_n \equiv -I_{nl}$ is the energy of the nth level in a Rydberg series, which is described by the one-electron equation (2.7) having an effective principal quantum number $n^* = n - \mu_l$, where μ_l is a quantum defect for the series and is weakly dependent on n. By expressing for large n the derivative

$$\frac{dE_n}{dn} \approx \frac{dE_n}{dn^*} = \frac{(Z^*)^2}{(n^*)^3} = \frac{|2E_n|^{3/2}}{Z^*}, \tag{5.41}$$

we find

$$\left.\frac{df}{d\varepsilon}\right|_{\varepsilon=0} = \lim_{n\to\infty} \frac{f_n Z^*}{|2E_n|^{3/2}} . \tag{5.42}$$

which allows us to derive the asymptotic relationship between integrals for a dipole transition and, in the general case, for a multipole transition

$$\lim_{n'\to\infty} \left(\frac{Z^*}{|2E_{n'}|^{3/2}}\right)^{1/2} \langle nl | r^t | n'\, l'\rangle = \langle nl | r^t | \varepsilon = 0\, l'\rangle . \tag{5.43}^*$$

In the dipole approximation, an electron from a shell having an orbital quantum number l can be excited into states having $l' = l \pm 1$. As a rule, the channel $l \to l + 1$ is dominant [26, 190]. This relationship can be explained, starting from a classical model: an energy--absorbing electron in a circular orbit must increase its orbital angular momentum. This result is partially valid for an elliptical orbit [26].

According to the hydrogenical model, in which the photoabsorption cross section has an algebraic expression [10], x-ray absorption by an inner shell increases sharply when the beam energy becomes equal to the electron binding energy, reaches a maximum, and decreases according to $\sigma \sim \omega^{-n}$ as the frequency increases, where n is of the order of 4. The asymptotic behavior of the cross section at high photon energies ($I_{nl} << \hbar\omega << mc^2$) is described by

$$\sigma_{nl}^{ion} \sim \omega^{-l-7/2} . \tag{5.44}$$

A small number of Rydberg photoexcitation lines usually adjoin to the low-energy side of the maximum in the spectrum of photionization from an inner subshell; the remaining lines in the series merge smoothly with the photoionization maximum, creating a photoabsorption edge. As the x-ray beam energy increases, absorption by electrons in the deeper shells becomes possible and new sawtooth-shaped maxima appear in the spectrum. Absorption for a neutral atom is given approximately by the semiempirical formula [1]

$$\sigma^{ph} \approx CZ^4 \omega^{-3} . \tag{5.45}$$

The constant C increases discontinuously at each ionization threshold.

When a photoelectron has a high energy because of oscillations in its wave function the area near the nucleus, where the Coulomb model is fairly realistic, makes the main contribution to the cross section.

With due regard for screening, this model can be used successfully for calculating σ_{nlj}^{ion} for photon energies above roughly 10 keV [26]. At low energies a nonrelativistic or a relativistic one-configuration approximation (depending on what kind of shell is being ionized) must be used. Near a threshold, when photoelectron energies are less than $Z^{2/3}$ Ry [27] and this approximation is inadequate, correlation effects and the interaction of not only discrete ion or atom states, but the various photoionization channels, begin to play an important role [27, 28, 190].

The traditional shape of a photoabsorption spectrum near an absorption edge can be significantly distorted by the effects due to a potential barrier and the collapse of a discrete or continuous spectrum wave function. As we showed in Sec. 3.3, the effective potential of an electron having an orbital angular momentum $l \geq 2$ usually looks like two potential wells separated by a potential barrier. If an electron in the $n'l + 1$ excited states of the $nl^{4l+1}n'l + 1$ Rydberg series is noncollapsing – its wave function is localized near an outer potential well far from the atomic core, the wave functions $P_{nl}(r)$ and $P_{n'l+1}(r)$ weakly overlap. This leads to the disappearance of the Rydberg series in the photoabsorption spectra from a *d*- or *f*-shell (excitation from the *p*-shell "through the barrier" is possible because of the low barrier for a *d*-electron [26]).

The existence of a potential barrier also leads to the suppression of absorption beyond the photoionization threshold. The function $P_{\varepsilon l+1}(r)$ for small ε barely penetrates the barrier and weakly overlaps $P_{nl}(r)$. If in an inner well there is a quasidiscrete level having a positive energy ε_r, a sudden penetration by the photoelectron wave function into the atomic core as $\varepsilon \to \varepsilon_r$ occurs, leading to the appearance of the so-called shape resonance in the absorption spectrum [147], which looks like a shifted absorption edge. Beyond it, a Cooper minimum can take place because the sign of the transition integral changes as ε increases [26, 190] (see Sec. 1.8).

When the wave function $P_{n'l+1}(r)$ collapses, the integral for the dipole transition $\langle nl \mid r \mid n'l + 1\rangle$ increases by several orders of magnitude, which makes possible a significant redistribution of the intensity in the x-ray absorption spectrum between ionization and excitation in support of the latter process. When $n = n'$ the strong overlap of the collapsing wave function $P_{nl+1}(r)$ with $P_{nl}(r)$ leads to the appearance of so-called giant photoexcitation maxima in the transition elements, which correspond to $nl^{4l+2}n(l + 1)^N \to nl^{4l+1}n(l + 1)^{N+1}$ $(l = p, d)$ transitions [29, 147]. Intense Rydberg series lines may appear in the spectrum during partial wave function collapse [194, 151].

Electrostatic exchange interaction plays a major role in configurations having two open shells $n_1l_1^{N_1}n_2l_2^{N_2}$, a special case of which are an atom's final configurations during photoabsorption, and determines the basic features of the energy spectrum if both shells belong to one outer layer ($n_1 = n_2$).

Summing the matrix elements of a multipole transition operator by the method of second quantization, it can be shown that the coefficient g_k ($k \neq 0$) at the exchange integral is expressed in terms of the submatrix elements of the transition operator [195]:

$$g_k(l_1^{N_1} l_2^{N_2} \gamma\gamma' J) =$$
$$= \Big\{ [J]^{-1} \langle n_1 l_1 | er^k | n_2 l_2 \rangle^{-2} \sum_{\gamma'' J''} \langle l_1^{N_1} l_2^{N_2} \gamma J \| O^{(k)} \| l_1^{N_1+1} l_2^{N_2-1} \gamma'' J'' \rangle \times$$
$$\times \langle l_1^{N_1} l_2^{N_2} \gamma' J \| O^{(k)} \| l_1^{N_1+1} l_2^{N_2-1} \gamma'' J'' \rangle - \delta(\gamma, \gamma') \frac{N_2}{2l_2+1} \langle l_1 \| C^{(k)} \| l_2 \rangle^2 \Big\}. \quad (5.46)$$

For a diagonal Coulomb matrix element the first term is positive and the second is negative. Thus, a larger value of the basic exchange coefficient g_1 corresponds to $l_1^{N_1} l_2^{N_2}$ configuration terms associated with $l_1^{N_1+1} l_2^{N_2-1}$ configuration terms by the matrix element of the $O^{(1)} \equiv D^{(1)}$ operator; therefore, they have higher energy. For example, for the $l^{4l_1+1} l_2$ configuration,

$$g_k(l_1^{4l_1+1} l_2 LS) = [l_2]^{-1} \langle l_1 \| C^{(k)} \| l_2 \rangle^2 \left[\frac{4l_2+2}{2k+1} \delta(S, 0) \delta(L, k) - 1 \right]. \quad (5.47)$$

The large positive exchange term for the ^1P term of the $l^{4l+1}(l + 1)$ configuration (in the dipole approximation it is mainly transitions into the $J = 1$ level of this term that occur during photoabsorption from a closed shell in the case of an LS coupling) significantly increases the effective potential barrier for an $(l + 1)$-electron in this many-electron state and "delays" its collapse in comparison with the lower $^3P_1$ and $^3D_1$ states. Because of this, when calculating the spectrum for photoabsorption from a subvalent shell having $l > 0$, especially near the critical Z value at which excited electron collapses, the HF-t wave functions must be used, because the HF-av. method may yield results that are qualitatively incorrect.

A similar division of terms into upper and lower groups by an exchange electrostatic interaction occurs in some other configurations of the $nl^{4l+1}n(l + 1)^N$ type when $N < 2l + 1$. According to Eq. (5.46), photoexcitation from a ground configuration is mainly into the upper group of terms (into the lower group only when these terms mix with the upper).

In practice, photoabsorption occurs at gas temperatures of $T \lesssim 5 \cdot 10^2$ K only from states in the ground term, and by atoms for most elements only from the lowest level of their ground configuration. In the case of a single open shell, the quantum numbers for the lowest term $\tilde{\gamma}\tilde{L}\tilde{S}$ are determined by Hund's rule and are given by the number of electrons in a shell and its orbital quantum number [196]:

$$\tilde{L} = \frac{1}{2} \mathcal{N}(2l+1-\mathcal{N}), \quad \tilde{S} = \mathcal{N}/2, \quad \tilde{\upsilon} = \mathcal{N}, \quad (5.48)$$

$$\tilde{J}=\begin{cases} N(2l-N)/2 & \text{for} \quad N<2l+1, \\ (4l+2-N)(N-2l)/2 & \text{for} \quad N\geqslant 2l+1, \end{cases}$$

$$\mathcal{N}=\begin{cases} N & \text{for} \quad N<2l+1, \\ 4l+2-N & \text{for} \quad N\geqslant 2l+1. \end{cases} \tag{5.49}$$

The expressions for the quantum numbers $\tilde{U}$ and $\tilde{W}$ of the f^N shell are somewhat more complicated. These relationships allow us to derive simple algebraic expressions for the electrostatic energy of the ground term and the spin– orbit energy of the ground level as polynomials in N for any l [196] (because the ground term is distant from the remaining terms, an almost pure *LS* coupling occurs for its level). The average energy and variance of the photoexcitation spectrum is given by the matrix elements of the ground level wave functions, and the corresponding formulas for the $l_1^{4l_1+2}l_2^{N_2} \rightarrow l_1^{4l_1+1}l_2^{N_2+1}$ transition are given in [197, 198].

5.3. Refined Methods for Describing Photoabsorption Spectra

Only refined methods that take the correlations between an atom's electrons and between these electrons and a photoelectron into consideration can completely describe photoabsorption spectra and be in good agreement with experiment, and only these methods can explain photoabsorption features caused by ionization and simultaneous excitation, by double ionization, by the existence of autoionization levels, etc. Particularly important are correlation effects that take place about 50 eV beyond the ionization threshold where, for atoms whose nuclear charge is near the critical value at which an excited electron collapses, the HF-av. method (for example, the *4d*-spectrum of Xe [26]) and even those methods such as the HF-t method and the random phase approximation with exchange (the *5p*-spectrum of Ba [138]) that consider some types of correlations can yield results that are not only quantitatively, but qualitatively, incorrect.

Correlation effects in an atom's initial state and in the final state of an ion (photoionization) or an atom (excitation) may be calculated by the many-configuration method [199]. When this is done, the admixed configurations associated with the ground or admixed configurations for another state by the major matrix elements of the transition operator make a major contribution to the cross section (although these configurations may be unimportant to refining the transition energy).

The method discussed in Secs. 2.4 and 5.1 is widely used (in a semiempirical, and less often, theoretical version) for calculating autoionization resonances in the photoabsorption cross section. Another means for taking the mixing of the different final states in an ion + electron system (different channels for the photoabsorption process) into account is the method of strong coupling, which has been borrowed from the theory of collisions between atoms and electrons (see Sec. 4.2). For photoionization, the wave function for the jth state of an atom + photoelectron system is the expansion

$$\Psi_j(x_1, \ldots, x_{N-1} x_N) = A \sum_i \Phi_i(x_1, \ldots, x_{N-1}) \varphi_{ij}(x_N). \tag{5.50}$$

Here, A is an antisymmetrization operator, Φ_i is the wave function of an ion in the ith state and is determined through a separate calculation, and φ_{ij} is the photoelectron's unknown wave function. As in the case where an atom scatters an electron, the function φ_{ij} in a spherically symmetric field is factorized into a standard spin-angular part and a radial function F_{ij}. To find this function the equations for strong coupling, Eq. (4.20), are obtained, starting from a variational principle. In order that Ψ_j be an ingoing spherical wave in the jth channel plus the outgoing spherical waves in all channels, $F_{ij}(r)$ must satisfy the asymptotic condition

$$F_{ij}(r) \underset{r\to\infty}{\simeq} k_i^{-1/2} (e^{-i\eta_i} \delta_{i,j} - S_{ij} e^{i\eta_i}), \tag{5.51}$$

where k_i is the wave number of an electron in the ith channel, S_{ij} is an element of the scattering matrix, and η_i is the phase, determined from Eq. (4.16).

A comparison of Eq. (5.51) with Eq. (4.14) with due regard to the relationship between the S- and K-matrices of Eq. (4.15) makes it possible to write Ψ_j in terms of the solutions of the equations for a strong coupling $\Psi_\beta{}'$ having the asymptote (4.14):

$$\Psi_j = i \sum_\beta (1 - iK)^{-1}_{\beta j} \Psi'_\beta. \tag{5.52}$$

Substituting Eq. (5.52) into the general formula for the photoionization cross section in a dipole approximation yields [200, 201]

$$\sigma^{ion}(\gamma \to j) = \frac{4\pi^2 \alpha \hbar \omega}{e^2} \sum_{\beta\beta'} \langle \Psi_\gamma | \hat{\epsilon}_{k\rho} \cdot \mathbf{D} | \Psi'_\beta \rangle \times$$
$$\times (1 - iK)^{-1}_{\beta j} (1 + iK)^{-1}_{j\beta'} \langle \Psi'_{\beta'} | \hat{\epsilon}_{k\rho} \cdot \mathbf{D} | \Psi_\gamma \rangle. \tag{5.53}$$

The most widely adopted of the many refined methods for calculating photoabsorption spectra have been the different versions of perturbation theory [27, 28, 128, 129, 138]. The main advantage of this approach is that correlations in the initial and final system states are taken into consideration when calculating the transition amplitude, circumventing a calculation of the refined wave functions. Another advantage is the clearness of the analysis

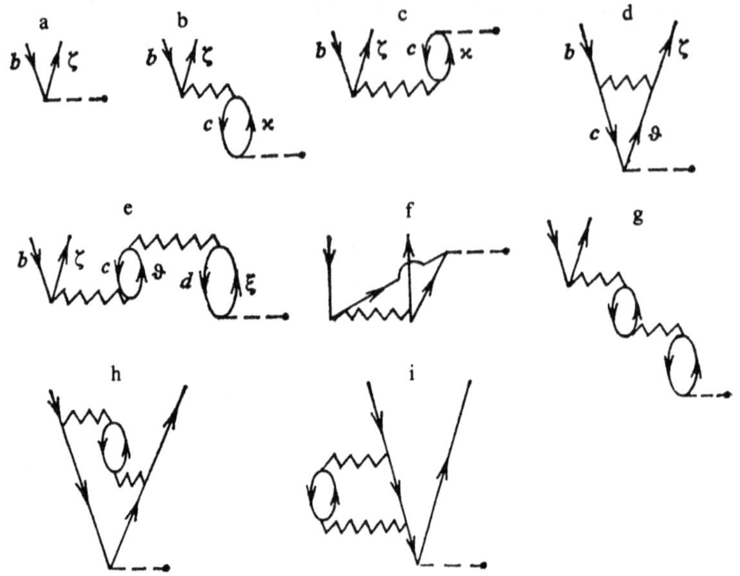

Fig. 5.1. Diagrams of a series from perturbation theory for the photoabsorption amplitude.

of the different correlations in the language of diagrams and the ease of performing different actions with them. Because these diagrams and the angular momentum diagrams are topologically equivalent, less complicated determinant wave functions may be used; only at the last stage is the transition made to the functions of coupled angular momenta. On the other hand, using perturbation theory raises a number of difficulties: computing time increases rapidly as the order of the terms increases, the problems of convergence in the series and of evaluating the importance of different diagrams have been little studied, and there is difficulty in generalizing the theory for atoms having open shells.

The foundation for a stationary perturbation theory was laid in Sec. 2.3 and its application to calculating the amplitudes of the radiative transitions were discussed. With the help of the wave operator Ω, which transforms an approximate model wave function to an exact wave function, Eq. (2.50), we can construct an effective transition operator O^{ef}, Eq. (2.89), whose matrix element for the model functions of an atom's initial and final states is equal to the matrix element of the transition operator O for the exact wave functions. Diagrams for the operator O^{ef} (and the transition amplitude diagrams that match them) are obtained by combining every possible method for wave operator diagrams for the initial and final states Ω_1 and Ω_2 with the diagram for the operator O in a manner similar to what is done for the effective energy operator (Sec. 2.3).

Figure 5.1 shows some lower-order diagrams that contribute to the photoabsorption amplitude of a closed shell. A solid line directed upward denotes an electron in an excited state; directed downward, it denotes a vacancy in a core state. A zigzag line denotes an electrostatic interaction, and a dashed line with a dot denotes an electron–photon interaction. Electrostatic interaction lines underneath the latter proceed from the diagrams for Ω_1 and show correlations in the initial state, and lines above it correspond to correlations in the final state. The rules for interpreting the diagrams, given in Sec. 2.3, make it possible to determine the contributions made by the individual diagrams. For example, the simplest diagram, a, shows the matrix element for the transition operator $\langle \zeta \mid d^{(1)} \mid b \rangle$, and the contributions from diagrams b and c are, respectively:

$$\sum_{c\varkappa} \frac{\langle \zeta c \mid h^e_{12} \mid b\varkappa \rangle \langle \varkappa \mid d^{(1)} \mid c \rangle}{\varepsilon_c - \varepsilon_\varkappa + \hbar\omega + i\eta}, \tag{5.54}$$

$$\sum_{c\varkappa b\zeta} \frac{\langle c \mid d^{(1)} \mid \varkappa \rangle \langle \zeta\varkappa \mid h^e_{12} \mid bc \rangle}{\varepsilon_b + \varepsilon_c - \varepsilon_\zeta - \varepsilon_\varkappa}. \tag{5.55}$$

The denominator of Eq. (5.54) may go to zero; therefore, the small imaginary quantity $i\eta$ is added to it.

Diagram b is interpreted as follows: a photon produces a vacancy in the state c and an electron in the excited state $\varkappa$, which interacts with and transmits the excitation energy to another electron, which produces a vacancy in the state b and a photoelectron ζ. If states b and c belong to the same shell, correlations described by diagram b are called innerchannel correlations (they are taken into consideration in the HF-t method); if they belong to different shells, we have interchannel correlations. The interpretation of diagram c, which describes double virtual excitations in the initial state, was given in Sec. 2.3.

As follows from the results from calculating photoionization cross sections (most of which were done for atoms of inert gases), these three types of "particle-hole" correlations are often more important to describing the process being examined [190]. It is worthwhile to take the lower and higher order correlations into consideration, which is typical in the random phase approximation with exchange (RPAE) that takes "electron–vacancy" excitations, shown in diagrams b, c, d, and e to infinite order, into consideration.

The sum of an infinite series of RPAE diagrams is the matrix element of an effective transition operator in this approximation and is designated $d(\omega)$. It considers every possible excitation of a given type and is shown by the very simple diagram with the shaded vertex (the first in Fig. 5.2). On the other hand, this infinite series of diagrams can be divided into the very simple diagram a and some infinite series each of which is written as a diagram containing as a part a fragment with an effective operator (Fig. 5.2). Writing the contributions

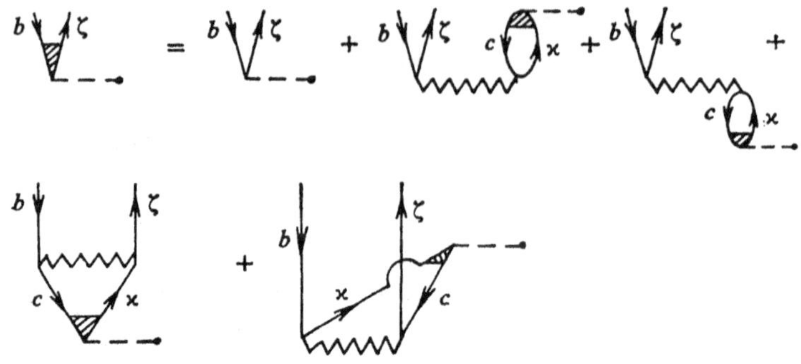

Fig. 5.2. A series of diagrams that describe photoabsorption in the random phase approximation with exchange and the composite diagram (on the left side of the equation).

of all diagrams for the right- and left-hand sides of the equation gives us an integral equation for finding the matrix element of the $d(\omega)$ operator:

$$
\begin{aligned}
&\langle\zeta\,|\,d(\omega)\,|\,b\rangle = \langle\zeta\,|\,d^{(1)}\,|\,b\rangle + \\
&+\sum_{\varkappa c}\left\{\frac{\langle c\,|\,d(\omega)\,|\,\varkappa\rangle[\langle\zeta\varkappa\,|\,h^{e}_{12}\,|\,bc\rangle-\langle\varkappa\zeta\,|\,h^{e}_{12}\,|\,bc\rangle]}{\varepsilon_c-\varepsilon_\varkappa-\hbar\omega+i\eta}+\right. \\
&\left.+\frac{[\langle\zeta c\,|\,h^{e}_{12}\,|\,b\varkappa\rangle-\langle c\zeta\,|\,h^{e}_{12}\,|\,b\varkappa\rangle]\langle\varkappa\,|\,d(\omega)\,|\,c\rangle}{\varepsilon_c-\varepsilon_\varkappa+\hbar\omega+i\eta}\right\}.
\end{aligned}
\tag{5.56}
$$

Only electrons or vacancies that take part in a transition or excitation are indicated in the matrix element notation. The sum over virtual states $\varkappa$ contains an integration over the continuous spectrum; in practice it is replaced by a sum over a finite number of continuum states [129].

The RPAE does not take relaxation effects into consideration (diagrams h and i in Fig. 5.1, etc.) – they can be incorporated into the RPAE approximately if the experimental values of electron binding energies and one-electron wave functions for the final state, calculated in the field of the ion [202], are used. Still, this method is inadequate for the strong correlation effects that are not taken into consideration in RPAE equations, for example, when a vacant state interacts with configurations of the type given by Eqs. (3.40) and (3.41). The contribution from the respective diagrams must be found individually [203].

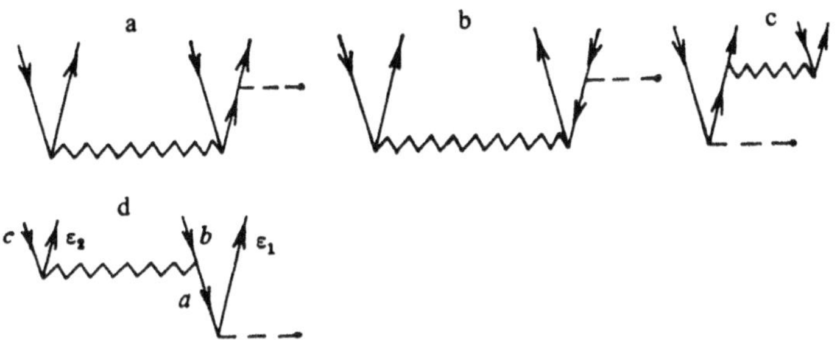

Fig. 5.3. Lower-order diagrams for double ionization.

Notice that photoionization cross sections are not distinguished in forms of distance and velocity in the random phase approximation [28].

The RPAE is used mainly for calculating photoabsorption spectra of atoms with closed shells (and for half-filled shells, which can be thought of as closed subshells whose electrons have the same direction of spin [204]). The RPAE equations were generalized in [124, 125] for atoms with open shells. A relativistic version of the RPAE has also been developed [205].

Perturbation theory is a consistent method for calculating the cross sections of many-particle processes, including two-electron photoionization or ionization with the excitation of another electron. Figure 5.3 shows lower-order diagrams that contribute to the amplitude of such a process; of these diagrams, a and b describe correlations in the initial state, and c and d are for the final state.

The excess energy transferred by a quantum of x rays to an atom may be distributed between both photoelectrons differently; therefore, the total two-electron photoionization cross section contains an additional integral with respect to the energy of one of these electrons.

We will also briefly describe a method for calculating an x-ray absorption spectrum. This method is based on using Brillouin–Wigner perturbations and the relationships between the photoabsorption cross section and the polarizability of an atom [138, 139] (we will examine an atom with closed shells).

The dynamic polarizability of an atom when acted upon by electromagnetic radiation having a frequency of ω is defined as the ratio of the dipole moment of the atom induced by the field to the electric field strength and is given as an infinite sum

$$\alpha_{\gamma J}(\omega) = \frac{e^2}{m} \sum_{\gamma' J'} \frac{f(\gamma J, \gamma' J')}{\omega^2_{\gamma' J', \gamma J} - \omega^2}, \quad (5.57)$$

where $f(\gamma J,\gamma' J')$ is the oscillator strength and $\hbar\omega_{\gamma' J',\gamma J}$ is equal to the energy of excitation into a discrete or continuous spectrum state $\gamma' J'$. In the one-electron approximation

$$\alpha(\omega)=\frac{e^2}{m}\sum_a\sum_\nu \frac{f_{\nu a}}{\omega_{\nu a}^2-\omega^2}, \tag{5.58}$$

where a is the core state, ν is an excited electron state, and $\hbar\omega_{\nu a} = \varepsilon_\nu - \varepsilon_a$.

We will rewrite Eq. (5.58) in another equivalent form:

$$\alpha(\omega)=\frac{1}{\hbar}\sum_a\sum_\nu |\langle\nu|d^{(1)}|a\rangle|^2\left[\frac{1}{\omega_{\nu a}+\omega}+\frac{1}{\omega_{\nu a}-\omega}\right]. \tag{5.59}$$

The second term close to $\omega_{\nu a} = \omega$ makes the main contribution to polarizability; therefore, we will keep only this term and add a small imaginary quantity $i\eta$ to its denominator. Using Eq. (2.84) to define the limit as $\eta \to 0$ and considering the fact that the square of the absolute value of the matrix element of a transition operator, summed over all initial and final states, is, to within a constant, equal to the total photoabsorption cross section, we obtain the relation (optical theorem)

$$\sigma^{ph}(\omega)=4\pi\frac{\omega}{c}\,\mathrm{Im}\,\alpha(\omega). \tag{5.60}$$

Correlations between electrons during photoabsorption can be accounted for by representing the vacancy being created as a quasiparticle whose energy differs from that of a "frozen" vacancy by the magnitude of the vacancy's self-energy Σ_a (Sec. 3.1). Replacing ε_a in the polarizability equation, Eq. (5.59), by $\varepsilon_a + \Sigma_a$ and using the relation $\hbar\omega = -E + \varepsilon_\nu$, where $-E$ is the energy of the state with the vacancy measured relative to the ground state energy, gives

$$\alpha(\omega)=\sum_a\sum_\nu \frac{|\langle\nu|d^{(1)}|a\rangle|^2}{E-\varepsilon_a-\Sigma_a(E)}. \tag{5.61}$$

Further, defining the imaginary part of $\alpha(\omega)$, substituting it into Eq. (5.60), and incorporating the spectral function, Eq. (3.14), gives us the following expression for the photoabsorption cross section:

$$\sigma_a^{ph}(\omega)=\frac{4\pi^2\omega}{c}\int |\langle\varepsilon|d^{(1)}|a\rangle|^2 A_a(\hbar\omega-\varepsilon)\,d\varepsilon. \tag{5.62}$$

The summation over excited ν states in discrete and continuous spectra is replaced here by an integration over the photoelectron energy ε (photoexcitation into an autoionizing state can be thought of as photoionization near a resonance).

The spectral function defines the shape of a photoabsorption spectrum containing autoionizing resonances. Their locations correspond to the maxima in the spectral function, Eq. (3.14), which occur at energies E_r that satisfy Dyson's equation

$$E - \varepsilon_a - \mathrm{Re}\,\Sigma_a(E) = 0. \tag{5.63}$$

which is solved by calculating the function $\mathrm{Re}\Sigma_a(E)$ and finding the points at which it intersects the straight line $E - \varepsilon_a$ [139].

If photoionization occurs at the ionization threshold of an inner shell, the time it takes a slow photoelectron to escape the atom, t_e, may be greater than the comparatively short lifetime of the vacancy, τ_v; i.e., the excited state decays before the photoelectron escapes the atom. Describing photoabsorption with due regard for its relationship to the decay process (dynamic relaxation of the atom) is a fairly complicated problem. The asymptotic case $t_e >> \tau_v$ has been examined in [206]. When this condition is satisfied, we can say that a photoelectron moves in the field of an ion that does not already contain a primary vacancy. The probability that an ion will be created in a state γ when a vacancy is filled is equal to the ratio of the partial width Γ_γ, which corresponds to a transition into this state, to the total width of the level containing the vacancy (Γ). In this manner we obtain the following simple formula for the total photoabsorption cross section:

$$\sigma^{ph} = \sum_\gamma \frac{\Gamma_\gamma}{\Gamma} \sigma_\gamma^{ph}, \tag{5.64}$$

where σ_γ^{ph} is the partial cross section of photoabsorption with the creation of an ion in the state γ when the vacancy decays (this cross section is given by the one-electron matrix element of the transition operator in which the photoelectron's wave function is calculated in the field of the relaxed ion [206]).

Chapter 6

PHOTOELECTRON SPECTRA

Spectra for x-ray absorption by gases and vapors yield data regarding the total photoionization and photoexcitation cross section of a free atom or molecule. Detecting the photoelectrons gives additional information about how atoms interact with photons: it lets us define partial cross sections that correspond to the removal of electrons from different shells, measure the electron binding energies in an atom, and study the geometry and dynamics of the process. X-ray photoelectron spectroscopy is a fundamental tool for studying an atom's electron shell structure and the correlation effects that occur in them. The widespread use of x-ray photoelectron spectroscopy for studying free atoms became possible through the use of synchrotron radiation as a source of x rays.

A photoelectron spectrum is generated by a beam of monochromatic x rays or ultraviolet light acting upon an atom. Since the same methods are used to theoretically describe both types of spectra, we will use the general term – photoelectron spectra – from here on.

In Sec. 6.1 we will examine the angular distribution of photoelectrons and how the method of angular momentum transfer is used to describe this distribution. One-configuration and correlation methods for calculating the integral photoelectron spectrum and the role played by correlation effects are discussed in Sec. 6.2. Section 6.3 discusses electron binding energies in atoms and ions.

6.1. The Angular Distribution of Photoelectrons

A photoabsorption cross section has a maximum near a threshold; therefore, radiation whose energy is somewhat greater than the electron binding energies in the atoms under consideration is used to obtain photoelectron spectra, which makes possible the use of a nonrelativistic dipole approximation to describe the angular distribution of those photoelectrons that escape the outer and intermediate shells.

The differential photoionization cross section of an atom having an electron escaping into a spatial angle Ω, $\Omega + d\Omega$ when a beam of monochromatic rays act upon the randomly oriented gas atoms is obtained by averaging the cross section, Eq. (5.17), over the projections of the total angular momentum of the initial state M and summing over the projections of the total angular momentum of the final state M' and the electron spin μ. In the dipole approximation we have [190, 207]

$$\frac{d\sigma}{d\Omega} = \frac{4\pi^2 mq\,\omega}{c\hbar^2}\,[J]^{-1} \sum_{MM'\mu} |\langle \gamma' J' M' \mathbf{q}\mu | \hat{e}_{\mathbf{k}\rho} \cdot \mathbf{D} | \gamma JM \rangle|^2. \tag{6.1}$$

In this section we will examine only photoionization cross sections; therefore, we can omit the subscript "ion" from the symbol σ; J' is the ion's total angular momentum.

A photoelectron is described by the wave function $\psi_{\mathbf{q}\mu}^{-}$, Eq. (4.40), which, far from the atom, is the superposition of plane and outgoing spherical waves. If the equation for ψ^{-} is substituted into Eq. (6.1), the ion and photoelectron angular momenta are recoupled into a common moment, and the Wigner–Eckart theorem is applied to the transition amplitude, we can do the summation over the projections. When we do this for randomly oriented atoms, only the terms with $p = 0$ and $m' = m$ are retained in the expansion of the product of the spherical harmonics

$$Y_{lm}(\hat{\mathbf{q}})\, Y^{*}_{l'm'}(\hat{\mathbf{q}}) = (-1)^m \frac{1}{\sqrt{4\pi}} \sum_{\nu p} [l, l', \nu]^{1/2} \times$$
$$\times \begin{pmatrix} l & l' & \nu \\ -m & m' & p \end{pmatrix} \begin{pmatrix} l & l' & \nu \\ 0 & 0 & 0 \end{pmatrix} Y_{\nu p}(\hat{\mathbf{q}}) \tag{6.2}$$

and the ν in the dipole approximation is limited to values of 0 and 2 [208]. The function $Y_{\nu 0}(\hat{\mathbf{q}})$ is, to within a constant, equal to a Legendre polynomial of degree ν

$$Y_{\nu 0}(\theta, \varphi) = \left[\frac{2\nu + 1}{4\pi}\right]^{1/2} P_\nu(\cos\theta) \tag{6.3}$$

and thus depends on only the polar angle θ. The Legendre polynomials of degree 0 and 2 that enter into Eq. (6.1) are equal to

$$P_0(\cos\theta) = 1, \quad P_2(\cos\theta) = \frac{3}{2}\cos^2\theta - \frac{1}{2}. \tag{6.4}$$

Linearly polarized beams are used most often for studying the angular photoelectron distribution and give maximum anisotropy in the photoelectrons. The z axis then coincides with the polarization vector. The differential cross section for the escape of a photoelectron from an nl shell is given in terms of the total photoionization cross section σ_{nl} and the anisotropy parameter β_{nl} [208, 190]:

$$\frac{d\sigma_{nl}}{d\Omega}=\frac{\sigma_{nl}}{4\pi}\,[1+\beta_{nl}\,P_2(\cos\theta)]. \tag{6.5}$$

Here, θ is the angle between the polarization vector and the direction in which the photoelectron escapes. In general, the β parameter is the sum of the products of submatrix elements of the transition operator, the $3nj$-symbols, and the phase factors [208]. When photoionization from a closed shell occurs, β has the form

$$\begin{aligned}\beta_{nl}=\{\,&l(l-1)\langle nl|r|\varepsilon l-1\rangle^2+(l+1)(l+2)\langle nl|r|\varepsilon l+1\rangle^2-\\&-6l(l+1)\langle nl|r|\varepsilon l+1\rangle\langle nl|r|\varepsilon l-1\rangle\cos[\vartheta_{l+1}(\varepsilon)-\vartheta_{l-1}(\varepsilon)]\,\}\times\\&\times(2l+1)^{-1}[l\langle nl|r|\varepsilon l-1\rangle^2+(l+1)\langle nl|r|\varepsilon l+1\rangle^2]^{-1}.\end{aligned} \tag{6.6}$$

From the condition that the cross section be nonnegative, it follows that β can have values $-1 \le \beta \le 2$ [208, 207]. A value of $\beta = 0$ corresponds to an isotropic photoelectron distribution, $\beta = 2$ corresponds to a $\cos^2\theta$ distribution having a maximum along the polarization vector, and $\beta = -1$ corresponds to a $\sin^2\theta$ dependence having a maximum perpendicular to the polarization vector. At an angle of 54°44 , called the magic angle, the $P_2(\cos\theta)$ polynomial goes to zero. The flux of photoelectrons escaping at this angle is independent of β, and this measurement allows us to find the total cross section.

The method of angular-momentum transfer was developed for studying relationships in the angular photoelectron distribution and for defining β and σ semiempirically [209, 210].

The angular momentum transferred to an atom during photoionization is equal to the difference between the total angular momenta $\mathbf{J}'$ of the final ion and the angular momenta $\mathbf{J}$ of the atom in the initial state. Adding the photoelectron spin s, which cannot be measured experimentally, to $\mathbf{J}' - \mathbf{J}$ yields a quantity called the transferred angular momentum $\mathbf{j}_t$:

$$\mathbf{j}_t=\mathbf{K}-\mathbf{J},\quad \mathbf{K}=\mathbf{J}'+\mathbf{s}. \tag{6.7}$$

It follows from the law of conservation of total angular momentum that

$$\mathbf{j}_t=\mathbf{j}_\omega-\mathbf{l}', \tag{6.8}$$

where $\mathbf{j}_\omega$ is the angular momentum of the absorbed photon (in the dipole approximation its quantum number is unity), and $\mathbf{l}'$ is the photoelectron's orbital angular momentum.

The main advantage to introducing $\mathbf{j}_t$ is that the differential cross section and β can be separated into contributions having different values of j_t, each of which has its own characteristic angular distribution [209, 210]:

$$\frac{d\sigma_{nl}}{d\Omega}=\sum_{j_t,K}\frac{d\sigma_{nl}(j_t)}{d\Omega}, \tag{6.9}$$

$$\frac{d\sigma_{nl}(j_t)}{d\Omega}=\frac{\sigma_{nl}(j_t)}{4\pi}\left[1+\beta_{nl}(j_t)P_2(\cos\theta)\right], \tag{6.10}$$

$$\sigma_{nl}=\sum_{j_t,K}\sigma_{nl}(j_t),\quad \beta_{nl}=\frac{1}{\sigma_{nl}}\sum_{j_t,K}\sigma_{nl}(j_t)\beta_{nl}(j_t). \tag{6.11}$$

For the sake of brevity, the dependence on different quantum numbers, including K, has been omitted from the notations $\sigma_{nl}(j_t)$ and $\beta_{nl}(j_t)$.

The quantity j_t is parity-unfavored if $PP' = -(-1)^{j_t}$ (P is the atom's parity, and P' the ion's); then l' takes on only the one value j_t, and σ and β are equal to [210]

$$\beta_{nl}(j_t)=-1, \tag{6.12}$$

$$\sigma_{nl}(j_t)=\frac{4}{3}\,\frac{\pi^2\alpha}{e^2}\,\hbar\omega\,\frac{2j_t+1}{2J+1}\,|D_0(j_t)|^2. \tag{6.13}$$

$D(j_t)$ is the sum of the submatrix elements of a dipole transition operator having weight

$$D(j_t)=\sum_{J''}(-1)^{J-J''-1}[J'']^{1/2}\begin{Bmatrix}K & l' & J''\\ 1 & J & j_t\end{Bmatrix}\times$$
$$\times\langle\gamma' J' s(K)\varepsilon l' J''\|D^{(1)}\|\gamma J\rangle. \tag{6.14}$$

The subscript 0 in the $D(j_t)$ in Eq. (6.13) means that the value $l' = j_t$ is used.

For a parity-favorable case, when $PP' = (-1)^{j_t}$, the partial contributions to β and σ take on the form of more complex expressions [210]:

$$\beta_{nl}(j_t) = \{ (j_t+2) | D_+(j_t) |^2 + (j_t - 1) | D_-(j_t) |^2 - 6[j_t(j_t+1)]^{1/2} \times$$
$$\times \operatorname{Re} [D_+(j_t) D_-^*(j_t)] \} (2j_t+1)^{-1} [| D_+(j_t) |^2 + | D_-(j_t) |^2]^{-1}, \tag{6.15}$$

$$\sigma_{nl}(j_t) = \frac{4}{3} \frac{\pi^2 \alpha}{e^2} \hbar\omega \frac{2j_t+1}{2J+1} [| D_+(j_t) |^2 + | D_-(j_t) |^2]. \tag{6.16}$$

The indices "$\pm$" on the $D(j_t)$ indicate that the matrix element is calculated with the photoelectron angular momentum's orbital quantum number, $l' = j_t \pm 1$.

The $D(j_t)$, which are independent of the projections, not only can be determined theoretically, but can also be thought of as semiempirical parameters.

The angular photoelectron distribution for unpolarized [211], partially polarized [212], and elliptically polarized [213] x rays are all expressed in terms of the same β parameter. For example, when unpolarized photons act on an atom the differential cross section becomes

$$\frac{d\sigma_{nl}}{d\Omega} = \frac{\sigma_{nl}}{4\pi} \left[1 - \frac{\beta_{nl}}{2} P_2(\cos\theta') \right]. \tag{6.17}$$

This time the z axis is aligned with the direction of the x-ray beam.

In the dipole approximation the equations for the differential cross section, Eqs. (6.5) and (6.17), retain the same form when correlation effects in the atom are taken into account, and even in relativistic theory [28, 214]. When a closed shell is ionized, the equation for β is written in terms of the matrix elements of an effective dipole operator [28].

As the energy of the x rays increases, the dipole approximation for the angular distribution of the electrons becomes inadequate (sooner than it does for the total cross section). Even at an energy of $\hbar\omega \approx 2$ keV a quadrupole term can make a significant correction to $d\sigma/d\Omega$. When this term is taken into account in Eqs. (6.5) and (6.17), the Legendre polynomials P_1 and P_3 also appear in addition to the polynomial P_2 [54].

A plane wave can be used as a photoelectron wave function if the energy of the x rays is much greater than the electron binding energy in the shell being examined. The matrix element of the general transition operator that has not been expanded into the multipoles may be integrated in quadratures in a Born approximation when a bound electron function is approximated by a hydrogenic function. For example, the differential cross section for photoionization from the K-shell becomes [10]

$$\frac{d\sigma_K}{d\Omega} = 2^{5/2} Z^5 \alpha^8 a_0^2 \left(\frac{\varepsilon}{mc^2} \right)^{-7/2} \frac{\sin^2\theta}{\left(1 - \frac{v}{c}\cos\theta\right)^4}, \tag{6.18}$$

when acted on by unpolarized photons ($I_K << \hbar\omega << mc^2$), where θ is the angle between the direction of beam propagation (along the z axis) and the photoelectron wave vector, v is its velocity, and ε is the photoelectron's kinetic energy.

According to Eq. (6.18), when $v << c$, photoelectrons escape mainly in a direction perpendicular to that of the primary beam of photons and the probability that they will escape forward or backward ($\theta = 0, \pi$) is zero. As the velocity of the photoelectrons increases, their preferred direction of motion approaches that of the x rays.

Measuring the total photoionization cross section and β, the most often used characteristics of the photoionization process, does not constitute a complete experiment that would let us define all matrix elements of a transition operator and phase shifts in the photoelectron's wave function. Photoionization in an atom with closed shells is described by three matrix elements in a dipole approximation (there are three possible sets of angular momenta in the final state $nlj^{-1}\varepsilon l'j'J = 1$ having a given ε) and two wave functions $P_{\varepsilon l\pm 1}(r)$; thus, five quantities must be measured before the photoionization process can be described completely [215–217]. Additional information can be obtained by measuring the extent to which the photoelectrons are spin-polarized:

$$P = \frac{I_{1/2} - I_{-1/2}}{I_{1/2} + I_{-1/2}} = \frac{W_{1/2} - W_{-1/2}}{W_{1/2} + W_{-1/2}}, \tag{6.19}$$

where I_μ is the flux intensity of photoelectrons having spin projection μ and escaping at some angle, and W_μ is the probability that such an electron will escape from the atom.

When an atom absorbs nonpolarized x rays, the degree of polarization assumes the form

$$P_{nl}(\theta) = \frac{2\xi_{nl} \sin\theta \cos\theta}{1 - \frac{\beta_{nl}}{2} P_2(\cos\theta)}. \tag{6.20}$$

Here, θ is the photoelectron escape angle with respect to the direction of the x-ray beam, β_{nl} is the anisotropy of the angular distribution of the photoelectrons, and ξ_{nl} is a parameter which, just like β_{nl}, is given in terms of the matrix elements of the transition operator and the scattering phases. The projection of the electron's spin is defined relative to the direction perpendicular to the plane of reaction.

When circularly polarized x rays are used, two more quantities needed for a complete experiment may be introduced – the total photoelectron spin polarization and its angular distribution parameter when such beams act upon an atom [215, 216].

6.2. The Energy Distribution of Photoelectrons. The Role of Correlation Effects

Recording the photoelectrons that escape from atoms being photoionized makes it possible to obtain two basic types of spectrum: 1) the distribution of photoelectrons over energy when monochromatic ultraviolet light or x rays acts on the atoms, and 2) the photoelectron intensity distribution as a function of photon energy when a given shell is ionized. The second type of spectrum, measured at two different angles, allows β to be defined as a function of ω and gives the photoabsorption spectrum when measured at the magic angle. A photoelectron spectrum in the narrow sense of the term is the first type of spectrum, measured at the magic angle. We will examine the theoretical description of this spectrum.

If in the initial state only an atom's ground level has been populated, the relative line intensities of the photoelectron spectrum are proportional to the cross section of photoionization from this level into the various final states, the cross section having been summed over the total momentum of the ion + photoelectron system and its projection, as well as over the photoelectron's quantum numbers. The photoelectron energy is found from the law of conservation of energy:

$$\varepsilon = \hbar\omega - [E^{i}(\gamma' J') - E^{a}(\gamma J)]. \tag{6.21}$$

The superscripts a and i denote the energies of the atom and the ion, respectively.

Substituting the equation for the oscillator strength density, Eq. (1.290), and the matrix elements of the transition operator, Eqs. (5.34) and (1.268), into the general formula for the photoionization cross section, Eq. (5.23), and summing over the total angular momenta of the photoelectron and the system in the final state gives us

$$\sigma(n_1 l_1^{N_1} \gamma_1 L_1 S_1 J_1 \to n_1 l_1^{N_1-1} \gamma_1' L_1' S_1' J_1' \varepsilon l) =$$
$$= \frac{4}{3}\pi^2 \alpha \hbar\omega N_1 [L_1, S_1, J_1'][l_1]^{-1} (l_1^{N_1} \gamma_1 L_1 S_1 \| l_1^{N_1-1} \gamma_1' L_1' S_1' l_1)^2 \times$$
$$\times \sum_y [y] \begin{Bmatrix} S_1' & \frac{1}{2} & S_1 \\ y & L_1' & J_1' \end{Bmatrix}^2 \tag{6.22}$$
$$\begin{Bmatrix} J_1 & l_1 & y \\ L_1' & S_1 & L_1 \end{Bmatrix}^2 \langle l_1 \| C^{(1)} \| l \rangle^2 \langle n_1 l_1 | r | \varepsilon l \rangle^2,$$

$$\sigma(n_1 l_1^{4l_1+2} n_2 l_2^{N_2} \gamma_2 L_2 S_2 J_2 \to n_1 l_1^{4l_1+1} j_1 n_2 l_2^{N_2} \gamma_2' L_2' S_2' J_2' J \varepsilon l) =$$

$$= \delta(\gamma_2 L_2 S_2 J_2, \gamma_2' L_2' S_2' J_2') \frac{4}{3} \pi^2 \alpha \hbar \omega [J] [l_1, J_2]^{-1} \times$$

$$\times \langle l_1 \| C^{(1)} \| l \rangle^2 \langle n_1 l_1 | r | \varepsilon l \rangle^2. \qquad (6.23)$$

The probability that a photoelectron will escape from an open shell is directly proportional to the square of the coefficient of fractional parentage that relates the states of the atom and the ion.

According to Hund's rule, the spin of the ground term has maximum value $\bar{S}$; therefore, when $N_1 \leq 2l_1 + 1$, S_1' can assume only the one value in a pure coupling, $\bar{S} - 1/2$, and a second value of $\bar{S} + 1/2$ for an almost filled shell. Since the terms of an open shell tend to be arranged according to the value of total spin (the higher the spin angular momentum, the lower the energy of the term [98]), for a coupling that is nearly an LS coupling, the photoelectron spectrum consists of a single group of lines when $N_1 \leq 2l_1 + 1$, and it consists of two groups of lines when $N_1 > 2l_1 + 1$. This relationship is seen, e.g., in the photoelectron spectra of the $4f^N$ shell of rare-earth elements [218, 219].

If an intermediate coupling is used to describe an ion's final configuration, the corresponding formula for the cross section is obtained by means of an expression for the matrix element of a transition operator in an intermediate coupling through a linear combination of the matrix elements of a pure coupling and substituting Eqs. (1.290) into Eq. (5.23); when this is done the summation over j and J can also be carried out.

In the relativistic and nonrelativistic theories the probability that photoelectrons will escape from a closed shell (subshell) is given by Eqs. (5.36) and (5.39), respectively.

Because the lifetime of a state of an ion having an inner vacancy is limited, a photoelectron line has some width. If the state decays in mainly a radiating manner (interacts weakly with a continuous spectrum), the photoelectron line looks like a Lorentz curve.

The correlations between electrons in initial and final states produce not only shifts in the photoelectron lines, changes in their shape, and a redistribution of intensities, but lead to the appearance of satellite lines that correspond to photoionization and the excitation of another electron.

Starting from multiconfiguration approximations, correlation effects in photoelectron spectra can be divided into several types [220]:

1) the interaction between configurations in the initial state;

2) the interaction between configurations in the final state. The latter are subdivided into correlations: a) in the final ion, b) in the final state of the continuum (interchannel coupling).

Fairly intense satellite lines that correspond to an interaction between the initial configuration and configurations in which two outer electrons have been excited (e.g., ns^2—np^2—$(n - 1)d^2$ in alkaline-earth and other elements of the second group [220, 221] or the ns^2np^6—$ns^2np^4n'd^2$ in inert gases [199, 222]) are seen in the photoelectron spectra of outer shells. $(s + d)^N$ mixing in an ion's initial configuration appears in the spectra of the vapors of transition elements having a d-shell that is filling and in the elements that precede them [223, 224].

If a state of ion having an inner vacancy is found in a super-Coster–Kronig or Coster–Kronig continuum (Sec. 3.4), the strong coupling with this continuum will lower the intensities and distort the shape of the photoelectron spectrum lines; however, wide double ionization bands are weakly discernible in it. On the other hand, if a vacancy is created in the inner shell of a layer and $n_1l_2^{-2}n_3l_3$ or $n_1l_2^{-1}n_3l_3^{-1}n_4l_4$ configurations of the discrete spectrum that strongly interacts with an ion state $n_1l_1j_1^{-1}$ are possible, this proves to be the cause of the intense satellite lines. This type of satellite has been observed in ns- or np-shell spectra [158, 113, 139, 203]. Satellite lines that correspond to the states of an ion having an inner vacancy and an electron that has been excited out of an outer shell also correspond to the photoelectron spectra of inert gases and other elements [222–226, 31].

Correlation effects such as interactions between configurations in the final state of a continuum are more important for photoelectron and photoabsorption spectra at low photoelectron energies and near a Cooper minimum [190]. The interaction between a final continuum and the autionizing states within it give rise to the so-called resonance satellites, the intensities of which are a function of photon energy [220, 138, 227]. These satellites are seen in photoelectron spectra of vapors of Ca, Ba [138, 228], Xe [229], and other elements that have been stimulated by photons whose energy is close to the binding energy in a subvalent shell. For example, in Ca the intensity of the $3p^63d$ satellite varies in a resonance fashion with respect to the $3p^64s$ main line when photoionization into the $3p^64s\varepsilon l$ continuum near the $3p^53d^3$ level occurs. These variations are the result of the mixing of these states, the interference between the photoexcitation and photoionization channels, and $(s + d)^2$ mixing in the initial state [138].

The method of sudden perturbation is used to describe the satellites that correspond to one-electron, monopole (the electron's orbital angular momentum is unchanged) excitations from an outer shell when a vacancy is created in an inner shell (a familiar example is the $1s$ photoelectron spectrum of Ne [225] or the $2p$ spectrum of Ar [226]). The multiconfiguration method is an effective tool for interpreting satellites in photoelectron spectra and is frequently applied in the configuration-interaction approximation. On the other hand, photoelectron spectra, which are simpler than Auger or radiation spectra, are a convenient tool for studying correlations between the electrons in atoms. In individual cases, for example, when satellites are caused only by the interaction between configurations in the final state of an ion, and the matrix elements of the transition operator between the initial and admixed configurations are zero (e.g., for $3s^23p^6 \to 3s3p^6 + 3s^23p^43d + 3s^23p^44s$ transitions), the intensity of the satellite lines make it possible to find the absolute values of the coefficients of the expansion of the multiconfiguration wave function with respect to the one-configuration functions.

Diagram methods in perturbation theory are used mainly for calculating the photoelectron spectra of atoms having closed and half-filled shells, where they make it possible to obtain good agreement with experiment [27, 28, 138].

Lower-order diagrams, which explain the two-electron transition when a photon is absorbed by an atom, are shown in Fig. 17. We will consider one of these, Fig. 17d, which represents the process as a single ionization and subsequent Auger transition, or shake-up, of the atom resulting from Coulomb interaction between the electrons [230]. Higher-order diagrams may be calculated by means of the spectral function, Eq. (3.14), and the vacancy's self-energy. This is tantamount to summing the contributions of the diagrams, which are obtained from the diagram of Fig. 17d by inserting into the vacancy line those fragments that correspond to possible intermediate excitations caused by an interaction between a vacancy and the surrounding electrons. A vacancy of this kind – a quasiparticle – is shown by a heavy line (Sec. 3.1). The amplitude of the effective operator that can be described by a diagram with a heavy line (Fig. 6.1) is found according to the general rules for interpreting the diagrams, except that the self-energy of the vacancy, Σ_a, is added to the energy of the zero approximation, ε_a.

If both electrons make a transition into the continuous spectrum, the amplitude has the form

$$M_a(\varepsilon_1, \varepsilon_2) = -\frac{\langle\varepsilon_1|d^{(1)}|a\rangle[\langle\varepsilon_2 a|h_{12}^e|cb\rangle - \langle\varepsilon_2 a|h_{12}^e|bc\rangle]}{-\varepsilon_1+\varepsilon_a+\hbar\omega+\Sigma_a(\varepsilon_1-\hbar\omega)}, \quad (6.24)$$

where ε_1 and ε_2 are the photoelectron energies. An exchange diagram having the same form, except that the designations b and c in it are replaced by spaces, is taken into consideration in Eq. (6.24).

Diagrams of the type shown in Fig. 17d usually make the main contribution to the two-electron photoionization cross section; then, using Eq. (6.24) and the definition of the spectral function $A_a(E)$, Eq. (3.14), the differential two-electron photoionization cross section when the energy of one electron is in the range ε_1, $\varepsilon_1 + d\varepsilon_1$ and that of the second is in the range ε_2, $\varepsilon_2 + d\varepsilon_2$ can be written [230]

$$\frac{d^2\sigma}{d\varepsilon_1 d\varepsilon_2} \sim \sum_{abc} |M_a(\varepsilon_1, \varepsilon_2)|^2 \delta(\varepsilon_1+\varepsilon_2-\varepsilon_b-\varepsilon_c-\hbar\omega) =$$

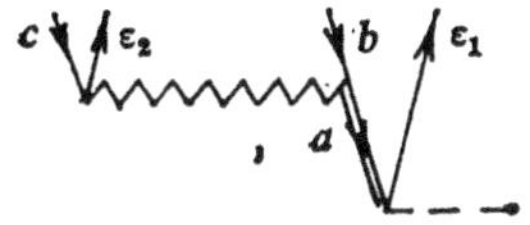

Fig. 6.1. Diagram, containing the quasiparticle line, of the amplitude of double photoionization (or photoionization and the subsequent Auger transition).

$$= \sum_{abc} |\langle \varepsilon_1 | d^{(1)} | a\rangle|^2 A_a(\varepsilon_1 - \hbar\omega) \frac{\pi |[\langle \varepsilon_2 a | h^e_{12} | cb\rangle - \langle \varepsilon_2 a | h^e_{12} | bc\rangle]|^2}{\mathrm{Im}\, \Sigma_a(\varepsilon_1 - \hbar\omega)} \times$$

$$\times\, \delta(\varepsilon_1 + \varepsilon_2 - \varepsilon_b - \varepsilon_c - \hbar\omega). \tag{6.25}$$

Equation (6.25) considers only those diagrams that can be represented by the total diagram of Fig. 6.1, in which vacancies in an ion's final state are said to be frozen.

6.3. Binding Energies. The Shifts in Electron and X-Ray Lines

If an atom is in its ground state before it interacts with a photon and an ion is created in the lowest energy state for the given vacancy, ζ^{-1}, the difference $E^i(\zeta^{-1}) - E^a$ in the photoelectron energy equation, Eq. (6.21), is the electron ζ binding energy in the atom. Equation (6.21) can then be rewritten

$$\varepsilon = \hbar\omega - I_\zeta. \tag{6.26}$$

X-ray photoelectron spectroscopy is direct and the most accurate method of finding the inner electron binding energy in free atoms (from emission and Auger spectra we can only find the differences in the binding energies of the different electrons). On the other hand, the x-ray photoelectron spectra of atoms whose electron binding energies are known fairly accurately are used to find the wavelengths of the x rays (x-ray photoelectron spectroscopy for analyzing x rays).

Sometimes an equivalent concept is used – the ionization energy of a shell or subshell (the energy needed to remove a single electron whose kinetic energy is zero) – instead of the binding energy. The ionization energy of an atom is the least amount of energy needed to remove a single electron from the atom and is expressed as the difference between the energies of the atom's and ion's ground states. If these energies differ only by the quantum numbers nl of only a single electron, it is equal to the binding energy of the electron that is least tightly bound to the atom. Because an electron has unit charge, then, according to the definition of the electron-volt, the binding energy given in electron-volts is numerically equal to the ionization potential in volts.

The electron binding energy in heavy atoms can reach magnitudes of $mc^2/4$ (m is an electron's rest mass). The accuracy obtained when theoretical calculations are done for the innermost shells (when one-configuration relativistic wave functions and other relativistic effects, including corrections for polarization of the vacuum and self-energy, are used) is within a few electron volts [160]. A similar absolute accuracy in calculating the binding energies of the electrons in intermediate shells is achieved with the Hartree–Fock–Pauli approximation when the relativistic corrections in first-order perturbation theory are calculated with nonrelativistic wave functions [79].

Correlation effects have a relatively minor influence on inner-shell energies for the following reasons: 1) correlations between the electrons in other layers, including the outer shells, are excluded when the atomic energy is subtracted from the ion's energy; 2) neighboring shells are closed; therefore, electrons cannot be excited into or between them; 3) although an ion is in an autoionizing state for many continua, an interaction between a final level and a continuum only slightly changes the binding energy if the level is remote from the continuum edge (Sec. 2.4).

The interaction between a final level and Coster–Kronig and super-Coster–Kronig continua (including the respective Rydberg series) makes a significant correction to binding energies, especially if the transitions to these continua are energetically allowed. For the ns- and np-electrons in different atoms this kind of a correction lowers binding energy by as much as 2–11 eV [160]. An exceptionally strong interaction with a super-Coster–Kronig continuum occurs for a $4p^{-1}$-vacancy when $Z \approx 48-58$ [138, 139, 160] and for a $4s^{-1}$-vacancy when $Z \approx 70-80$ [160]. For example, in Xe, for which the double ionization edge $4d^{-2}$ almost coincides with the binding energy of a $4p$-electron, the $4p_{3/2}$-photoelectron line is spread over the $4d^{-2}\varepsilon f$ continuum and the Rydberg series corresponding to it, and the $4p_{1/2}$-line is shifted almost 11 eV in comparison with the Hartree–Fock value [142]. The significant mixing of $2s^{-1}-2p^{-1}3d^{-1}n(\varepsilon)f$ configurations for $21 < Z < 60$, of $3p^{-1}-3d^{-2}n(\varepsilon)f$ for $Z \approx 30$, and of $3s^{-1}-3p^{-1}4f^{-1}n(\varepsilon)g$ for $Z > 70$ must be noted [162].

The possibility of super-Coster–Kronig and Coster–Kronig transitions is limited and eliminated when a vacancy's orbital angular momentum increases. This explains the correlation correction to electron binding energy as a function of its quantum number l: for a given n this correction tends to increase as l decreases [79]. Another important type of correlation, which is fundamental for an I_{nl} having $l = n - 1$, is caused by a reduction in the number of pairs when a vacancy is created. Because the energy of a pair correlation is negative, this correction raises the binding energy (it has a magnitude of from 1 to 3 eV and is weakly dependent on Z [160, 162]).

In concluding this section, we will examine the change in binding energies and the energy of the diagram x-ray lines as an atom is ionized.

The shift in an ion's binding energy in comparison with that of an atom is the difference of the differences in total energies; therefore, if the electron being examined and the electrons being removed from an atom are in different layers, the correlation, relativistic, and relaxation effects compensate one another for the most part and we can use a fairly simply model. In the nonrelativistic theory we will define the shift in terms of the average energies of the atom (E^a) and ion (E^i):

$$\delta I_{nl}\,(\text{ион} - \text{атом}) = [E^i(nl^{-1}) - E^i] - [E^a(nl^{-1}) - E^a]. \tag{6.27}$$

Let, in comparison with an atom, the electrons $n_1 l_1^{q_1} n_2 l_2^{q_2}$ be absent from an ion. Using the generalized Koopmans theorem

$$E^a = E^{i'} + q_1\,\varepsilon_{n_1 l_1} + q_2\,\varepsilon_{n_2 l_2} - \bar{E}^e\,(n_1\,l_1^{q_1}\,n_2\,l_2^{q_2}), \tag{6.28}$$

we can write an atom's energy in terms of the ion energy that was calculated with the atom's frozen wave functions (this is indicated by a prime), where ε_{nl} is the one-electron energy. The last term in Eq. (6.28) is the average energy of a Coulomb interaction between electrons that are removed from an atom. Substituting Eq. (6.28) and an analogous relationship between $E^a(nl^{-1})$ and $E^{i'}(nl^{-1})$ into Eq. (6.27), using Eq. (1.144) for $\bar{E}^e$, and assuming that relaxation corrections approximately compensate one another, we find

$$\begin{aligned}
&\delta I_{nl}\,(n_1\,l_1^{N_1-q_1}\,n_2\,l_2^{N_2-q_2} - n_1\,l_1^{N_1}\,n_2\,l_2^{N_2}) = \\
&= q_1\,\Delta\varepsilon_{n_1 l_1}(nl^{-1}) + q_2\,\Delta\varepsilon_{n_2 l_2}(nl^{-1}) + q_1\,(q_1 - 1)\,A_{n_1 l_1}(nl^{-1}) + \\
&+ q_2\,(q_2 - 1)\,A_{n_2 l_2}(nl^{-1}) + q_1\,q_2\,B_{n_1 l_1, n_2 l_2}(nl^{-1}).
\end{aligned} \tag{6.29}$$

Here, the notations

$$\begin{aligned}
&A_{n_1 l_1}(nl^{-1}) = \\
&= -\frac{1}{2}\,\Delta F^0(n_1 l_1,\ n_1 l_1) + \sum_{k>0} \frac{\langle l_1 \| C^{(k)} \| l_1 \rangle^2}{(4l_1+1)\,(4l_1+2)}\,\Delta F^k\,(n_1 l_1,\ n_1 l_1),
\end{aligned} \tag{6.30}$$

$$B_{n_1 l_1,\, n_2 l_2}(nl^{-1}) =$$
$$= -\Delta F^0(n_1 l_1,\; n_2 l_2) + \sum_k \frac{\langle l_1 \| C^{(k)} \| l_2 \rangle^2}{2(2l_1+1)(2l_2+1)} \Delta G^k(n_1 l_1,\; n_2 l_2), \tag{6.31}$$

$$\Delta X = X - X(nl^{-1}), \text{ where } X = \varepsilon_{nl},\; F^k,\; G^k. \tag{6.32}$$

are introduced. The quantity $\Delta\varepsilon_{n_1 l_1}(nl^{-1})$ is the positive change in the energy of an $n_1 l_1$ electron when an nl^{-1} vacancy is created. The parameters A and B are roughly an order of magnitude smaller than $\Delta\varepsilon$ [141]; therefore, the shift increases almost linearly as the degree of ionization increases. The nonlinear A and B parameters also much weaker than $\Delta\varepsilon$ depend on the configuration of the atom by whose wave function they are calculated.

Comparing the results obtained from calculating δI_{nl} according to Eqs. (6.27) and (6.29) shows [141] that Eq. (6.29) is a good approximation of the shift in binding energy. The advantage of Eq. (6.29) is that it contains in explicit form the dependence of the shift on the number of electrons removed and requires only that the wave functions of an atom with and without an nl^{-1} vacancy be calculated.

The energy of a diagram x-ray line is equal to the difference between the binding energies of the jumping electron in its initial and final states; therefore, the equation for the radiation line shift $n_3 l_3 \to n_4 l_4$ as $n_1 l_1{}^{q_1} n_2 l_2{}^{q_2}$ electrons are removed from the atom follows from Eq. (6.29) [231]:

$$\delta E_{n_3 l_3 \to n_4 l_4}(n_1 l_1^{-q_1} n_2 l_2^{-q_2}) = \delta I_{n_4 l_4} - \delta I_{n_3 l_3} = q_1 C_{n_1 l_1} + q_2 C_{n_2 l_2} +$$
$$+ q_1(q_1 - 1) A'_{n_1 l_1} + q_2(q_2 - 1) A'_{n_2 l_2} + q_1 q_2 B'_{n_1 l_1,\, n_2 l_2}. \tag{6.33}$$

The parameters A', B', and C are all given in terms of the parameters of the binding energy shift:

$$C_{n_1 l_1} = \Delta\varepsilon_{n_1 l_1}(n_4 l_4^{-1}) - \Delta\varepsilon_{n_1 l_1}(n_3 l_3^{-1}), \tag{6.34}$$

$$A'_{n_1 l_1} = A_{n_1 l_1}(n_4 l_4^{-1}) - A_{n_1 l_1}(n_3 l_3^{-1}), \tag{6.35}$$

$$B'_{n_1 l_1,\, n_2 l_2} = B_{n_1 l_1,\, n_2 l_2}(n_4 l_4^{-1}) - B_{n_1 l_1,\, n_2 l_2}(n_3 l_3^{-1}). \tag{6.36}$$

The C_{nl} parameter has the sense of a shift in the energy of an x-ray line as an *nl* electron is removed from an atom and, in contrast to $\Delta\varepsilon_{nl}$, may be positive (*s*- and *p*-electrons) as well as negative (*d*- and *f*-electrons) [232]. The squared terms are also more important in an x-ray line shift [231].

Equations (6.29) and (6.33) remain valid in relativistic theory if the quantum numbers *nl* in these equations are replaced by *nlj* and the definitions

$$A_{n_1 l_1 j_1}(nlj^{-1}) = = -\frac{1}{2}\Delta F^0(nlj,\ nlj) + \frac{2j+1}{4j}\sum_{k>0}\begin{pmatrix} j & j & k \\ -\frac{1}{2} & \frac{1}{2} & 0 \end{pmatrix}^2 \Delta F^k(nlj,\ nlj), \tag{6.37}$$

$$B_{n_1 l_1 j_1,\ n_2 l_2 j_2} = = -\Delta F^0(n_1 l_1 j_1,\ n_2 l_2 j_2) + \sum_k \begin{pmatrix} j_1 & j_2 & k \\ \frac{1}{2} & -\frac{1}{2} & 0 \end{pmatrix}^2 \Delta G_k(n_1 l_1 j_1,\ n_2 l_2 j_2). \tag{6.38}$$

are used for the *A* and *B* parameters. Here, F^k and G^k are the relativistic integrals of Eqs. (1.186) and (1.195).

X-ray line shifts that occur when an outer electron is removed are only a few tenths or hundredths of a electron-volt [233]. Shifts in the binding energies δI_{nl} are several electron-volts [141] and, because of the relationship between them and δE (6.33) and the fact that δE is small, are only weak functions of *nl*.

Thus, we see that the binding energy of the various electrons increases by almost the same amount when vacancies are created – multiple ionization satellites are seen to be roughly the same distance from the base lines (if ionization does not lead to their further splitting).

Chapter 7

CHARACTERISTIC EMISSION SPECTRA

X rays are generated by acceleration of moving charged particles, by plasma heated to temperatures of the order of a million degrees, and by highly excited atoms having inner vacancies due to the radiative transitions that occur in them. During spontaneous emission of atoms with vacancies, characteristic x-ray spectra that reflect the structure of the energy levels in the excited atoms arise. We will examine these spectra in this chapter. The x-ray spectra of multiply charged ions – the main component of a plasma – are more similar to optical spectra [46, 48] and will not be considered here.

We will begin our examination of x-ray characteristic spectra from the very simple spectra of atoms having a single inner vacancy. These spectra comprise the so-called diagram lines (Sec. 7.1), the study of which has played a major role in quantum mechanics. The foundations for a theoretical description of the x-ray spectra of atoms in a one-configuration approximation will be discussed in Sec. 7.2, with allowances made for correlations between the electrons.

7.1. Fundamental Characteristics of Emission Spectra. Diagram and Satellite Lines

An atom having an inner vacancy can be spontaneously deexcited, whereby it emits a photon whose energy is equal to the difference between the energies of the atom in its initial and final states:

$$\hbar\omega = E_1 - E_2, \tag{7.1}$$

When this happens the vacancy is filled by an electron from the shell having the smaller binding energy. The characteristic emission spectrum for a given element is the set of all the possible x-ray transition lines.

Lines that correspond to different transitions from initial states having a vacancy in the same layer constitute a spectral series, for example, *K*-, *L*-, *M*-series, etc. An electron can make a transition from any other layer to the innermost layer, the *K*-layer, which consists of only one shell. The x-ray spectrum of an element whose atomic number is less than 11 can contain only the *K*-series: as the number of shells and layers in an atom increases, new series, along with their lines, appear.

When transitions are made into deep inner shells, the energy interval of one series is generally several times narrower than the distances between adjacent series. The lines of different series converge and even overlap one another if there are primary vacancies in the intermediate and subvalent shells.

The main diagram lines correspond to dipole transitions between subshells in the different layers. Quadrupole transition lines are also related to diagram lines. Their relative intensity increases with decreasing wavelength and when this wavelength is a few angstroms or less, the quadrupole lines are no longer the very weak forbidden lines as found in optical spectra.

Line intensity decreases rapidly as the difference between the principal quantum numbers for both states of an electron making a transition increases. The selection rules for dipole transitions $\Delta l = 1, \Delta j = 0, \pm 1$ and the one-particle nature of the energy levels in an atom with closed electron shells and a single inner vacancy (Sec. 3.1) are due mainly to the doublet form of the spectra, similar to the optical spectra of atoms of the alkali elements [55]. The x-ray spectrum becomes even simpler because its series, in contrast to an optical series, contains only a small number of lines – fewer than the number of subshells in the atom.

The basic x-ray emission spectral lines, which correspond to transitions in the inner shells, maintain their nature even for atoms with open outer shells, since they only slightly affect the properties of the inner shells.

X-ray emission spectral lines defined by this simple model are called diagram lines and their designations (Table 2) have gradually emerged as new lines have been discovered; therefore, there is no simple rule for interpreting them. Along with these traditional designations, designations that indicate the initial and final states of an atom are often used (especially for transitions in which open outer shells participate).

Because different x-ray lines may correspond to the same initial and final states, there is a combinational principle for the transition energies or frequencies; e.g.,

$$E(K^{\mathrm{I}}_{\beta_5}) = E(K_{\alpha_1}) + E(L_{\alpha_1}), \tag{7.2}$$

$$E(K^{II}_{\beta_5}) = E(K^{II}_{\beta_5}) + E(M_{\zeta_5}). \tag{7.3}$$

Similar relationships allow us to find the energies of excited states with vacancies from experimental values of the line energies.

If the expression for the energy of an atom with a vacancy, Eq. (3.12), is substituted into Eq. (7.1) and only basic terms that are quadratic in Z are retained, we obtain Moseley's law, discovered semiempirically, for the diagram line:

$$\sqrt{\frac{\nu}{cR}} = A(Z - \sigma'). \tag{7.4}$$

Here,

$$A = \left(\frac{1}{n_1^2} - \frac{1}{n_2^2}\right)^{\frac{1}{2}}, \tag{7.5}$$

where ν is the line frequency and σ' is the average value of the total screening constant, given by

$$(Z - \sigma')^2 \left[\frac{1}{n_1^2} - \frac{1}{n_2^2}\right] = \frac{(Z - \sigma_1')^2}{n_1^2} - \frac{(Z - \sigma_2')^2}{n_2^2}. \tag{7.6}$$

Moseley's law is not satisfied for transitions with $n_1 = n_2 = n$. The term that is quadratic in Z goes to zero and the frequency ν is only linearly dependent on Z:

$$\frac{\nu}{cR} = \frac{2(\sigma_2' - \sigma_1')}{n^2}\left[Z - \frac{\sigma_1' + \sigma_2'}{2}\right]. \tag{7.7}$$

Transitions such as these have comparatively low energy even in the heavy elements. Because of this and another dependence on Z, they usually are not classified to x-ray diagram lines.

Spin-doublets and screening doublets are characteristic for an x-ray emission spectrum [1, 10].

According to Eq. (3.12), the distance between two levels having the same quantum numbers n and l and having $j = l \pm 1/2$ (neglecting the dependence of σ_{nlj} on j) is

$$\Delta E = Rhc\alpha^2 \frac{(Z - \sigma_{nl})^4}{n^3 l(l+1)}. \tag{7.8}$$

These levels are called a spin-doublet. Two emission spectrum lines are also called spin-doublet if they correspond to the same initial or final level, and the second transition levels are a spin-doublet (e.g., $K_{\alpha 1}$- and $K_{\alpha 2}$-lines). Using the relation

$$|\Delta\lambda| = \frac{c\Delta\nu}{\nu^2} = \frac{c\Delta E}{h\nu^2}, \tag{7.9}$$

expressing the ν^2 from Eq. (7.4) and ΔE from Eq. (7.8), and taking the approximation $\sigma \approx \sigma'$ into consideration, we find that the difference between the wavelengths of a spin-doublet is independent of Z (for this reason they are sometimes called regular doublets):

$$|\Delta\lambda| \approx \frac{\alpha^2}{A^4 n^3 l(l+1) R} = \text{const.} \tag{7.10}$$

As a matter of fact, this quantity increases slowly with increasing Z.

There is another dependence on Z, which is characteristic of the distance between pairs of levels having the same nj, but with an l that is not unity. A vacancy's energy is not degenerate in l as is that of an electron in a one-electron atom: it depends on the orbital angular momentum through the screening constants. Leaving only the ground term that is quadratic in Z in Eq. (3.12), we find that

$$\left(\frac{E_l - E_{l+1}}{Rhc}\right)^{1/2} \approx \frac{\sigma'_{l+1} - \sigma'_l}{n}. \tag{7.11}$$

Screening doublets are lines that correspond to a common level, and the other pair of levels with some nj values but different l values $l = j \pm 1/2$. The difference in their energies is only a linear function of Z. Because the levels of a doublet have different parity, both lines cannot correspond to electric dipole transitions. Usually, one of them corresponds to an electric quadrupole transition. For example, a screening doublet comprises the lines $L_{\beta 3}$ and $L_{\beta 10}$.

Equations (7.8) and (7.11) allow us to find the total and inner screening constants from experimental data [10].

Emission spectral lines that cannot be explained by the single-vacancy model have historically been called satellite lines. Their appearance is due to several factors:

1) transitions in two or more highly ionized atoms that appear when an atom interacts with an incident particle or during the radiationless decay of an excited state, and in atoms having a vacancy and an excited outer electron;

2) an interaction between a vacancy in an initial or final state and an open outer shell or the participation of this shell in the transition, and

3) correlation effects in an initial or final state.

There are many x-ray spectra, especially in the long-wave spectrum (which corresponds to transitions in the subvalent and valent shells), for which the factors noted above so dramatically alter the spectrum that the satellite line intensities become comparable to, or even exceed, the diagram line intensities. When this happens, dividing a spectrum into diagram and satellite lines becomes meaningless.

If in an initial state a subshell has two vacancies the one-electron radiative transition lines that appear are called hypersatellites [235, 236, 162]. Because effective nuclear charge increases as an atom is ionized, hypersatellites, as well as the satellites from multiple ionization, are shifted from the main lines toward the higher energies.

Although emission is the reverse of absorption, emission spectra are usually more complex than photoexcitation spectra: electrons in closed shells can take part in radiative transitions, but excitation into closed shells is impossible; excitation proceeds mainly from one level in the initial configuration, whereas many levels for both configurations appear in emission spectra.

According to their origin, x-ray spectra can be subdivided into primary spectra, obtained by collisions between atoms and electrons or other particles, and secondary – or fluorescent spectra – when stimulated by x rays.

Among the recorded x-ray lines for atoms, one of the longest in wavelength – the M_2-N_1 line emitted by a potassium atom – is 69.2 nm long and has an energy of only 17.9 eV; on the other hand, the $K_{\beta 4}(K - N_{4,5})$ line from uranium correspond to a wavelength of 0.0108 nm and a transition energy of 115.01 keV [234]. The characteristic x-ray radiation from atoms ranges roughly between these values.

7.2. A Theoretical Description of Emission Spectra

An atom's excited state is not a stationary state in an electromagnetic field + atom system; therefore, the state decays in a radiating manner even when no external field is present — the atom's spontaneous emission [237]. Because the electromagnetic interaction is weak, the use of first-order perturbation theory is justified.

An equation for the radiation probability is obtained in much the same way that the absorption equation was, starting from Fermi's rule, Eq. (1.251), when the operator for an interaction between the atom and electromagnetic field, Eq. (1.238), or the first term in Eq. (1.240) is used. If there are $n_{\mathbf{k}\rho}$ photons in the initial state, each having a wave vector $\mathbf{k}$ and polarization vector $\hat{\epsilon}_{\mathbf{k}\rho}$ and their number is increased by one in the final state, the photon component of the transition amplitude is then determined by the matrix element for the $a_{\mathbf{k}\rho}{}^*$, Eq. (1.243). The probability per unit time that an atom will make a transition from the state γ into a state γ' by emitting a photon having the polarization $\hat{\epsilon}_{\mathbf{k}\rho}$ into an element of the solid angle $d\Omega$ is

in the nonrelativistic theory,

$$dW_{\mathbf{k}\rho}(\gamma \rightarrow \gamma') = \frac{e^2 \omega}{2\pi \hbar c^3 m^2} \left| \hat{\epsilon}_{\mathbf{k}\rho} \cdot \langle \gamma' | \sum_j \mathbf{p}_j e^{-i\mathbf{k}\cdot\mathbf{r}_j} | \gamma \rangle \right|^2 (n_{\mathbf{k}\rho} + 1)\, d\Omega, \quad (7.12)$$

and in the relativistic theory,

$$dW^{rel}_{\mathbf{k}\rho}(\gamma \to \gamma') = \frac{e^2 \omega}{2\pi \hbar c} \left| \hat{\epsilon}_{\mathbf{k}\rho} \cdot (\gamma' | \sum_j \boldsymbol{\alpha}_j e^{-i\mathbf{k}\cdot\mathbf{r}_j} | \gamma) \right|^2 (n_{\mathbf{k}\rho} + 1)\, d\Omega. \qquad (7.13)$$

Here, α_j is a Dirac matrix, ω is the cyclic frequency of the photon, equal to $(E_\gamma - E_{\gamma'})/\hbar$, and $\mathbf{p}_j$ is the momentum of the jth electron. The summation is carried out over the coordinates of all electrons in the atom.

The transition probability contains two terms. One is proportional to the number of quanta in the initial state and describes the atom's electromagnetic field-induced radiation; the second term, independent of the external field, describes the atom's spontaneous emission. From now on, we will be examining spontaneous emission and will say that $n_{\mathbf{k}\rho}$ is zero.

We can use Eq. (1.275) to transform the nonrelativistic operator for a radiative transition into the more widely used r-form:

$$dW_{\mathbf{k}\rho}(\gamma \to \gamma') = \frac{e^2 \omega^3}{2\pi \hbar c^3} \left| \hat{\epsilon}_{\mathbf{k}\rho} \cdot \langle \gamma' | \sum_j \mathbf{r}_j e^{-i\mathbf{k}\cdot\mathbf{r}_j} | \gamma \rangle \right|^2 d\Omega. \qquad (7.14)$$

The probability of principal transitions in which the orbital angular momentum of the electron making the transition changes by one, even at x-ray wavelengths, is essentially determined by the dipole transition operator obtained when all terms of order higher than the first term in the expansion of $e^{-i\mathbf{k}\cdot\mathbf{r}_j}$ — unity — are neglected (see Sec. 1.8). The operator $-e\Sigma\mathbf{r}_j$ is the operator for an atomic dipole moment $\mathbf{D}$.

Substituting the expansion of the scalar product, Eq. (5.19), into Eq. (7.14), specifying the final and initial states, and using the Wigner–Eckart theorem, in the dipole approximation we find that

$$dW_{\mathbf{k}\rho}(\gamma J M \to \gamma' J' M') = \frac{\omega^3}{2\pi \hbar c^3} \left| \sum_q (-1)^q C^{(1)}_{-q}(\hat{\epsilon}_{\mathbf{k}\rho})(-1)^{J'-M'} \times \right.$$

$$\left. \times \begin{pmatrix} J' & 1 & J \\ -M' & q & M \end{pmatrix} \langle \gamma' J' \| D^{(1)} \| \gamma J \rangle \right|^2 d\Omega. \qquad (7.15)$$

Because of the parity of the operator, dipole transitions can only be made between configurations with different parities. The conditions under which the Wigner symbol does not go to zero are given by the transition selection rules:

$$|J - J'| \leqslant 1 \leqslant J + J', \quad \Delta M = M - M' = -q. \qquad (7.16)$$

According to the second condition for transitions having $\Delta M = 0$, only the term having the polarization vector component $\epsilon_0^{(1)} = \epsilon_z$ does not go to zero in the summation over q; i.e., x rays are polarized along the axis of quantization z (isolated by some kind of external perturbation), and if the projection of the total angular momentum changes by one ($\Delta M = \pm 1$), then the polarization vector is, according to Eq. (1.53), equal to

$$\epsilon_{\mp 1}^{(1)} = \pm \frac{1}{\sqrt{2}} (\epsilon_x \mp i\epsilon_y) \tag{7.17}$$

and right-hand or left-hand polarization of the radiation in the xy plane occurs. The radiation angular distribution is determined by the spherical function $C_{-q}^{(1)}(\hat{\epsilon})$. Multiplying the transition probability by the number of excited atoms in the state γ per unit volume [the population of the initial state $\mathcal{N}(\gamma)$] and the photon energy gives us the radiation intensity:

$$dI_{\mathbf{k}\rho}(\gamma \to \gamma') = \mathcal{N}(\gamma)\, dW_{\mathbf{k}\rho}(\gamma \to \gamma')\, \hbar\omega. \tag{7.18}$$

If the population is independent of the projection of the atom's total angular momentum – all states in the initial level are populated identically – the equation for the probability of a transition from a level γJ to a level $\gamma' J'$ is then found by averaging Eq. (7.15) over M and summing over M'. When this is done, Eq. (5.21) becomes, according to Eq. (1.13), independent of both the direction of the polarization vector and of the photon's escape angle. Doing the summation over $\rho = 1, 2$ and integrating with respect to the angles, we can write the total transition probability in terms of the line strength $S(\gamma J, \gamma' J')$:

$$W(\gamma J \to \gamma' J') = \frac{4}{3\hbar^4 c^3 (2J+1)} \Delta E(\gamma J, \gamma' J')^3 S(\gamma J, \gamma' J'). \tag{7.19}$$

The numerical coefficient in Eq. (7.19) is $5.181248 \cdot 10^{-7}$ if all quantities are defined in atomic units, and has a value of $2.141998 \cdot 10^{10}$ if the energy and line strength are in a.u. and the probability is in $\sec^{-1}$.

When the population $\mathcal{N}(\gamma JM)$ is dependent on M (the atoms are aligned with respect to the axis of quantization, the direction of the stimulating beam), this dependence then affects the angular distribution and the polarization of the subsequent x-ray radiation [208, 238].

An equation for the probability of spontaneous multipolar radiation is obtained by using the expansion, Eq. (1.256), for the exponent $e^{i\mathbf{k}\cdot\mathbf{r}}$ in H'. In the nonrelativistic theory the total probability of an electric multipole transition between two levels becomes [39]

$$W^t(\gamma J \to \gamma' J') = \frac{2(2t+1)(t+1)}{[(2t+1)!!]^2 t} \frac{1}{\hbar} \left(\frac{\omega}{c}\right)^{2t+1} \frac{1}{2J+1} S_t(\gamma J, \gamma' J'). \quad (7.20)$$

Comparing Eq. (7.20) when $t = 2$ with Eq. (7.19) gives us the following estimate for the relative magnitude of the probability (or intensity) of electric quadrupole radiation:

$$\frac{W_{\text{эл. квадр}}}{W_{\text{эл. дип}}} \approx \left(\frac{\omega}{c}\right)^2 \frac{\langle | r^2 | \rangle^2}{\langle | r | \rangle^2} \approx \frac{1}{\lambda^2} \frac{(a^2)^2}{a^2} = \left(\frac{a}{\lambda}\right)^2, \quad (7.21)$$

where a is the radius of the atom's radiating area.

For soft x rays, $\lambda \approx 10$ nm and $a \approx 0.1$ nm (valence electrons can take part in the transition); thus, $(a/\lambda)^2 \approx 10^{-4}$. With hard x rays, $\lambda \approx 0.1$ nm, the size of the radiating area also decreases ($a \approx 10^{-2}$ nm) and the ratio increases but remains much smaller than one: $(a/\lambda)^2 \approx 10^{-2}$. Consequently, in comparison with the optical spectrum, the intensity of electric quadrupole transitions in the x-ray spectrum increases by several orders of magnitude, although they are still much weaker than dipole transitions.

The nonrelativistic theory does not explain x-ray magnetic dipole transitions – their probability is zero (the operator does not act on radial variables; therefore, its matrix element goes to zero because radial wave functions are orthogonal).

In the relativistic theory the probabilities of electric and magnetic multipole transitions are given, much like Eq. (7.20), in terms of the matrix elements of the ${}_eO^{(t)}$ [Eq. (1.277)] and ${}_mO^{(t)}$ [Eq. (1.278)] operators:

$$W^t(\gamma J \to \gamma' J') = \frac{2(2t+1)(t+1)}{[(2t+1)!!]^2 t} \left(\frac{\omega}{c}\right)^{2t+1} \frac{1}{\hbar(2J+1)} |(\gamma' J' \| O^{(t)} \| \gamma J)|^2. \quad (7.22)$$

The submatrix elements of the ${}_eO^{(t)}$ and ${}_mO^{(t)}$ operators are found from Eqs. (1.279)–(1.281).

The equations for the radiative transition probabilities presented above correspond to the total probabilities of a transition between the given levels. Because the upper, and sometimes both levels, have a finite lifetime and, in connection with this, a certain width, the intensity of the line is distributed over a certain energy interval. If a state decays in only a radiating manner, the density of the states $\rho(E)$ in Eq. (1.246) and the shape of the line are described by a Lorentz function (Sec. 9.1):

$$I(E) = \frac{J\Gamma}{2\pi} \frac{1}{(E-E_0)^2 + (\Gamma/2)^2}, \quad (7.23)$$

where I is the total intensity of the line, Γ is the line width, equal to the sum of the widths of the initial and final levels, and E_0 is the location of an intensity maximum.

When a transition is made from an autoionizing level, the autoionizing level wave function, which looks like the superposition of continuous and discrete spectra functions (Sec. 2.4), must be used in the expression for the radiative transition probability to find the shape of a line. Because an excited state interacts with a continuum, a radiation line may become asymmetric [239, 240].

The fields of the various multipoles interfere with one another; however, the total radiation intensity is equal to the sum of the contributions from the various multipoles when integrating over the angles because spherical harmonics are orthogonal [88, 39].

The intensity of multipole, as well as dipole, radiation is given in terms of the population of the upper level, the probability, and the energy of transition:

$$I(\gamma J \to \gamma' J') = \mathcal{N}(\gamma J) W(\gamma J \to \gamma' J') \Delta E(\gamma J, \gamma' J'). \tag{7.24}$$

For $n_1 l_1 j_1^{-1} \to n_2 l_2 j_2^{-1}$ transitions we can derive simple expressions for the relative intensities of a multiplet of lines having different values of j_1 and j_2, assuming that the populations of the $j_1 = l_1 \pm 1/2$ levels are proportional to their statistical weights and that the transition energy is independent of j_1 and j_2,. By summing the equation for the line intensity that follows from Eqs. (7.24), (7.19), and (1.271),

$$I(n_1 l_1 j_1^{-1} \to n_2 l_2 j_2^{-1}) \sim \Delta E(n_1 l_1^{-1}, n_2 l_2^{-1})^4 \times$$
$$\times [j_1, j_2] \begin{Bmatrix} l_1 & j_1 & \frac{1}{2} \\ j_2 & l_2 & 1 \end{Bmatrix}^2 \langle l_1 \| C^{(1)} \| l_2 \rangle^2 \langle n_1 l_1 | r | n_2 l_2 \rangle^2, \tag{7.25}$$

and using Eq. (A2.7) over j_1 and j_2, we find

$$\sum_{j_1} I(n_1 l_1 j_1^{-1} \to n_2 l_2 j_2^{-1}) \sim \frac{2j_2+1}{2l_2+1};$$
$$\sum_{j_2} I(n_1 l_1 j_1^{-1} \to n_2 l_2 j_2^{-1}) = \frac{2j_1+1}{2l_1+1}. \tag{7.26}$$

Depending on the quantum numbers of both vacancies, these sums will contain one or two terms. The relative intensities when the total angular momentum of the other vacancy is fixed are

$$\frac{\sum\limits_{j_1} I(n_1 l_1 j_1^{-1} \to n_2 l_2 \bar{j}_2^{-1})}{\sum\limits_{j_1} I(n_1 l_1 j_1^{-1} \to n_2 l_2 \bar{\bar{j}}_2^{-1})} = \frac{2\bar{j}_2+1}{2\bar{\bar{j}}_2+1}, \quad \frac{\sum\limits_{j_2} I(n_1 l_1 \bar{j}_1^{-1} \to n_2 l_2 j_2^{-1})}{\sum\limits_{j_2} I(n_1 l_1 \bar{\bar{j}}_1^{-1} \to n_2 l_2 j_2^{-1})} = \frac{2\bar{j}_1+1}{2\bar{\bar{j}}_1+1}. \qquad (7.27)$$

Equations (7.27) correspond to the familiar empirical Burger–Dorgelo rule [1, 10]. Applying these equations to, for example, the L_{β_1}, L_{α_1}, and L_{α_2} lines we obtain the system (g is the statistical weight of the level)

$$\frac{I(L_{\alpha_2})+I(L_{\beta_1})}{I(L_{\alpha_1})} = \frac{g(M_4)}{g(M_5)} = \frac{2}{3}, \quad \frac{I(L_{\alpha_1})+I(L_{\alpha_2})}{I(L_{\beta_1})} = \frac{g(L_3)}{g(L_2)} = 2, \qquad (7.28)$$

the solution of which yields

$$I(L_{\alpha_2}):I(L_{\alpha_1}):I(L_{\beta_1}) = 1:9:5. \qquad (7.29)$$

An open outer shell in an atom affects the emission spectrum, even if it does not take part in the transitions being examined. An interaction between a vacancy and an open shell, where the vacancy is produced in a not too deep shell, splits the levels of the core. Depending on the width of the levels and the strength of the different interactions in the atom, this multiplet splitting leads to a widening of the diagram lines, their asymmetry, the appearance of satellites, or division of the spectrum into a set of lines [241–244, 112, 113, 161, 175].

It was shown in Sec. 4.2 that, in an approximation, we can say that the electrons in an open outer shell retain their term when an inner vacancy is created in an atom in the ground state. Let then the radiative transition

$$n_1 l_1^{4l_1+1} n_2 l_2^{4l_2+2} n_3 l_3^{N_3} \to n_1 l_1^{4l_1+2} n_2 l_2^{4l_2+1} n_3 l_3^{N_3}.$$

of an electron from a closed shell into a shell with a vacancy occur. The emission spectrum is divided into two groups of lines according to the quantum number j_1 of the first shell, and the splitting of the lines in the groups when $n_2 = n_3$ is governed mainly by electrostatic interaction between the $l_2^{4l_2+1}$ and $l_3^{N_3}$ shells [241, 242, 175, 195]. For the final configuration, the ground term approximation loses validity but, because a configuration of this type has an inherent tendency to separate the terms according to their spin, the spin of the lowest term in an open shell $\tilde{S}_3 = N_3/2$ is maintained during radiation and transitions occur mainly in the two groups of levels that correspond to total spin values of $\tilde{S}_3 + 1/2$ and $\tilde{S}_3 - 1/2$. If we ignore the multiplet splitting of the initial configuration caused by an electrostatic interaction, and the energy of an exchange electrostatic interaction $l_2^{4l_2+1}l_3^{N_3}$ in the final configuration is averaged over all but the spin quantum numbers, the average energy difference of two groups of transition lines from a given level j_1 is then found to be [241, 195]

$$E\left(S=\tilde{S}_3-\frac{1}{2}\right)-E\left(S=\tilde{S}_3+\frac{1}{2}\right)= =(2\tilde{S}_3+1)\sum_k \frac{\langle l_2 \| C^{(k)} \| l_3\rangle^2}{(2l_2+1)(2l_3+1)}\, G^k(n_2 l_2,\ n_3 l_3), \tag{7.30}$$

i.e., is proportional to the number of unpaired electrons in the $l_3^{N_3}$ shell $(2\tilde{S}_3 + 1 = N_3 + 1)$. The satellites that satisfy this relationship are seen in the x-ray $K_{\beta_{1,3}}$-spectra $(1s^{-1} \rightarrow 3p^{-1})$ of elements with a $3d^N$-shell that is being closed [241, 175, 244], in the L_{γ_1} and $L_{\beta_{2,15}}$-spectra $(2p^{-1} \rightarrow 4d^{-1})$ of the lanthanides [112, 244], etc.

If the initial vacancy is created in a layer containing the outer shell and a spectrum is emitted during the transition of an electron from the outer shell $n_1l_1^{4l_1+1}n_1l_2^{N_2} \rightarrow n_1l_1^{4l_1+2}n_1l_2^{N_2-1}$, then, assuming that the maximum spin of the $l_2^{N_2}$ shell is maintained when the atom is ionized and that the total spin is maintained during a radiative transition (its operators do not act on the spin variables), we have the following regularity: when $N_2 \leq 2l_2 + 1$, the spectrum will contain one group of lines having $S = \tilde{S}_2 - 1/2$, and when $N_2 > 2l_2 + 1$, also a second group of lines having $S = \tilde{S}_2 + 1/2$.

An important characteristic of an emission spectrum is its average energy. If we assume that all states in an initial configuration are populated equally, then the average energy depends on only the atom's characteristics and is defined by

$$\bar{E}_e(K-K')=\frac{\sum\limits_{\beta\gamma}[\langle\beta|H|\beta\rangle-\langle\gamma|H|\gamma\rangle]W_{\beta\gamma}}{\sum\limits_{\beta\gamma}W_{\beta\gamma}} \quad (\beta\in K,\ \gamma\in K'), \tag{7.31}$$

where H is the Hamiltonian for the atom and $W_{\beta\gamma}$ is the probability of the radiative transition $\beta \rightarrow \gamma$.

The average energy of a radiation spectrum is distinguished from the difference in the average energies of a final and an initial configuration by a quantity called the spectrum shift:

$$\delta E_e(K-K') = \bar{E}_e(K-K') - [\bar{E}(K) - \bar{E}(K')]. \tag{7.32}$$

In the HF-av. approximation, if we replace the energy term in $W_{\beta\gamma}$ by its average value for specific K and K' configurations, the summation over β and γ in Eq. (7.31) can generally be done by using the method of second quantization [40]. The shift in the spectrum is determined only by electrostatic interaction between the shells that participate in the transition:

$$\begin{aligned}
&\delta E_e(K_0 n_1 l_1^{N_1} n_2 l_2^{N_2} - K_0 n_1 l_1^{N_1+1} n_2 l_2^{N_2-1}) = \\
&= \frac{1}{(4l_1+1)(4l_2+1)} \sum_{k>0} \begin{Bmatrix} l_1 & l_1 & k \\ l_2 & l_2 & t \end{Bmatrix} \langle l_1 \| C^{(k)} \| l_1 \rangle \langle l_2 \| C^{(k)} \| l_2 \rangle \times \\
&\times [-N_1(4l_2+2-N_2) F_i^k(n_1 l_1, n_2 l_2) + (N_2-1)(4l_1+1-N_1) F_f^k(n_1 l_1, n_2 l_2)] + \\
&+ \frac{1}{(4l_1+1)(4l_2+1)} \sum_{k} \left[\frac{2}{3}\,\delta(k, t) - \frac{1}{2(2l_1+1)(2l_2+1)}\right] \langle l_1 \| C^{(k)} \| l_2 \rangle^2 \times \\
&\times [N_1(4l_2+2-N_2) G_i^k(n_1 l_1, n_2 l_2) - (N_2-1)(4l_1+1-N_1) G_f^k(n_1 l_1, n_2 l_2)].
\end{aligned} \tag{7.33}$$

The subscripts i and f on the integrals denote the fact that they are calculated with the wave functions of the initial and final configurations, respectively. The K_0 are any shells that do not participate in the transition and t is the rank of the transition operator. The shift is zero only when $N_1 = 4l_1 + 1$ and $N_2 = 4l_2 + 2$.

Radiative transitions in which electrons from the deep inner shells participate are highly dependent on relativistic effects and affect the emission spectrum thus [245, 162]:

— an increase in the transition energy that appears to the fourth power in the expression for the line intensity;

— a change in the wave functions for which the matrix element of the transition operator is calculated;

— a change in the transition operator and the atom's Hamiltonian (this, in particular, explains the lines that are forbidden in the nonrelativistic theory).

The x-ray emission that arises in deep inner shells having a single vacancy is fairly well explained in first-order perturbation theory by the one-configuration Dirac–Fock method when relativistic and higher-order quantum electrodynamic effects of the transition energy are considered [246, 247, 162].

Questions relating to correlation effects in x-ray spectra, whose role increases particularly for transitions in which outer and subvalent shell electrons participate, have been resolved less satisfactorily.

Operators for radiative transitions are operators of the one-electron type, but, because of correlations between electrons, two-electron – one-photon transitions are possible. For example, the strong mixing of configurations having a vacancy with discrete-spectrum configurations of the type given by Eq. (3.40) induces intense groups of satellite lines in the spectra of inert gases [248, 249], of the transition elements [113, 116], and of other elements.

Two-electron – one-photon transitions are divided into several types.

Radiation Auger effect: a photon and an electron escape from an atom having a vacancy when its excited state decays [250]. These transitions from autoionizing levels are possible due to mixing of configurations from the discrete and continuous spectra or the shaking off of an atom during a radiative transition. Sometimes the transitions which excite an electron into a discrete state are related to a radiation Auger effect, although such transitions are more often called two-electron radiative transitions.

$K_{\alpha\alpha}(KK - LL)$, $K_{\alpha\beta}(KK - LM)$, and similar transitions: two vacancies in an inner shell are filled by two electrons and one photon escapes [251, 252].

KL^N are transitions of the two-electron radiative rearrangement $1s^{-1}2p^{-q} \to 2s^{-2}2p^{-q+1}$ which are possible due to mixing of the $2p^{-q-1}$ and $2s^{-2}2p^{-q+1}$ configurations [154].

In a radiation Auger effect, the transition energy – equal to the energy of the respective Auger transition – may be distributed between the photon and electron in a different manner:

$$\hbar\omega + \varepsilon = E(a^{-1}) - E(b^{-1}c^{-1}) = E_A(a^{-1} \to b^{-1}c^{-1}), \tag{7.34}$$

therefore, the satellite that arises is a broad band from the low-energy side of the lines in the principal spectrum. The shortwave edge of a satellite corresponds to a photon energy of $\hbar\omega_{max} = E_A$. The probability of a two-electron transition which will not change the orbital quantum number of one of the electrons making the transition is nonzero when one-configuration wave functions of both states are used, because the wave functions are not orthogonal to one another. This corresponds to the shake-up model: the radiative transition of one electron causes a sudden perturbation of the atomic field, which causes another electron to be excited into the discrete or continuous spectrum [250, 186, 235]. A perturbation may be considered sudden only when a transition occurs in the inner shells and an electron is excited from an outer layer. The shake-up model was successfully used to interpret multiple ionization satellites of the K-series in light atoms [253, 236]. These satellites are explained by transitions in ions containing an additional vacancy in the L-layer, which is caused by the sudden perturbation of an atom when the first vacancy is created.

The method of configuration interaction is the fundamental method used for calculating the probabilities and energies of two-electron—one-photon transitions in atoms with open shells [250, 113, 161]. Different versions of perturbation theory are used for simpler configurations [254, 255].

The diagram method of perturbation theory, which uses a spectral function and the self-energy of a vacancy, is useful for calculating the shape of an emission spectrum line. This method also makes it possible to generalize the description of the spectrum with due regard for the relationship that exists between the processes of the production of an excited state and its radiative decay, i.e., to examine photoionization and radiation from an atom as a unified process [256]. Such a refinement may be significant if photoionization occurs near a threshold and a slow photoelectron affects the radiating atom, or the relaxation time of the ion is of the same order of magnitude as the lifetime of the excited state; it is also useful in describing resonant fluorescence when an atom is excited by quasimonochromatic rays whose energy is nearly equal to that of the transition [257].

According to Secs. 3.1 and 6.2, correlations in the states of an atom with vacancies can be accounted for by replacing the vacancy line in the process under examination with a heavy quasiparticle line. The self-energy of the vacancy $\Sigma_a(E)$ is found by summing the series of diagrams that are obtained when fragments that describe the various excitations are inserted into the vacancy line. Only the part of self-energy produced by radiationless excitation or fluctuations was examined in Sec. 3. For a deep inner vacancy the radiative part of the self-energy, $\Sigma_a{}^r(E)$, which corresponds to the emission and subsequent absorption of a photon, must also be taken into consideration. This energy is shown in its lowest order by the diagram of Fig. 7.1a, whose contribution is [230]

$$\Sigma_a^r(E) = -\sum_b \frac{|g_{ab}(\omega_x)|^2}{-E-\hbar\omega_x+\varepsilon_b+i\eta}, \tag{7.35}$$

where $g_{ab}(\omega_x)$ is a quantity that describes the electron—photon interaction. In the dipole approximation this quantity is

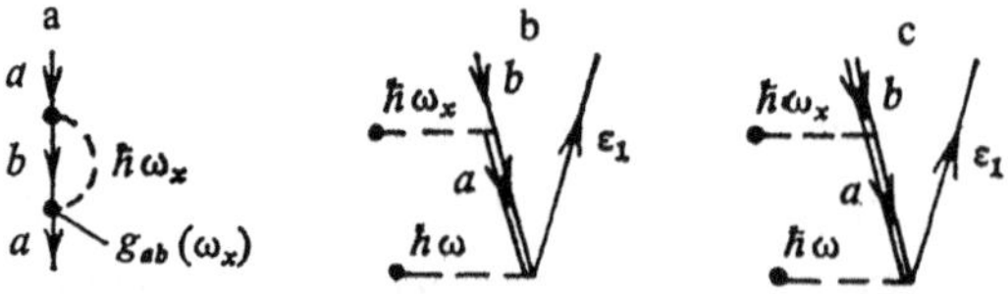

Fig. 7.1. Lower-order diagrams of the radiative part in a vacancy's self-energy (a), and of the amplitude of the unified process of photoionization and emission of photon $\hbar\omega_x$ when one (b) or two (c) quasiparticle lines are used.

$$g_{ab}(\omega_x) \sim (\omega_x)^{1/2} \langle a | d^{(1)} | b \rangle . \tag{7.36}$$

The unified process of photoionizing an atom and its spontaneous emission is described by the generalized photoionization cross section:

$$\frac{d\sigma_{a,b}(\omega_x, \omega)}{d\omega_x} \sim \int | M_{ab}(\varepsilon_1, \omega_x, \omega) |^2 \delta(\varepsilon_1 + \hbar\omega_x - \varepsilon_b - \hbar\omega)\, d\varepsilon_1, \tag{7.37}$$

where ω is the frequency of the absorbed photon, ω_x is the frequency of the emitted photon, ε_1 is the photoelectron's energy, and M_{ab} is the process amplitude. When the first vacancy is said to be interacting, M_{ab} is shown by Fig. 7.1b and is written

$$M_{ab}(\varepsilon_1, \omega_x, \omega) = -\frac{\langle a | d^{(1)} | b \rangle \langle \varepsilon_1 | d^{(1)} | a \rangle}{-\varepsilon_1 + \varepsilon_a + \hbar\omega + \Sigma_a(\varepsilon_1 - \hbar\omega)} . \tag{7.38}$$

When Eq. (7.38) is substituted into Eq. (7.37) and the spectral function $A_a(E)$, Eq. (3.14), is incorporated, the differential cross section for the absorption of a photon $\hbar\omega$ by an electron in a state a and subsequent emission of the photon in the energy band $\hbar\omega_x$, $\hbar(\omega_x + d\omega_x)$ and creation of a vacancy in a state b becomes [230]

$$\frac{d\sigma_{ab}(\omega_x, \omega)}{d\omega_x} \sim | \langle \varepsilon_b - \hbar\omega_x + \hbar\omega | d^{(1)} | a \rangle |^2 A_a(\varepsilon_b - \hbar\omega_x) \frac{\pi | \langle a | d^{(1)} | b \rangle |^2}{\operatorname{Im} \Sigma_a(\varepsilon_b - \hbar\omega_x)} . \tag{7.39}$$

In general, if the final vacancy can also be considered to be interacting, as was the initial vacancy, the δ-function in Eq. (7.37) must then, according to Eq. (3.15), be replaced by the second spectral function, and the cross section is the convolution of these:

$$\frac{d\sigma_{ab}(\omega_x, \omega)}{d\omega_x} \sim \int | \langle \varepsilon_1 | d^{(1)} | a \rangle |^2 \frac{\pi | \langle a | d^{(1)} | b \rangle |^2}{\operatorname{Im} \Sigma_a(\varepsilon_1 - \hbar\omega)} \times$$
$$\times A_a(\varepsilon_1 - \hbar\omega) A_b(\varepsilon_1 - \hbar\omega + \hbar\omega_x)\, d\varepsilon_1. \tag{7.40}$$

Extending this method to configurations with open shells entails some theoretical and computational difficulties.

The complex satellite structure of the x-ray spectra of atoms, even for inert gases with closed shells [239, 249], is caused not only by interelectron correlations, but by the various radiation and radiationless processes that affect the population of excited levels. Thus, a detailed interpretation of the spectrum requires that the entire cascade of processes that take place when an atom interacts with an incident particle and the subsequent rearrangement of the atom be examined [258, 248].

Chapter 8

AUGER SPECTRA

Auger transitions are autoionizing transitions from the highly excited states of atoms containing vacancies and are the fundamental process in spontaneous deexcitation of atoms having a vacancy, excluding the deepest K- and L-vacancies in heavy atoms, where the total probability of an Auger decay of a state is less than that of a radiative decay. With vacancies in light atoms and in the subvalent shells of heavy atoms (when the transition energy is less than $5-10$ keV [9, 259]) Auger transitions become the dominant channel for spontaneous deexcitation of the atoms.

A general description of Auger transitions in atoms and their classification will be given in Sec. 8.1. The widely used approximation of describing Auger spectra in first-order perturbation theory is examined (Sec. 8.2). The possibilities for refining this approximation by taking correlation effects into consideration, and the connection between an Auger transition and the process of creating an excited state that precedes it are discussed in Sec. 8.3.

8.1. Auger Transitions in Atoms

According to Sec. 2.4, the wave function of an autoionizing state is the superposition of discrete and continuous spectra wave functions. Because of the admixture of a continuum function having the same energy, an excited state is not localized – as time passes it makes a spontaneous, radiationless transition into a continuum state. An interaction between a discrete state and a continuum or their mixing is determined by the matrix element of the Hamiltonian for the wave functions of these states. Since a state with a vacancy is distinguished from the final state of an ion + free electron system by the quantum numbers of two electrons, Auger transitions are usually defined by a basic two-electron interaction

– the electrostatic interaction between electrons. One electron fills the vacancy and this interaction transfers the transition energy to the other electron which then escapes the atom as a free Auger electron. This process is not, however, an internal photoeffect – the energy is transferred by a virtual, not an actual photon [4]. Because the wave function is asymmetric it is, in principle, impossible to establish the kind of shell or subshell from which an electron has gone into a free state and what kind of electron has escaped the atom.

In Auger transition notation the subshell containing the first vacancy is shown first, followed by the subshells in which vacancies in the final state were created (for example, KL_1L_2, $M_{4,5}O_{2,3}O_{2,3}$, or $M_{4,5}O_{2,3}{}^2$). An individual spectral line is characterized by the term of a final state that has two vacancies ($M_4O_1O_{2,3}{}^3P_1$). For Auger transitions in which an atom's open shells take part, the initial and final states are frequently indicated instead of the traditional notations. The group of Auger lines that correspond to the same three vacancies is called a spectral group (often shortened to spectrum: a KL_1L_2-Auger spectrum), and those corresponding to the first vacancy in the same layer are called an Auger spectral series (K-series, L-series, etc.).

In the special case in which the first vacancy and one of the second vacancies belong to the same layer, Auger transitions are called Coster—Kronig transitions ($L_1L_2M_{2,3}$).

If a transition occurs in a single layer (all three subshells have the same principal quantum number), we have a super Coster–Kronig transition ($N_{2,3}N_{4,5}{}^2$). Because wave functions having the same n overlap more strongly, the probability of super-Coster–Kronig transitions is higher than for Coster–Kronig transitions, which are more probable than the usual Auger transitions.

The energy of an Auger electron is independent of the energy of the particle that created the first vacancy and, according to the law of conservation of energy, is equal to the difference in the energies of the ion in the initial and final states:

$$\varepsilon_A = E(a^{-1}) - E(b^{-1}c^{-1}), \tag{8.1}$$

where a^{-1}, b^{-1}, and c^{-1} are any vacancies for which an Auger transition is energetically allowed, i.e., $\varepsilon_A > 0$.

The energy of an Auger electron can be written in terms of the binding energies of the electrons taking part in the transition:

$$\varepsilon_A = [E(a^{-1}) - E] - [E(b^{-1}) - E] - [E(b^{-1}c^{-1}) - E(b^{-1})] =$$

$$= I_a - I_b - I_c(b^{-1}) = I_a - I_c - I_b(c^{-1}). \tag{8.2}$$

Here, $I_c(b^{-1})$ is the binding energy of an electron in the c subshell when a b^{-1} vacancy is present. If b^{-1} is an inner vacancy and c^{-1} an outer (but not in a valence shell, see Sec. 3.1) vacancy, $I_c(b^{-1})$ can, in an approximation, be replaced by the binding energy of an electron in the c subshell of an atom having nuclear charge $Z + 1$:

$$\varepsilon_A \approx I_a(Z) - I_b(Z) - I_c(Z+1). \tag{8.3}$$

More accurate semiempirical formulas [260, 262] are often used instead of Eq. (8.3). For example, the equation

$$\varepsilon_A = I_a(Z) - \frac{1}{2}\,[I_b(Z) + I_b(Z+1)] - \frac{1}{2}\,[I_c(Z) + I_c(Z+1)]. \tag{8.4}$$

was used in [261] for composing tables of Auger line energies.

Relaxational and relativistic corrections can be incorporated into the semiempirical equation for ε_A [262].

The energy of an Auger transition between different layers increases with increasing nuclear charge, just like the energy of an x-ray transition. However, because Eq. (8.1) contains the energy of a state with two vacancies, ε_A is not as simple a function of Z as is the dependence according to Moseley's law. In fact, ε_A is determined by the effective nuclear charge and its growth for a given transition gradually slows down in a number of atoms [261]. For the case of Coster–Kronig transitions – when the principal quantum numbers of the electrons are equal – ε_A begins to decrease at a specific Z, which leads to their being energetically forbidden in the heavier atoms [10, 8].

An Auger transition, just as a radiative transition, can be described in the language of vacancies: in the initial state there is a vacancy in an inner shell and in the continuous spectrum (it is then filled by an Auger electron); and in the final state, there are two other vacancies. An electrostatic interaction between two vacancies is equivalent to an interaction between two electrons; this allows us to use a simple model of two vacancies or two electrons taking part in an interaction to describe Auger transitions in the inner closed shells. Auger spectral lines described by this model are called diagram, or normal, lines. Violation of the two-particle model and the appearance of satellite lines in the spectra of free atoms are determined by the following basic reasons:

1. The participation of open-shell electrons in an Auger transition or their strong interaction with electrons involved in the transition.

2. The supplemental ionization or excitation of atoms when an initial vacancy is created.

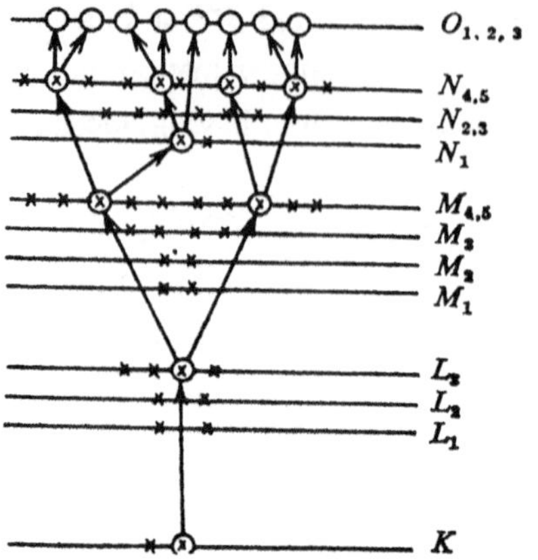

Fig. 8.1. A cascade of Auger transitions that appear when a K-vacancy is created in Xe. ×) Electrons; ○) vacancies; ⊗) vacancies that are subsequently filled by electrons.

3. Correlation effects in the initial or final state.

Many-particle Auger transitions are possible because of correlation effects.

A three-electron Auger transition (a hypersatellite, a double, or a half Auger effect) in the initial state has two vacancies that are filled by two electrons, and the transition energy is transferred to one Auger electron [251, 135, 263].

A double Auger effect is one in which one electron makes the transition into a free state and two Auger electrons escape the atom (or one electron escapes and the other is excited) [250, 159].

A radiation Auger transition occurs when an electron and a photon escape the atom during the decay of a state with a vacancy (see Sec. 7.2).

Because of the radiationless filling of a vacancy, electrons in neighboring shells that interact most strongly with a vacancy have the highest probability of being removed (if this is not energetically forbidden). Consequently, an ion that results from an Auger transition is often found in an excited state – it contains two vacancies in the shells having lesser binding energy – and the atom can be deexcited by subsequent Auger or radiative transitions. The series of radiationless transitions that occur after a vacancy has been created in a deep shell is called an Auger cascade (Fig. 8.1). When this happens, an atom's degree of ionization increases rapidly because of an ordinary, as well as a double, Auger effect or shake-up of the atom during Auger transitions. Thus, a many-electron atom containing a vacancy may turn out to be ionized a factor of 20 times or more [9, 264].

8.2. The Auger Spectra of Many-Electron Atoms

If an atom has a quasidiscrete state γ that interacts with the continuum in which it is found, the probability that this state will decay per unit time in a radiationless process into an ion in a state γ' and a free electron having energy ε is, according to Eq. (2.104), equal to

$$W_A(\gamma \to \gamma' \varepsilon\gamma'') = \frac{2\pi}{\hbar} \, |\langle \gamma' \varepsilon\gamma'' | H | \gamma \rangle|^2, \tag{8.5}$$

where γ'' denotes the remaining quantum numbers of the electron, as well as the overall system.

A transition can only be made between system states having the same energy; therefore, ε is defined by the relation

$$\varepsilon = E(\gamma) - E(\gamma'). \tag{8.6}$$

The wave function of an Auger electron in Eq. (8.5) has been normalized over the energy scale and in this case the density of the continuous spectrum states is unity [135]; if another normalization is used, Eq. (8.5) must be multiplied by $\rho(\varepsilon)$ – the number of states in an interval equal to the unit of energy by which the ε in $P_{\varepsilon l}(r)$ is measured. Additionally, if independent variations are used to find the wave functions of the atom and the ion, for example, the solutions of the Hartree–Fock equations for the respective configurations, the more general formula for the probability of an Auger transition, Eq. (2.105), must be used. Thus,

$$W_A(\gamma \to \gamma' \varepsilon\gamma'') = \frac{2\pi}{\hbar} \, \langle \gamma' \varepsilon\gamma'' | (H - E) | \gamma \rangle|^2 \rho(\varepsilon), \tag{8.7}$$

where E is the energy of the initial autoionizing level. This equation is based on the following basic assumptions:

1. An Auger transition and the creation of an excited state are independent processes.

2. An autoionizing level is remote from other levels having the same symmetry and its interaction with the continuous spectrum can be thought of as an isolated resonance.

3. The spectroscopic factor for the initial vacancy, Eq. (3.18), is approximately equal to unity.

The possibilities for improving this model will be examined in Sec. 8.3.

In practice, some additional simplifications are made when Auger spectra are calculated: the term containing E in Eq. (8.7) is omitted, as are the small corrections to the matrix element of the H operator, due to the nonorthogonality of the wave functions and only a two-electron type of operator remains in the Hamiltonian. In the relativistic theory only the operator for an electrostatic interaction between electrons is ordinarily taken into account, since the basic part of the spin–orbit interaction has a one-electron nature. The two-electron "spin–other orbit" and "spin–spin" interactions become significant even for heavy atoms only when Auger transitions are forbidden because of electrostatic interaction.

It is convenient to normalize the wave function of an Auger electron to unity electron flux density, Eq. (1.169). Then, $\rho(\varepsilon) = (2\pi)^{-1}$ [135] and, taking the previously mentioned simplifications into account, the probability of an Auger transition becomes

$$W_A(\gamma \to \gamma' \varepsilon\gamma'') = \hbar^{-1} |\langle \gamma' \varepsilon\gamma'' | H^e | \gamma \rangle|^2. \tag{8.8}$$

This equation (Wentzel's formula) can also be obtained in first-order perturbation theory, starting from Fermi's golden rule if autoionization is thought of as a transition that occurs due to the action of a constant perturbation [56]. A perturbation operator is the difference between the total Hamiltonian for a many-electron atom and the model Hamiltonian for a zero approximation, H_0. Ordinarily, H_0 is a one-electron type of operator (as, for example, a Hartree–Fock Hamiltonian), and it, just as the remaining one-electron component H, can be omitted from the transition amplitude; then, taking only the fundamental electrostatic interaction between electrons into consideration, we arrive at Eq. (8.8). It must be noted that the contribution from the Hamiltonian for a zero approximation in Eq. (8.7) is compensated by the corresponding part of the term in E.

An expression similar to Eq. (8.8) for the probability of an Auger transition in the relativistic theory (Mueller's formula) is obtained when a relativistic two-electron operator, Eq. (1.178), or its approximation as the sum of Coulomb and Breit operators, Eq. (1.179), is used:

$$W_A(\gamma \to \gamma' \varepsilon\gamma'') = \hbar^{-1} |(\gamma' \varepsilon\gamma'' | (H^e + H^B) | \gamma)|^2. \tag{8.9}$$

Expressions (8.8) and (8.9) are fundamental approximations and are widely used for calculating the probabilities of Auger transitions.

The matrix elements in the right-hand sides of Eqs. (8.8) and (8.9) define the selection rules for Auger transitions. Because a transition operator is a scalar (in contrast to a radiative transition operator), an Auger transition is possible only when the total angular momentum of the system, J, is conserved and the probability is independent of the projection of this angular momentum. The operator is even; therefore, system parity must be preserved. Then the orbital angular momenta of the atom's l_1, l_2, l_3 electrons taking part in the transition and the Auger electron, l, satisfy the condition that

$$-1)^{l_1+l_2+l_3+l} = 1. \tag{8.10}$$

In the nonrelativistic theory, the values of the orbital and total angular momenta of a free electron are limited by the conditions that one-electron submatrix elements of the spherical harmonic and of the $3nj$-symbols not go to zero. These conditions are not very restrictive; therefore, Auger transitions due to electrostatic interaction are forbidden in only fairly rare situations, mainly in total system spin. A similar situation occurs in the relativistic theory.

The presence of an open outer shell and its interaction with the vacancies that participate in a transition are more important for transitions in the subvalent and outer shells where the nonrelativistic theory is useful; therefore, we will use Eq. (8.8) to describe Auger transitions between complex configurations having open shells.

Because the matrix element of the H^e operator is diagonal for the projections of the total angular momentum of the initial and final states and is independent of them, the probability of a transition between states is then equal to the probability of a transition between levels. Taking the scalar and Hermitian natures of the operator into account, we have

$$W_A(K\gamma J \rightarrow K'\gamma' J' \varepsilon ljJ) = W_A(K\gamma JM \rightarrow K'\gamma' J' \varepsilon ljJM) = \\ = \hbar^{-1} |\langle K'\gamma' J' \varepsilon ljJ | H^e | K\gamma J\rangle|^2 = \hbar^{-1} |\langle K\gamma J | H^e | K'\gamma' J' \varepsilon ljJ\rangle|^2. \quad (8.11)$$

The angular momenta of an ion and an Auger electron are coupled by a Jj-coupling, because this coupling lets us use for the ion the wave functions of an intermediate coupling.

The probability of creating a $K'\gamma' J'$ ion as a result of an Auger transition is obtained by summing Eq. (8.11) over the l and j angular momenta of the Auger electron:

$$W_A(K\gamma J \rightarrow K'\gamma' J' \varepsilon) = \hbar^{-1} \sum_{jl} |\langle K\gamma J | H^e | K'\gamma' J' \varepsilon ljJ\rangle|^2. \quad (8.12)$$

If there is only one vacancy in the initial configuration and the remaining shells are closed, the interconfiguration matrix element for the H^e operator is, to within the accuracy of simple multipliers, equal to the two-electron matrix element, and the probability is

$$W_A(n_1 l_1 j_1^{-1} \rightarrow n_2 l_2 j_2^{-1} n_3 l_3 j_3^{-1} J_0 \varepsilon lj) = \\ = \hbar^{-1} [J_0] [j_1]^{-1} |\langle n_2 l_2 j_2 n_3 l_3 j_3 J_0 | h^e | n_1 l_1 j_1 \varepsilon lj J_0\rangle|^2. \quad (8.13)$$

Let there be an LS coupling between the vacancies in the final configuration. Then the probability, averaged over the states of the initial vacancy and summed over the j of the free electron, can also be given in terms of the two-electron matrix element in the LS coupling:

$$W_A(n_1 l_1^{-1} \to n_2 l_2^{-1} n_3 l_3^{-1} L_0 S_0 \varepsilon l) = \\ = \hbar^{-1} \frac{(2L_0+1)(2S_0+1)}{4l_1+2} |\langle n_2 l_2 n_3 l_3 L_0 S_0 | h^e | n_1 l_1 \varepsilon l L_0 S_0 \rangle|^2. \tag{8.14}$$

Formulas for the probability of an Auger transition when the electron configurations are complex can be derived by the procedure discussed in Chapter 1. General expressions for configurations having any number of shells (vacancies, not electrons, are considered) were presented in [265]. We will confine ourselves here to the practically more important cases of transitions from an initial configuration containing a single vacancy and only one open shell [266].

Not only is the absolute value of the matrix element important in calculating the probabilities of transitions in an intermediate coupling, but its phase as well. Therefore, we will present formulas not for the probabilities, but for the matrix elements.

The simplest form convenient for calculations in an intermediate coupling is acquired by a matrix element in the LS coupling (from here on, we will omit principal quantum numbers for the sake of brevity):

1) an open shell that takes part in an Auger transition:

$$\langle l_1^{4l_1+1} l_2^{4l_2+2} l_3^{N_3} \gamma_3 L_3 S_3 LS | H^e | l_1^{4l_1+2} l_2^{4l_2+1} l_3^{N_3-1} \gamma_3' L_3' S_3' (L_0 S_0) \varepsilon l LS \rangle = \\ = (-1)^{l_2+l_3} \sqrt{N_3} [L_3, S_3, L_0, S_0]^{1/2} (l_3^{N_3-1} \gamma_3' L_3' S_3' l_3 \| l_3^{N_3} \gamma_3 L_3 S_3) \times \\ \times \sum_{L_0' S_0'} (-1)^{L_0'+S_0'} [L_0', S_0'] \langle l_2 l_3 L_0' S_0' | h^e | l_1 \varepsilon l L_0' S_0' \rangle \times \\ \times \begin{Bmatrix} l_1 & L_3 & L \\ L_0 & l & L_0' \end{Bmatrix} \begin{Bmatrix} \frac{1}{2} & S_3 & S \\ S_0 & \frac{1}{2} & S_0' \end{Bmatrix} \begin{Bmatrix} L_3' & l_3 & L_3 \\ L_0' & L_0 & l_2 \end{Bmatrix} \begin{Bmatrix} S_3' & \frac{1}{2} & S_3 \\ S_0' & S_0 & \frac{1}{2} \end{Bmatrix}, \tag{8.15}$$

$$\langle l_1^{4l_1+1} l_2^{N_2} \gamma_2 L_2 S_2 LS | H^e | l_1^{4l_1+2} l_2^{N_2-2} \gamma_2' L_2' S_2' \varepsilon l LS \rangle = \\ = (-1)^{L_2+S_2+L_2'+S_2'+N_2} \left[\frac{N_2(N_2-1)}{2} \right]^{1/2} \sum_{L_0 S_0} [L_0, S_0, L_2, S_2]^{1/2} \times \\ \times (l_2^{N_2-2} \gamma_2' L_2' S_2' l_2^2 L_0 S_0 \| l_2^{N_2} \gamma_2 L_2 S_2) \begin{Bmatrix} L_2 & L_0 & L_2' \\ l & L & l_1 \end{Bmatrix} \begin{Bmatrix} S_2 & S_0 & S_2' \\ \frac{1}{2} & S & \frac{1}{2} \end{Bmatrix} \times \\ \times \langle l_2^2 L_0 S_0 | h^e | l_1 \varepsilon l L_0 S_0 \rangle; \tag{8.16}$$

2) a passive open shell:

$$\langle l_1^{4l_1+1} l_2^{4l_2+2} l_3^{N_3}\gamma_3 L_3 S_3 LS \,|\, H^e \,|\, l_1^{4l_1+2} l_2^{4l_2}\gamma_2 L_2 S_2 l_3^{N_3}\gamma_3' L_3' S_3' (L_0 S_0)\,\varepsilon lLS\rangle =$$
$$= \delta(\gamma_3 L_3 S_3,\ \gamma_3' L_3' S_3')(-1)^{L_0-S_0+L_2-S_2+N_3} \times$$
$$\times \sqrt{4l_2+1}\,[l_2,\ L_0,\ S_0]^{1/2} (l_2^{4l_2}\gamma_2 L_2 S_2 l_2^2 L_2 S_2 \| l_2^{4l_2+2}) \times$$
$$\times \begin{Bmatrix} l_1 & L & L_3 \\ L_0 & L_2 & l \end{Bmatrix} \begin{Bmatrix} \frac{1}{2} & S & S_3 \\ S_0 & S_2 & \frac{1}{2} \end{Bmatrix} \langle l_2^2 L_2 S_2 | h^e | l_1 \varepsilon l L_2 S_2\rangle, \tag{8.17}$$

$$\langle l_1^{4l_1+1} l_2^{4l_2+2} l_3^{4l_3+2} l_4^{N_4}\gamma_4 L_4 S_4 LS \,|\, H^e \,|$$
$$| l_1^{4l_1+2} l_2^{4l_2+1} l_3^{4l_3+1} (L_0 S_0) l_4^{N_4}\gamma_4' L_4' S_4' (L_0' S_0')\,\varepsilon lLS\rangle =$$
$$= \delta(\gamma_4 L_4 S_4,\ \gamma_4' L_4' S_4')(-1)^{L_0+S_0+L_0'+S_0'+L_4+S_4+N_4+1} \times$$
$$\times [L_0,\ S_0,\ L_0',\ S_0']^{1/2} \begin{Bmatrix} l_1 & L & L_4 \\ L_0' & L_0 & l \end{Bmatrix} \begin{Bmatrix} \frac{1}{2} & S & S_4 \\ S_0' & S_0 & \frac{1}{2} \end{Bmatrix} \times$$
$$\times \langle l_2 l_3 L_0 S_0 | h^e | l_1 \varepsilon l L_0 S_0\rangle. \tag{8.18}$$

A transition in a final configuration to a Jj coupling between an ion and an Auger electron is accomplished by multiplying the matrix element by the transformation matrix:

$$\langle K\gamma LSJ \,|\, H^e \,|\, K'\gamma' L' S' J' \varepsilon ljJ\rangle =$$
$$= [L,\ S,\ j,\ J']^{1/2} \begin{Bmatrix} L' & l & L \\ S' & \frac{1}{2} & S \\ J' & j & J \end{Bmatrix} \langle K\gamma LSJ \,|\, H^e \,|\, K'\gamma' L' S' \varepsilon lLSJ\rangle. \tag{8.19}$$

The summation over LS disappears because of the diagonality of the matrix element relative to these quantum numbers.

For a JJ coupling Eqs. (8.15)—(8.18) are replaced by

$$\langle l_1^{4l_1+1} j_1 l_2^{4l_2+2} l_3^{N_3}\gamma_3 L_3 S_3 J_3 J \,|\, H^e \,|\, l_1^{4l_1+2} l_2^{4l_2+1} j_2 l_3^{N_3-1}\gamma_3' L_3' S_3' J_3' (J_0)\,\varepsilon ljJ\rangle =$$
$$= \sum_{j_3} (-1)^{J_3+J_0+1} \sqrt{N_3}\,[j_3,\ L_3,\ S_3,\ J_3,\ J_3',\ J_0]^{1/2} \times$$
$$\times (l_3^{N_3}\gamma_3 L_3 S_3 \| l_3^{N_3-1}\gamma_3' L_3' S_3' l_3) \begin{Bmatrix} L_3' & l_3 & L_3 \\ S_3' & \frac{1}{2} & S_3 \\ J_3' & j_3 & J_3 \end{Bmatrix} \sum_{J_0'} (-1)^{J_0'} [J_0'] \times$$

$$\times \begin{Bmatrix} J_3' & j_3 & J_3 \\ J_0' & J_0 & j_2 \end{Bmatrix} \begin{Bmatrix} j_1 & J_3 & J \\ J_0 & j & J_0' \end{Bmatrix} \langle l_2 j_2 l_3 j_3 J_0' \,|\, h^e \,|\, l_1 j_1 \varepsilon l j J_0' \rangle, \tag{8.20}$$

$$\langle l_1^{4l_1+1} j_1 l_2^{N_2} \gamma_2 L_2 S_2 J_2 J \,|\, H^e \,|\, l_1^{4l_1+2} l_2^{N_2-2} \gamma_2' L_2' S_2' J_2' \varepsilon l j J \rangle =$$

$$= (-1)^{-J_2+J_2'+N_2} \left[\frac{N_2 (N_2-1)}{2} \right]^{1/2} [L_2,\ S_2,\ J_2,\ J_2']^{\frac{1}{2}} \sum_{L_0 S_0 J_0} (-1)^{J_0} [J_0] \times$$

$$\times (l_2^{N_2} \gamma_2 L_2 S_2 \,\|\, l_2^{N_2-2} \gamma_2' L_2' S_2' l_2^2 L_0 S_0) \begin{Bmatrix} L_2' & L_0 & L_2 \\ S_2' & S_0 & S_2 \\ J_2' & J_0 & J_2 \end{Bmatrix} \times$$

$$\times \begin{Bmatrix} J_2' & J_0 & J_2 \\ j_1 & J & j \end{Bmatrix} \langle l_2^2 L_0 S_0 J_0 \,|\, h^e \,|\, l_1 j_1 \varepsilon l j J_0 \rangle, \tag{8.21}$$

$$\langle l_1^{4l_1+1} j_1 l_2^{4l_2+2} l_3^{4l_3+2} l_4^{N_4} \gamma_4 L_4 S_4 J_4 J \,|\, H^e \,|$$

$$|\, l_1^{4l_1+2} l_2^{4l_2+1} j_2 l_3^{4l_3+1} j_3 (J_0) l_4^{N_4} \gamma_4' L_4' S_4' J_4' (J_0') \varepsilon l j J \rangle =$$

$$= \delta (\gamma_4 L_4 S_4 J_4,\ \gamma_4' L_4' S_4' J_4') (-1)^{J_0+J_0'+J_4+N_4+1} [J_0,\ J_0']^{1/2} \times$$

$$\times \begin{Bmatrix} j_1 & J_0 & j \\ J_0' & J & J_4 \end{Bmatrix} \langle l_2 j_2 l_3 j_3 J_0 \,|\, h^e \,|\, l_1 j_1 \varepsilon l j J_0 \rangle, \tag{8.22}$$

$$\langle l_1^{4l_1+1} j_1 l_2^{4l_2+2} l_3^{N_3} \gamma_3 L_3 S_3 J_3 J \,|\, H^e \,|\, l_1^{4l_1+2} l_2^{4l_2} \gamma_2 L_2 S_2 J_2 l_3^{N_3} \gamma_3 L_3 S_3 J_3 (J_0) \varepsilon l j J \rangle =$$

$$= (-1)^{J_2-J_3-J_0+N_3} [L_2,\ S_2]^{-1/2} \sqrt{4l_2+1}\, [J_2,\ J_0,\ l_2]^{1/2} \times$$

$$\times (l_2^{4l_2} \gamma_2 L_2 S_2 l_2^2 L_2 S_2 \,\|\, l_2^{4l_2+2}) \begin{Bmatrix} j_1 & J & J_3 \\ J_0 & J_2 & j \end{Bmatrix} \langle l_2^2 L_2 S_2 J_2 \,|\, h^e \,|\, l_1 j_1 \varepsilon l j J_2 \rangle. \tag{8.23}$$

The two-electron matrix element in a jj coupling is [266]

$$\langle n_2 l_2 j_2 n_3 l_3 j_3 J_0 \,|\, h^e \,|\, n_1 l_1 j_1 \varepsilon l j J_0 \rangle = \sqrt{2}\, N_{n_2 l_2 j_2,\, n_3 l_3 j_3} (-1)^{J+J_0} [j_1,\ j_2,\ j_3,\ j]^{1/2} \times$$

$$\times \sum_k \left[(-1)^{J_0} \begin{pmatrix} j_2 & k & j_1 \\ -\frac{1}{2} & 0 & \frac{1}{2} \end{pmatrix} \begin{pmatrix} j_3 & k & j \\ -\frac{1}{2} & 0 & \frac{1}{2} \end{pmatrix} \begin{Bmatrix} j_1 & j_2 & k \\ j_3 & j & J_0 \end{Bmatrix} \times \right.$$

$$\times (l_2 k l_1)(l_3 k l) R^k (n_2 l_2 n_3 l_3, n_1 l_1 \varepsilon l) + \begin{pmatrix} j_2 & k & j \\ -\frac{1}{2} & 0 & \frac{1}{2} \end{pmatrix} \begin{pmatrix} j_3 & k & j_1 \\ -\frac{1}{2} & 0 & \frac{1}{2} \end{pmatrix} \times$$

$$\left. \times \begin{Bmatrix} j_2 & j & k \\ j_1 & j_3 & J_0 \end{Bmatrix} (l_2 k l)(l_3 k l_1) R^k (n_2 l_2 n_3 l_3, \varepsilon l n_1 l_1) \right]. \tag{8.24}$$

The designations mean the same thing as in Eq. (2.14); the functions for both states must have the same parity.

An expression for the probability of an Auger transition in an intermediate coupling – which contains not only quadratic, but interference terms – is obtained by substituting expansions of the wave functions for initial and final states in terms of functions of a pure coupling into Eq. (8.11). When this happens a redistribution of the probabilities between transitions occurs, but the total probability of a transition between given configurations is conserved. For Coster–Kronig, and especially super-Coster–Kronig transitions, the dependence of an Auger electron's wave function on a term may be substantial. This dependence is also determined by the form of the effective potential of the field in which the electron is moving. Specifically, when an Auger electron with a large orbital angular momentum ($l \geq 2$) escapes from an atom that no longer contains electrons with a given l in the ground configuration, the effective potential for small positive energies may decay very slowly; the result is that a small change in the Auger electron's energy produces a large change in the extent to which its wave function can penetrate the core area [267]. A significant dependence of $P_{\varepsilon l}(r)$ on the term can also be expected for an effective potential having a large positive barrier; the suppression of some energetically allowed super Coster–Kronig transitions is possible (similar to the suppression of the photoionization cross section at an ionization limit when a positive barrier to a photoelectron is present).

The Auger spectrum is often calculated by using the relaxed wave function of the final state, but when this is done the partially compensating corrections for overlap and exchange [135] because of the difference between the radial functions for both states are frequently ignored. The method of a transition operator [90, 135], in which the wave functions of initial and final states are optimized simultaneously and one set of one-electron orbitals is found, does not significantly improve the results [135].

In order to compare the different channels of Auger decay, as well as verify the results of calculating Auger spectra, it is worthwhile to introduce the concept of the Auger transitions [analogous to the intensity of a radiative transition group, Eq. (1.287)]. This quantity divided by $\hbar$ is the total probability of Auger transitions between all levels of K and K' configurations with the escape of an εl (in the nonrelativistic theory) or an εlj (in relativistic theory) Auger electron. In the nonrelativistic case we have

$$S_A(K, K'\varepsilon l) = \hbar \sum_{\gamma JM\gamma' J'j} W_A(K\gamma J \to K'\gamma' J' \varepsilon ljJ) = \\ = \sum_{\gamma JM\gamma' J'j} |\langle K\gamma J | H^e | K'\gamma' J' \varepsilon ljJ\rangle|^2. \tag{8.25}$$

If we can consider the radial integrals to be independent of the quantum numbers γJ and $\gamma' J'j$, the summation over spin-angular coefficients can then be done algebraically, as it was for radiative transitions, Eq. (1.287), and the following general equation is obtained [266]:

$$S_A(K, K'\varepsilon l) = g(K, K')\, s_A(n_2l_2n_3l_3,\ n_1l_1\varepsilon l). \tag{8.26}$$

Here, n_1l_1, εl and n_2l_2, n_3l_3 are, respectively, the quantum numbers of the electrons making the transitions in the final and initial states, and $g(K, K')$ is a statistical factor that is derived from the sums over the quantum numbers of the passive electrons:

$$g(K, K') = g(K_0) \prod_i \binom{4l_i + 2 - m_i}{\min N_i}, \tag{8.27}$$

where $g(K_0)$ is the statistical weight of the passive shells; m_i is the number of electrons in the shell n_il_i taking part in the transition; min N_i is the least number of electrons in the shell n_il_i before and after the transition, and the subscript i covers the shells taking part in the transition.

The last term in Eq. (8.26) is the total line strength of Auger transitions in the two-electron model and, just as $S_A(K, K'\varepsilon l)$, is independent of the type of coupling:

$$s_A(n_2l_2n_3l_3,\ n_1l_1\varepsilon l) = \hbar \sum_{L_0S_0} W_A(n_2l_2n_3l_3L_0S_0 \to n_1l_1\varepsilon lL_0S_0) =$$

$$= \sum_{L_0 S_0} [L_0, \; S_0] \, | \langle n_2 l_2 n_3 l_3 L_0 S_0 \, | \, h^e \, | \, n_1 l_1 \varepsilon l L_0 S_0 \rangle \, |^2 =$$

$$= 8N^2_{n_2 l_2, \, n_3 l_3} \sum_k \Big[[k]^{-1} \langle l_1 \| C^{(k)} \| l_2 \rangle^2 \langle l \| C^{(k)} \| l_3 \rangle^2 R^k (n_2 l_2 n_3 l_3, \; n_1 l_1 \varepsilon l)^2 +$$

$$+ [k]^{-1} \langle l_2 \| C^{(k)} \| l \rangle^2 \langle l_1 \| C^{(k)} \| l_3 \rangle^2 R^k (n_2 l_2 n_3 l_3, \; \varepsilon l n_1 l_1)^2 -$$

$$- \sum_{k'} (-1)^{k+k'} \langle l_2 \| C^{(k)} \| l_1 \rangle \langle l_3 \| C^{(k)} \| l \rangle \langle l_2 \| C^{(k')} \| l \rangle \langle l_3 \| C^{(k')} \| l_1 \rangle \times$$

$$\times \begin{Bmatrix} l_1 & l_2 & k \\ l & l_3 & k' \end{Bmatrix} R^k (n_2 l_2 n_3 l_3, \; n_1 l_1 \varepsilon l) R^{k'} (n_2 l_2 n_3 l_3, \; \varepsilon l n_1 l_1) \Big]. \tag{8.28}$$

Here, $N_{a,b}$ is the normalizing factor of Eq. (1.107) and R^k is the general integral for an electrostatic interaction, Eq. (2.15) (its algebraic form for hydrogenic functions is given in [268]).

Note that the intensity of a transition group has such a simple dependence on N_i only because the quantum numbers nl of the two electrons change during the transition [269] [thus, Eq. (8.26) does not cover the situation $n_1 l_1 j_1{}^{-1} \to n_1 l_1 j_2{}^{-1} n_3 l_3 j_3{}^{-1}$].

In the relativistic theory Eq. (8.26) is replaced by

$$S_A^{rel} (K, \; K' \varepsilon lj) = g (K, \; K') \, s_A^{rel} (n_2 l_2 j_2 n_3 l_3 j_3, \; n_1 l_1 j_1 \varepsilon lj), \tag{8.29}$$

where

$$g (K, \; K') = g (K_0) \prod_i \binom{2j_i + 1 - m_i}{\min N_i}, \tag{8.30}$$

$$s_A^{rel} (n_2 l_2 j_2 \;\; n_3 l_3 j_3, \; n_1 l_1 j_1 \;\; \varepsilon lj) = \hbar \sum_{J_0} W_A (n_2 l_2 j_2 \;\; n_3 l_3 j_3 J_0 \to n_1 l_1 j_1 \;\; \varepsilon lj J_0) =$$

$$= \sum_{J_0} [J_0] \, | (n_2 l_2 j_2 \; n_3 l_3 j_3 J_0 \, | \, (h^e + h^B) \, | \, n_1 l_1 j_1 \varepsilon lj J_0) \, |^2 . \tag{8.31}$$

Here, $g(K_0)$, m_i, and N_i have the same meaning as in Eq. (8.27).

If we divide the total Auger transition probability by the statistical weight of the initial configuration we obtain the average partial Auger width of the initial level or, if we divide by $\hbar$, we obtain the average probability of the respective Auger transition from the initial level [266, 135, 270].

The relativistic two-electron model is widely used to calculate the Auger spectra that arise during transitions in the inner shells (due to the large width of levels, an interaction between a vacancy and an open outer shell only leads to line broadening). The probability of an Auger transition is determined by the two-electron matrix element of the h^{rel} operator, Eq. (1.178) (or its approximation, $h^e + h^B$), for the four-component wave functions [271]:

$$
\begin{aligned}
&W_A(n_1 l_1 j_1^{-1} \rightarrow n_2 l_2 j_2^{-1} n_3 l_3 j_3^{-1} J_0 \varepsilon l j) = \\
&= \hbar^{-1} [J_0] [j_1]^{-1} |(n_2 l_2 j_2 \, n_3 l_3 j_3 J_0 | h^{rel} | n_1 l_1 j_1 \varepsilon l j J_0)|^2 .
\end{aligned}
\qquad (8.32)
$$

The following relationships have been established for most elements in the periodic system from systematic calculations in this approximation (using Dirac–Fock–Slater wave functions) of the Auger spectra for the *K*-, *L*-, and *M*-series [245, 162].

Relativistic corrections to the probabilities of Auger transitions (when the transition energy is correct) are determined by three fundamental factors: 1) the difference between the relativistic and nonrelativistic wave functions; 2) a retardation of the Coulomb interaction; and 3) a magnetic "current–current" interaction. Depending on their relative contributions and mutual compensation, the magnitude of relativistic effects varies markedly for different lines. The retardation and magnetic interactions are significant for transitions in the deep inner shells (e.g., KL_1L_1 and KL_3L_3) and the effect from relativistic wave functions dominates if outer shells ($L_2L_3N_2$) take part in the transition. As in the other refinements, the probabilities of weak lines are more sensitive to relativistic corrections than are the principal lines.

The influence of relativistic effects on the energies of Auger transitions is more important for Coster–Kronig and super-Coster–Kronig transitions having low transition energies [79, 162].

In general, the intensities of Auger spectrum lines are determined by not only the probabilities of the Auger transitions, but by the population densities of the levels as well. This, in particular, gives rise to the resonance Auger effect – a strong dependence of the distribution of spectral intensities on the energy of monochromatic x rays when a state with a vacancy is created by exciting an electron into a discrete spectrum state [272].

During nonresonance population of initial states with one vacancy nlj^{-1}, we can usually assume that the population densities are proportional to $2j + 1$ (the ionization cross section of a subshell is directly proportional to the number of electrons in it, and the width of the level is independent of j). The relative intensity of an Auger line is then equal to the probability of a transition multiplied by the statistical weight $2j + 1$.

A symmetric Lorentz curve is typical of most individual Auger lines. This property is used in a very simple method for calculating the shape of an Auger spectrum – the total intensity of each line is distributed by this function according to the results from calculating its position and width, and summation over all spectral lines is done. A more accurate calculation of the shape of the lines, which takes into account interference effects when discrete and continuous spectrum states merge, is possible using resonance scattering theory or perturbation theory.

Until now we have examined the total probabilities of Auger transitions that are independent of the escape angle of an Auger electron. This angle can then be described by a partial spherical wave having a certain angular momentum. In studying the angular distribution of the electrons, the wave function of the Auger electron must be used as a nonrelativistic, Eq. (4.40), or relativistic, Eq. (5.24), function $\psi_{\mathbf{q}\mu}^{-}$ that describes the motion of an electron having a wave vector $\mathbf{q}$ and spin projection μ. An expression for the relative probability of an Auger electron escaping at a spatial angle $d\Omega$ in the direction $\mathbf{q}$ is obtained by multiplying the probability of an Auger transition by the relative population density of the initial level. In addition to this, for randomly oriented gas atoms this expression must be averaged over the projections of the total angular momentum of an ion in the initial state with a vacancy and summed over the projections of the angular momenta in the final state,

$$W(\hat{\mathbf{q}})\,d\Omega = [J]^{-1} \sum_{MM'\mu} \mathcal{N}(\gamma JM)\, dW_A(\gamma JM \to \gamma' J' M' \mathbf{q}\mu), \tag{8.33}$$

where $W(\mathbf{q})d\Omega$ is the probability that the Auger electron will escape into the spatial angle $d\Omega$ in the direction $\hat{\mathbf{q}}$ during the given transition, and $\mathcal{N}(\gamma JM)$ is the relative population density of the γJM level. The quantization axis z coincides with the direction of the beam of primary particles that ionize the atom.

Substituting the expansion for the function $\psi_{\mathbf{q}\mu}^{-}$, Eq. (4.40), into the expression for dW_A [see Eq. (1.244)], recombining the angular momenta $\mathbf{l}$ and $\mathbf{s}$ of the electron into the total angular momentum $\mathbf{j}$, combining the latter with the ion's angular momentum $\mathbf{J}'$, and considering the fact that the matrix element of the scalar operator in this coupling is independent of the projection, we obtain

$$W(\hat{\mathbf{q}}) = \frac{2\pi}{\hbar} \sum_{MM'\mu} \mathcal{N}(\gamma JM) \left| \sum_{lm_l jm_j} i^{-l} e^{i\vartheta_l} Y_{lm_l}(\hat{\mathbf{q}}) [j]^{1/2} \times \right.$$

$$\left. \times (-1)^{l-s+m_j+J'-j+M} \begin{pmatrix} l & s & j \\ m_l & \mu & -m_j \end{pmatrix} \begin{pmatrix} J' & j & J \\ M' & m_j & -M \end{pmatrix} \langle \gamma' J' \varepsilon ljJ | H^e | \gamma J \rangle \right|^2 . \quad (8.34)$$

The probability that an Auger electron will escape depends only on the angle θ, measured relative to the direction of the ionizing particles. If a unique value of the quantum number l is possible, $W(\hat{\mathbf{q}})$ also becomes independent of the partial scattering phase ϑ_l and is given in terms of the total probability of an Auger transition. When summing over projections, it turns out that the distribution of Auger electrons is isotropic when there is no alignment of the atoms during the ionization process [if $\mathcal{N}(\gamma JM)$ is independent of M], and even for aligned ions, if $J \leq 1/2$ [273].

For unpolarized atoms and the particles that ionize them, the angular distribution function for Auger electrons $W(\theta)$ assumes the form [238, 274]

$$W(\theta) = \frac{W}{4\pi} \left[1 + \sum_{\nu=2} \alpha_\nu \mathcal{A}_{\nu 0} P_\nu (\cos\theta) \right], \quad \nu \leqslant 2J'. \quad (8.35)$$

Here, W is the probability of an Auger transition with the escape of an electron into the solid angle 4π; P_ν is a Legendre polynomial of order ν; α_ν are parameters that are functions of the amplitudes of Auger decay and the scattering phase of the Auger electron in the different channels, and the $\mathcal{A}_{\nu 0}$ are alignment parameters that describe the initial aligned state and are given in terms of the ionization cross section of the atom [238, 274]. For p^{-1} and $d_{3/2}{}^{-1}$ vacancies the asymmetry of the angular distribution of Auger electrons depends on the single parameter $\alpha_2 \mathcal{A}_{20}$, whose values, found semiempirically and theoretically for Auger transitions stimulated by photons or electrons, are small (<0.1) [135, 238]. The angular distribution of the electrons may be even more asymmetric in an ion collision [30]. Notice that at the angle $\theta = 54°44'$, $P_2(\cos\theta) = 0$; therefore, the Auger spectrum measured at this angle is preferred for comparison with the integral theoretical spectrum.

Auger electrons that escape from randomly oriented gas atoms may also be oriented in the direction of their spin. It follows, starting from the symmetry of the process, that the spin polarization vector must be perpendicular to the plane of the reaction in which an incident beam of unpolarized particles and the Auger electron move. The degree of polarization is determined according to Eq. (6.19) and is given in the following form [275, 274]:

$$P(\theta) = \sum_{\nu=2} \beta_\nu \mathscr{A}_{\nu 0} P_\nu^1 (\cos\theta) \Big/ \Big[1 + \sum_{\nu=2} \alpha_\nu \mathscr{A}_{\nu 0} P_\nu (\cos\theta)\Big]. \tag{8.36}$$

The α_ν and $\mathscr{A}_{\nu 0}$ parameters mean the same here as in Eq. (8.35); β_ν is a polarization parameter containing interference terms between the various channels of Auger decay [274], and P_ν^1 is an adjoint, first-order Legendre polynomial.

8.3. More Accurate Methods for Calculating Auger Spectra

Correlation effects in Auger spectra and in photoelectron spectra can be classified as an interaction between configurations in the initial state and in the final state of an ion + electron system. The latter can be further subdivided into correlations in the ion state and in the continuum state [159]. The fundamental types of strong mixing configurations for states with vacancies were examined in Sec. 3.4 and they are significant in Auger spectra [30, 159, 259, 276]. We note the mixing $sp^N - s^2p^{N-2}d - s^2p^{N-2}s$ that leads to satellite lines which are comparable in intensity to the basic lines in the $L_{2,3}MM$-spectrum of Ar [277], the $M_{4,5}N_1N_{2,3}$-spectrum of Kr, and the $N_{4,5}O_1O_{2,3}$-spectrum of Xe [278, 279], and the resonance-increasing $(s + d)^N$ mixing during the collapse of a d-electron for configurations having an inner vacancy [29, 120] that appears in various Auger spectra for Ca and Ba [159, 259].

Strong correlation effects occur in Auger spectra when the electrons in the open outer shells take part in the transitions and the numbers of electrons in these shells are different in strongly mixing configurations [265]. The interaction between a configuration having two vacancies in the final state of an ion and, primarily, quasidegenerate configurations [30, 280, 259], especially if they are close to one another in energy, is characteristic of Auger spectra. For example, the strong mixing of quasidegenerate $3d^{-2}$ and $3s^{-1}3d^{-1}$ configurations near $Z = 60$ sharply increases the intensity of $L_1M_3M_3$ Auger transitions by one-and-a-half orders of magnitude [281].

Configuration mixing may open additional channels for the Auger decay of an excited state and must be considered when admixed configurations are chosen. When the admixed configuration for the expansion of another state's wave function is taken into consideration, separating the correlations into an interaction between initial-state and final-state configurations becomes a conditional matter.

The effect on Auger spectra of an interchannel interaction in the final state of a system has as yet been little studied. Calculations done according to methods from perturbation theory [282] and the scattering channel coupling [135] show that this interaction can be as important as an interaction between discrete spectrum configurations in a final ion. For example, $(1s^22p^6\varepsilon s - 1s^22s2p^5\varepsilon' p)^2S$ mixing significantly redistributes the intensities of the *KLL*-spectrum in neon, as in the familiar superposition of $1s^22p^6 - 1s^22s^22p^4$ configurations in the double ion of neon [282].

Correlation effects for Auger transitions in the inner shells are tightly interwoven with relativistic effects – their simultaneous calculation requires that the relativistic one-configuration method [280, 281] be used.

Satellites in Auger spectra can be caused by many-electron effects not only while the Auger transition is taking place, but even as an initial vacancy is being created due to the supplemental ionization or excitation of the atom.

The perturbation theory [230, 283, 284] is effective for studying correlation effects in diagram Auger spectra, especially anomalies in the shape and locations of lines and satellites that correspond to many-particle transitions. As was shown in Sec. 2.3, the probability of an Auger transition is defined as the imaginary part of the contributions from the diagrams for energy (Fig. 6b–e). To describe an initial state with a vacancy, it is worthwhile to introduce the self-energy of a vacancy and its spectral function, Eq. (3.14). Using this method to perform systematic calculations of the Auger level widths and, in particular, the probabilities of transitions shows that the spectroscopic factor, Eq. (3.18), is nearly unity in most cases; even for $4d^{-1}$ vacancies that interact with the $4f^{-2}\varepsilon l$ super Coster–Kronig continuum, it is approximately 0.97 [284]. It is only when the energies of one- and two-vacancy levels almost match one another, $E(n_1l_1^{-1}) \gtrsim E(n_2l_2^{-1}n_3l_3^{-1})$ as, for example, for $4s^{-1}$ and $4p^{-1}$ vacancies in Xe, that the spectroscopic factor becomes much less than unity and, in general, Eq. (3.17) becomes invalid.

Perturbation theory makes it also possible to extend the description of Auger transitions by considering their relationship with the preceding process of creating a vacancy in the atom.

Photoionization and the subsequent Auger transition are shown by the very same diagrams as in double photoionization (see Fig. 5.3). Ordinarily, diagram *d* makes the basic contribution to the unified process amplitude. If the vacancy a^{-1} is thought of as a quasiparticle and shown by a heavy line (see Fig. 6.1), the amplitude is then, with due regard for only this diagram, defined by an expression that matches Eq. (6.24), and the double differential cross section for the observation of an electron in the energy range ε_1, $\varepsilon_1 + d\varepsilon_1$ and for another electron having an energy between ε_2 and $\varepsilon_2 + d\varepsilon_2$ has the form given by Eq. (6.25). Integrating this expression with respect to the photoelectron energy ε_1 gives us a formula that describes the Auger spectrum:

$$\frac{d\sigma_A(\omega)}{d\varepsilon} \sim \sum_{abc} |\langle \varepsilon_b + \varepsilon_c - \varepsilon + \hbar\omega | d^{(1)} | a \rangle|^2 A_a(\varepsilon_b + \varepsilon_c - \varepsilon) \times$$

$$\times \pi |\langle \varepsilon a | h_{12}^e | cb \rangle - \langle \varepsilon a | h_{12}^e | bc \rangle|^2 [\mathrm{Im}\, \Sigma_a (\varepsilon_b + \varepsilon_c - \varepsilon)]^{-1}. \quad (8.37)$$

The last two terms are the partial and total Auger widths. If we also consider the other unified process diagrams that describe photoelectron scattering (Fig. 5.3c), or correlations in the initial state (Figs. 5.3a,b), the differential cross section does not reduce to so simple a form.

Another general approach that allows us to think of the creation and radiationless decay of an excited state as a unified process was developed in [131, 135, 137]. The theory of autoionizing levels and their respective resonances in the cross sections of the scattering processes, discussed briefly in Sec. 2.3, forms the basis of this approach. The fundamental concept of the method is the following. The creation of a primary vacancy a^{-1} as a particle x is scattered by an atom and the subsequent decay of this vacancy can be thought of as the double ionization of the atom by the x particles when the autoionization level a^{-1} is present (for example, the Auger effect as a resonance in the double ionization cross section). The amplitude of the process near a level consists of a monotonically varying term that corresponds to direct double ionization and a resonance term that describes the excitation of an atom into a quasistationary state and its subsequent Auger decay. These two processes – direct and two-stage – have the same initial and final states; therefore, they interfere with each other. If the interference terms can be neglected, the expression for the Auger line intensity can be factored into the population density of the initial level and the probability of an Auger transition. In the general case the intensity is given in terms of the cross section of the unified process.

If the distance between initial autoionizing levels having the same symmetry is greater than the sum of their widths,

$$\Delta E_{12} > \Gamma_1 + \Gamma_2, \quad (8.38)$$

which is often the case for inner vacancies, then the examination of Auger transitions from each level can be done individually as the creation and decay of an isolated, quasidiscrete state that interacts with one or more continua. The cross section near a resonance is found from Eq. (2.114) and the probability of Auger decay of the initial state γ is proportional to the resonance width and consists of the additive contributions from all Auger decay channels β:

$$W(\gamma) = \hbar^{-1} \sum_{\beta} |\langle \beta | H^e | \gamma \rangle|^2. \quad (8.39)$$

Each term in the sum represents the probability of the respective Auger transition and is defined according to Eq. (8.8), but a basis of wave functions in which the energy matrix for the ion + Auger electron system is diagonal is now used to describe the continua [131].

If Eq. (8.38) is not satisfied – the resonances overlap one another – the coupling between adjacent resonances must be considered and the more complex problem of multichannel scattering when proximate quasidiscrete levels are present must be solved [131, 135]. The resonance scattering method for describing the Auger effect must also be used for a strong interaction between a vacant state and a continuum.

A unified description of the creation and decay of excited states can be generalized by including channels for not only Auger, but radiative decay as well [135–137]. Terms corresponding to the electromagnetic field energy and its interaction with an atom are then added to the Hamiltonian, and the wave function contains a photon function as well. When this is done the fundamental equations retain their form; however, certain problems arise. These problems are associated with the absence of a photon wave function in coordinate representation and with the fact that the resonance energy shift caused by an electron—photon interaction between a discrete state and a continuum is infinite and must therefore be renormalized [135]. Because an atom can make a transition into the same final state when an autoionizing state decays in a radiative or radiationless process (the Auger electron recombines with the ion, emitting a photon; the photon emitted during the radiative transition is again absorbed by the atom, which emits an electron), they can interfere with each other when the probability of both decay channels is similar, which appears in the structure of Auger or emission spectra [136].

The relationship between the processes of the creation and decay of an excited state becomes important if the time required to create the vacancy is comparable to its decay time. When a shell is ionized by a photon whose energy is nearly equal to the shell's ionization energy, a slow electron escapes from the atom but is not removed from the atom during the time that the Auger transition takes place, and a secondary Auger electron escapes. Three electrons can interact when an electron collision occurs near an atom. This so-called post-collision interaction leads to a change, seen in experiments, in the locations and shapes of Auger lines that can be stimulated by particles whose energy is near the threshold energy.

A post-collision interaction is described by the method of multichannel scattering and perturbation theory [131, 285]; simpler approximations [286, 287] can also be used. In the manner of [287], we will give a qualitative, semiclassical description of this effect during photoionization, which will help in understanding its basic features.

Let an atom absorb a quantum of x rays whose energy exceeds the binding energy of an electron in the shell being ionized by an amount ΔE, called the energy excess. A photoelectron will have just such energy when it is removed to infinity. If the electron is located a distance $R(t)$ from an ion whose field is a Coulomb field of unity charge, the photoelectron's kinetic energy is

$$\varepsilon_1(t) = \Delta E + \frac{e^2}{R(t)}. \tag{8.40}$$

At time $t = t^*$ a fast Auger electron escapes from the atom and the ion's field is suddenly changed to a field of double charge, which leads to a change

$$\Delta\varepsilon_1(t^*) = \frac{e^2}{R(t^*)} - \frac{2e^2}{R(t^*)} = -\frac{e^2}{R(t^*)}. \tag{8.41}$$

in the photoelectron's energy.

According to the energy balance, an Auger electron acquires this much extra energy (the photoelectron partially screens the ion's field; therefore, the Auger electron experiences a smaller attractive potential):

$$\Delta\varepsilon_A \equiv \varepsilon = \frac{e^2}{R(t^*)}. \tag{8.42}$$

The distribution function for Auger electrons in a post-collision shift can be found from the law of the decay of an excited state:

$$\mathcal{N}(t) = \mathcal{N}(0)\, e^{-t/\tau}, \tag{8.43}$$

where τ is the lifetime of the state, and $\mathcal{N}(t)$ is the number of excited electrons at time t.

Taking the distribution function

$$P(\varepsilon) = \frac{d\mathcal{N}}{d\varepsilon} \tag{8.44}$$

from Eq. (8.43) and using the equation

$$\frac{d\mathcal{N}}{dt} = \frac{d\mathcal{N}}{d\varepsilon}\,\frac{d\varepsilon}{dR}\,\frac{dR}{dt}, \tag{8.45}$$

and the classical relation

$$\frac{dR}{dt} = v(t), \tag{8.46}$$

we find, that when $t = t^*$,

$$P(\varepsilon)=\frac{\mathcal{N}(0)\,e^{2}\exp(-t^{*}/\tau)}{\tau\varepsilon^{2}v(t^{*})}. \tag{8.47}$$

The Auger decay time t^* can be found from

$$t^{*}=\int_{0}^{t^{*}}dt=\int_{R(0)}^{R(t^{*})}\frac{dR}{v(R)}. \tag{8.48}$$

The most probable value of Auger line shift follows from the condition that

$$\frac{dP(\varepsilon)}{d\varepsilon}=0. \tag{8.49}$$

An explicit expression is found for t^* in a quasiclassical approximation; using this [and neglecting, as in Eq. (8.47), the direct, double photoionization into the final state of an Auger transition as well as the interference between this and the two-stage process], we find the equation [286]

$$\Gamma[2(\Delta E+\varepsilon)]^{1/2}-4\varepsilon(\Delta E+\varepsilon)-\varepsilon^{2}=0, \tag{8.50*}$$

where Γ is the natural width of a level in the atom having the vacancy.

When $\Delta E >> \varepsilon$ (in practice, for $\Delta E > 0.2$ a.u.), the solution of this equation is closely approximated by the formula

$$\varepsilon=\Gamma(8\Delta E)^{-1/2}. \tag{8.51*}$$

Consequently, the post-collision shift in the Auger line increases as the lifetime of the excited state and the energy excess decrease. The magnitude of the shift at the threshold is about 10^{-1} eV. The shift may be as much as 1 eV in a two-well effective potential having a positive barrier which impedes the removal of a photoelectron [288]. A post-collision interaction also leads to Auger line broadening.

Describing a post-collision interaction when an atom is ionized by electrons is a more complex theoretical problem because two slow electrons whose motion is highly correlated are present. This problem can only be solved for asymptotic cases – equal to zero and when the energy excess is very large [287].

Chapter 9

THE WIDTH AND SHAPE OF THE LINES. THE FLUORESCENCE YIELD

The shape and width of the different spectral lines – x-ray, emission, Auger, photoelectron – are all determined by the same fundamental principles, which makes it worthwhile to combine our examination of these questions into a single chapter. We will discuss the various factors involved in line broadening in Sec. 9.1: spontaneous transitions, distortions caused by the equipment, Doppler broadening, etc. A quantity closely related to natural linewidth – the fluorescence yield, which is characteristic of the relative role of a radiation and Auger decay of a state with a vacancy – is the topic of Sec. 9.2. The natural linewidth and fluorescence yield as a function of the nuclear charge, one- and many-electron quantum numbers, and the approximate invariance of these quantities that occurs for some configurations is studied in Sec. 9.3.

9.1. Factors that Determine the Shape of X-ray and Electron Lines

Let the population density of an excited level k be $\mathcal{N}_k$ at time t. During the time interval t, $t + dt$ the population density decreases an amount

$$-d\mathcal{N}_{k\to i} = \mathcal{N}_k W_{ki}\, dt. \tag{9.1}$$

because of transitions into the level i (spontaneous transitions and those due to external factors) that occur with total probability W_{ki}. Summing Eq. (9.1) over all possible decay channels

$$\frac{d\mathcal{N}_k}{dt} = -\mathcal{N}_k \sum_i W_{ki} \tag{9.2}$$

and integrating this differential equation (W_{ki} is assumed to be time-independent), we obtain the exponential decay law:

$$\mathcal{N}_k(t) = \mathcal{N}_k(0)\, e^{-\sum_i W_{ki} t}, \tag{9.3}$$

where $\mathcal{N}_k(0)$ is the population density of the level at the initial moment in time.

The average lifetime (often called simply the lifetime) of a level is defined as the time during which the population density of that level decreases by a factor of e. As a result,

$$\tau_k = \frac{1}{\sum_i W_{ki}}. \tag{9.4}$$

On the other hand, according to the uncertainty principle, there is a relationship

$$\Gamma_k \approx \hbar/\tau_k. \tag{9.5}$$

between the lifetime τ_k of a level and its width Γ_k. Hereafter, we will replace the "$\approx$" symbol with "$=$." Comparing Eqs. (9.4) and (9.5) gives us an equation for Γ_k in terms of the total probability that the level k will be deexcited:

$$\Gamma_k = \hbar \sum_i W_{ki}. \tag{9.6}$$

The width of a spectral line is equal to the sum of the widths of the initial and final levels.

Because of the short lifetime of highly excited states having an inner vacancy, they are characterized by a large natural width caused by spontaneous transitions. Distortions caused by the equipment, Doppler and collision broadening, etc., which usually cause only small and sometimes insignificant corrections to the natural linewidth, are other causes of x-ray and electron line broadening.

An isolated atom may be in its ground state for an indefinite period of time – the natural width of the ground level is zero. A lifetime on the order of $10^{-14} - 10^{-16}$ sec is typical of an excited state with an inner vacancy, which corresponds to a natural width of $0.1 - 10$ eV. The closer a vacancy is to the nucleus, the greater are the various possibilities of its being filled and the broader the level becomes; therefore, $\Gamma_K > \Gamma_L > \Gamma_M$.

We will examine how the total line intensity is distributed over its natural width. We will begin from a classical oscillator – an electron oscillating about its equilibrium position. Being accelerated, the electron radiates electromagnetic waves and loses energy; therefore, its radiation cannot be monochromatic. The equation for the electric field created by the charge contains, in addition to a retarded wave, a factor $e^{-\gamma t/2}$ that describes the attenuation:

$$E(t)=E_0\,e^{-\frac{\gamma t}{2}}\,e^{i\omega_0(t-r/c)}, \tag{9.7}$$

where γ is the constant of radiation attenuation, ω_0 is the frequency of the radiation in the absence of attenuation, and r is the distance from the center of oscillation to the point at which the electric field $E(t)$ is measured.

Using the Fourier transform, $E(t)$ can be expanded with respect to monochromatic waves:

$$E(t)=\int_0^\infty E(\omega)\,e^{i\omega t}\,d\omega, \tag{9.8}$$

$$E(\omega)=\frac{1}{\pi}\int_0^\infty E(t)\,e^{-i\omega t}\,dt \quad \big(E(t)=0 \quad \text{for} \quad -\infty<t<0\big). \tag{9.9}$$

Integrating Eq. (9.9) with due regard for Eq. (9.7)

$$E(\omega)=\frac{E_0}{\pi}\,\frac{-i(\omega-\omega_0)+\frac{1}{2}\gamma}{(\omega-\omega_0)^2+\frac{1}{4}\gamma^2}\,e^{-i\omega_0\frac{r}{c}}. \tag{9.10}$$

and using a familiar relationship between the amplitude of the electric field and the radiation intensity we obtain the law of intensity distribution as a function of frequency:

$$I(\omega)=\frac{c}{4\pi}\,|E(\omega)|^2=\frac{c}{4\pi^3}\,\frac{E_0^2}{(\omega-\omega_0)^2+\frac{1}{4}\gamma^2}. \tag{9.11}$$

The spectral intensity reaches a maximum when $\omega = \omega_0$

$$I(\omega_0) = \frac{cE_0^2}{\pi^3 \gamma^2} \tag{9.12}$$

and decreases symmetrically on both sides of this value.

The intensity $I(\omega)$, given in terms of $I(\omega_0)$, assumes its familiar form, called the Lorentz (natural) shape of the line (the dispersion formula):

$$I(\omega) = I(\omega_0) \frac{(\gamma/2)^2}{(\omega - \omega_0)^2 + (\gamma/2)^2}, \tag{9.13}$$

where γ is equal to the linewidth at its half-height.

A quantum-mechanics examination of the spontaneous radiation by an atom leads to the very same intensity distribution function [289]. If k is an excited level having an energy width Γ and i is the ground level, the probability that a photon with energy in the range E, $E + dE$ will be emitted during a $k \to i$ transition is given by

$$W_{ki}(E)\,dE = \frac{W_{ki}}{\pi} \frac{\Gamma/2}{(E - E_0)^2 + (\Gamma/2)^2} dE, \quad W_{ki} = \int W_{ki}(E)\,dE. \tag{9.14}$$

The transition probability distribution (9.14) corresponds to the intensity distribution relative to E_0, the energy of the line without regard to its broadening:

$$\frac{I_{ki}(E)\,dE}{I_{ki}} = \frac{W_{ki}(E)}{W_{ki}} dE, \quad I_{ki}(E) = I_0 \frac{(\Gamma/2)^2}{(E - E_0)^2 + (\Gamma/2)^2}, \tag{9.15}$$

where I_0 is the height of the line at maximum $E = E_0$, expressed in terms of the total line intensity:

$$I_0 = \frac{2I_{ki}}{\pi\Gamma}. \tag{9.16}$$

Equation (9.15) also describes the shape of a photoexcitation spectral line, which corresponds to transitions from the atom's ground state.

Let both levels that take part in a transition have a nonzero width. If the energy of the upper level is somewhere between E and $E + dE$, and that of the lower level is somewhere between E' and $E' + dE'$, then the photon energy is somewhere between $E - E' - dE'$ and $E - E' + dE$. Thus,

$$\hbar\,d\omega = dE + dE'. \tag{9.17}$$

The probability of a radiative transition having a photon energy in this interval is equal to

$$W_k(E)\,W_i(E')\,dE\,dE', \tag{9.18}$$

where $W_k(E)$ and $W_i(E')$ are found from Eq. (9.14).

Integrating this expression with respect to dE and considering the relation $E' = E - \hbar\omega$ yields [1, 10]

$$I_{ki}(E)=\frac{I_{ki}}{2\pi}\,\frac{\Gamma_k+\Gamma_i}{[E-(E_k-E_i)]^2+\left(\dfrac{\Gamma_k+\Gamma_i}{2}\right)^2}, \tag{9.19}$$

where E_i is the energy of the maximum in the distribution for the ith level.

Thus, in the general case, a linewidth equal to the sum of the widths of the initial and final levels must be used in the dispersion formula.

If one or both levels taking part in a transition are autoionizing levels, the shape of a line also depends on the peculiarities of an interaction between discrete states and continua and may become asymmetric (Sec. 2.4). All in all, such asymmetry is rarely encountered in x-ray and Auger spectra – the majority of the individual lines have a symmetric Lorentz shape. On the other hand, the possibility of the Auger decay of a state leads to additional broadening of the level, and its total natural width, equal to the sum of the radiation (Γ_r) and the Auger (Γ_A) widths,

$$\Gamma=\Gamma_r+\Gamma_A. \tag{9.20}$$

must be used in Eqs. (9.15) and (9.19). The quantities Γ_A and Γ_r are given in terms of the sum of the probabilities of the respective transitions, Eq. (9.6).

We will briefly examine some other reasons for x-ray and electron line broadening.

Doppler broadening of a radiation line at its half height (the Doppler width), caused by the random movement of gas atoms when they have a Maxwell distribution in velocity, is described by the equation [39, 66]

$$\delta E_D=2\sqrt{2\ln 2}\,\frac{E_0}{c}\sqrt{\frac{kT}{M}}, \tag{9.21}$$

where E_0 is the energy at the maximum in the line, T is the absolute temperature, k is Boltzmann's constant, and M is the mass of the atom. If M is measured in a.m.u. (atomic mass units; one a.m.u. is equal to 1/12 the mass of a ^{12}C nucleus), we have

At a temperature of $T = 500$ K for $K_{\alpha 1}$ – the line for the lightest boron atom containing a $2p$-electron – this ratio is $5 \cdot 10^{-6}$ and the absolute value of δE_D is only 10^{-3} eV. For heavy atoms the ratio $\delta E_D/E_0$ decreases, but because the transition energy increases as Z^2, δE_D at these temperatures is on the order of 10^{-1} eV.

Photoelectron or Auger line broadening due to the thermal movement of the gas atoms is given by [9]

$$\delta\varepsilon_D = 7{,}5 \cdot 10^{-4} \sqrt{\frac{\varepsilon T}{M}}, \tag{9.23}$$

in which the Doppler width $\delta\varepsilon_D$ and the photoelectron energy ε are measured in eV, and M is measured in a.m.u.

For light atoms $\delta\varepsilon_D$ reaches magnitudes of about 0.01–0.1 eV at $T = 500$ K but, in contrast to Doppler broadening, decreases with increasing atomic number. This is determined by the increase in the atomic mass, since the spectra of the different elements are usually registered at low values of photoelectron energy.

If a beam of atoms is used in an experiment, a Doppler shift that is linear with respect to velocity is easily eliminated by recording the spectrum in a direction perpendicular to the velocity of the beam.

The electron spectra stimulated during ion—atom collisions are functions of the kinematics of the collision – the velocities acquired by the particles during the collision. If the collisions are accompanied by a noticeable scattering of the particles, the Doppler shift and electron line broadening that are caused by this can significantly distort the spectrum [290]. The complex dependence of the shape of the spectrum on the distribution of the atoms with respect to their velocities and scattering angles makes it difficult to establish a connection between the recorded lines and the true transition energies.

Another possible reason for line broadening is the influence of the electric fields of the scattered particles. In spectra stimulated by electrons, protons, or photons when the atoms are ionized to a low degree this broadening can, as a rule, be neglected.

The effects due to self-absorption in the spectra for gases and vapors are, in contrast to the spectra for solids, usually insignificant.

The shape of the lines is distorted to some extent by the equipment: for example, when recording x rays, by diffraction effects, inexact focusing, etc. [2], depending on the quality of the devices being used. We will define an instrumental response function as the shape that a line acquires when a very narrow delta-shaped line is recorded. If the actual shape of the line is described by the function $I_n(E)$, then, because of equipment distortions, it is transformed into the convolution of the instrumental response $I_a(E - E')$ and $I_n(E)$:

$$I(E) = \int_{-\infty}^{\infty} I_n(E') I_a(E - E')\, dE'. \tag{9.24}$$

In addition, a detector records not the intensity distribution itself, but the integral intensity that impinges through a slit in the detector:

$$\bar{I}(E) = \int_{E-s/2}^{E+s/2} I(E')\, dE', \tag{9.25}$$

where s is the energy width of the slit.

The instrumental response for x-ray and electron lines is ordinarily very well approximated by a Gaussian function [291, 292]:

$$I_a(E - E') = \frac{2\sqrt{\ln 2}}{\sigma\sqrt{\pi}} \exp\left[-4\ln 2\left(\frac{E - E'}{\sigma}\right)^2\right], \quad \int_{-\infty}^{\infty} I_a(E - E')\, dE' = 1, \tag{9.26}$$

where σ is the width of the instrumental response.

Substituting Eq. (9.26) and the Lorentz function, Eq. (9.15), for $I_n(E)$ into Eqs. (9.24) and (9.25) yields

$$\bar{I}(E) = \frac{I\Gamma\sqrt{\ln 2}}{\sigma\pi\sqrt{\pi}} \int_{E-s/2}^{E+s/2} dE' \int_{-\infty}^{\infty} dE'' \frac{\exp\left[-4\ln 2\left(\frac{E' - E''}{\sigma}\right)^2\right]}{(E'' - E_0)^2 + (\Gamma/2)^2}. \tag{9.27}$$

Changing the variables of integration converts this expression into another, equivalent form [291]:

$$\frac{\delta E_D}{E_0} = 7{,}16 \cdot 10^{-7} \sqrt{\frac{T}{M}}. \tag{9.22}$$

$$I(E)=\frac{I\Gamma\sqrt{\ln 2}}{\sigma\pi\sqrt{\pi}}\int_{-\infty}^{\infty}d\beta\int_{E-s/2}^{E+s/2}dE'\,\frac{\exp[-4\ln 2(\beta/\sigma)^2]}{(E'-E_0+\beta)^2+(\Gamma/2)^2}. \tag{9.28}$$

According to Eq. (9.27), the height of a line when $E = E_0$ is

$$I(E_0)=\frac{I\Gamma\sqrt{\ln 2}}{\sigma\pi\sqrt{\pi}}\int_{E_0-s/2}^{E_0+s/2}dE'\int_{-\infty}^{\infty}dE''\,\frac{\exp\left[-4\ln 2\left(\frac{E'-E''}{\sigma}\right)^2\right]}{(E''-E_0)^2+(\Gamma/2)^2}. \tag{9.29}$$

The integration shown in Eq. (9.29) is carried out in squares only in the asymptotic cases. If $s < \sigma$ and $\sigma << \Gamma$, the quantities $| E' - E'' | < \sigma$ make the main contribution to the integral in Eq. (9.29). Considering the fact that

$$|E''-E_0| < \sigma + s \ll \Gamma, \tag{9.30}$$

only the term $(\Gamma/2)^2$ remains in the denominator of the subintegral expression. Then,

$$I(E_0)\approx\frac{2Is}{\pi\Gamma} \tag{9.31}$$

and the total theoretical intensities divided by the natural linewidths correspond to the line heights observed in an experimental spectrum.

Because of the large width of the levels in configurations having inner vacancies, the transition lines in atoms having open shells often overlap one another, which is a specific feature of x-ray and Auger spectra. If there is a passive outer shell that interacts weakly with the inner electrons taking part in a transition, the small multiplet splitting leads only to broadening of the diagram lines and asymmetry in them (for example, in transition elements having an open *d*- or *f*-shell). The asymmetry of a line can be described by an asymmetry index [1]

$$a=\frac{\alpha}{\beta},\quad \alpha+\beta=\Gamma, \tag{9.32}$$

where α and β are the longwave and shortwave half widths of a line, respectively, and are measured from the ordinate corresponding to the line's maximum to its edge.

9.2. The Fluorescence Yield

The fluorescence yield is a quantity that describes the capability of an atom having an inner vacancy to deexcite itself in a radiating manner:

$$\omega = \frac{W_r}{W} = \frac{W_r}{W_r + W_A}, \tag{9.33}$$

where W_r and W_A are the total probabilities that an excited state will decay spontaneously via radiative and Auger transitions, respectively. Since the total probabilities are independent of the projections of the angular momenta, the fluorescence yields from the states and the levels are the same.

The relationship between the total probability and the width of a level, Eq. (9.6), makes it possible for us to put ω into another, equivalent form:

$$\omega = \frac{\Gamma_r}{\Gamma_r + \Gamma_A}, \tag{9.34}$$

where Γ_r and Γ_A are the radiative and Auger widths of the level.

The Auger electron yield,

$$\omega_A = \frac{\Gamma_A}{\Gamma_r + \Gamma_A}, \tag{9.35}$$

which is uniquely related to the fluorescence yield,

$$\omega + \omega_A = 1. \tag{9.36}$$

is sometimes used. The yield ω can be given in terms of quantities that can be measured experimentally: $\mathcal{N}$ is the number of excited atoms per unit volume in the state being examined, and $\mathcal{I}$ is the number of photons emitted by an atom as a result of spontaneous radiative transitions out of this state:

$$\omega = \frac{\mathcal{I}}{\mathcal{N}}. \tag{9.37}$$

The fluorescence yield of a subshell, a shell, or a layer is defined as the average probability that a vacancy in the given subshell, shell, or corresponding layer will be filled by a radiative process. The fluorescence yield of a subshell i in a layer X thus depends not only on the fluorescence yields of the levels $\omega(\gamma J)$, but on their population densities $\mathcal{N}'(\gamma J)$:

$$\omega_i^X = \frac{\sum_{\gamma J} \mathcal{N}(\gamma J)\,\omega(\gamma J)}{\sum_{\gamma J} \mathcal{N}(\gamma J)} \tag{9.38}$$

as well. Here, the summation over γJ covers all levels of a configuration that has this vacancy. The quantities in the denominator and numerator of Eq. (9.38) have the sense of the number of vacancies in a subshell $\mathcal{N}_i^X$ and the number of photons emitted as these vacancies are filled, $\mathcal{G}_i^X$,

$$\mathcal{N}_i^X = \sum_{\gamma J} \mathcal{N}(\gamma J), \quad \mathcal{G}_i^X = \sum_{\gamma J} \mathcal{N}(\gamma J)\,\omega(\gamma J). \tag{9.39}$$

As a result, Eq. (9.38) may be rewritten in a form that is analogous to Eq. (9.37):

$$\omega_i^X = \frac{\mathcal{G}_i^X}{\mathcal{N}_i^X}, \tag{9.40}$$

and ω_i^X becomes a purely atomic quantity only when all states are populated equally:

$$\omega_i^X = \frac{\sum_{\gamma J} (2J+1)\,\omega(\gamma J)}{\sum_{\gamma J} (2J+1)}. \tag{9.41}$$

Equation (9.41) is not equivalent to the equation for the fluorescence yield of a subshell in terms of the average probabilities of filling a vacancy in a radiation (${}_r\bar{W}_i^X$) and any ($\bar{W}_i^X$) manner,

$$\bar{\omega}_i^X = \frac{{}_r\bar{W}_i^X}{\bar{W}_i^X}, \tag{9.42}$$

and using it leads to substantial differences between theoretical results and experimental data [294, 162].

The fluorescence yield of a layer is given in terms of the fluorescence yields of the subshells; the simplest relationship is obtained when Coster–Kronig transitions are forbidden (vacancies are not rearranged in the subshells of the layer being examined):

$$\omega_X = \frac{\sum_i \mathcal{N}_i^X \omega_i^X}{\sum_i \mathcal{N}_i^X} \quad (i \in X). \tag{9.43}$$

Equation (9.43) can also be transformed to the form of Eq. (9.37):

$$\omega_X = \frac{\mathcal{G}^X}{\mathcal{N}^X}, \tag{9.44}$$

where $\mathcal{N}^X$ is the total number of atoms having primary vacancies in a layer X and $\mathcal{G}^X$ is the number of photons emitted by these atoms.

If Coster–Kronig transitions can change the primary distribution of vacancies in a layer, then, starting from an expression similar to Eq. (9.38), and using f_{ij}^X to denote the probability of a radiationless transition of a vacancy from a subshell i into a subshell j in the same layer X, instead of Eq. (9.43), we obtain [293]

$$\omega_X = \frac{\sum_i V_i^X \omega_i^X}{\sum_i V_i^X}. \tag{9.45}$$

Here, the V_i^X are secondary population densities with due regard for the change made in them by Coster–Kronig transitions:

$$\begin{aligned} V_1^X &= \mathcal{N}_1^X, \\ V_2^X &= \mathcal{N}_2^X + \mathcal{N}_1^X f_{12}^X, \\ V_3^X &= \mathcal{N}_3^X + \mathcal{N}_2^X f_{23}^X + \mathcal{N}_1^X (f_{13}^X + f_{12}^X f_{23}^X). \\ &\text{-- -- -- -- -- -- -- -- -- -- -- -- --} \end{aligned} \tag{9.46}$$

It is sometimes more convenient to use primary population densities when calculating ω_X, but, instead of the ω_i^X, coefficients ν_i^X are introduced, which have the sense of the number of x-ray quanta emitted due to the creation of a vacancy in a subshell X_i (but not only when a vacancy in this subshell is filled in a radiative manner) [293]:

$$\omega_X = \frac{\sum_i \mathcal{N}_i^X \nu_i^X}{\sum_i \mathcal{N}_i^X}. \tag{9.47}$$

The equations for the ν_i^X follow from comparing Eq. (9.47) with Eqs. (9.45) and (9.46).

9.3. The Natural Width of the Levels and the Fluorescence Yield as a Function of the Characteristics of an Atom

The radiative width of an atom's level is, according to Eqs. (9.6) and (7.19),

$$\Gamma_r(\gamma) \sim \sum_{\gamma'} E(\gamma, \gamma')^3 S(\gamma, \gamma'). \tag{9.48}$$

Transition energy as a function of the effective nuclear charge can be established via Moseley's law as $(Z - \sigma)^2$, and the line strength of a dipole transition varies according to $(Z - \sigma)^{-2}$ (σ is approximately equal to the screening constant for an electron making a transition in an initial state). Thus,

$$\Gamma_r \sim (Z - \sigma)^4. \tag{9.49}$$

The radiative width of the deepest K-vacancy in heavy atoms is approximately equal to the total width, which is described very well by the semiempirical formula [10]

$$\Gamma_K = 1.73\, Z^{3,93}\, 10^{-6} \text{ eV}. \tag{9.50}$$

For large Z the quantity Γ_K reaches magnitudes of tens and even hundreds of electron volts.

The Auger width of a level is given in terms of the squares of the matrix elements for an electrostatic interaction operator. The radial integral of an electrostatic interaction for wave functions of a discrete spectrum is proportional to $(Z - \sigma)$. When one function in it is replaced by the function of a free electron, the integral becomes, according to Eq. (1.170), independent of $Z - \sigma$ [46]. Consequently, the Auger width can, in an approximation, be considered independent of the nuclear charge:

$$\Gamma_A \approx \text{const.} \tag{9.51}$$

Coster–Kronig transitions make the fundamental contribution to the Auger width; if they are forbidden, the Auger width is only a few tenths of an electron-volt. If they are allowed, the Auger width is several electron-volts wide and reaches ten or more electron volts when super-Coster–Kronig transitions are present.

According to Eqs. (9.33), (9.49), and (9.51) the fluorescence yield when $Z >> \sigma$ can be approximated by the formula

$$\omega = \frac{BZ^4}{C+BZ^4} = \frac{Z^4}{A+Z^4}, \tag{9.52}$$

where $A = C/B$ is an empirical constant that can be found by comparing Eq. (9.52) with experimental data. As $Z \to \infty$, $\omega \to 1$ (Fig. 9.1).

Equation (9.52) can be made more accurate by incorporating additional parameters that take screening and relativistic effects into consideration. For example,

$$\omega_K = \frac{(A+BZ+CZ^3)^4}{1+(A+BZ+CZ^3)^4}. \tag{9.53}$$

is used for the fluorescence yield of the K-layer [8, 293]; $\omega > 1/2$ only for the K-layer, starting from $Z \gtrsim 30$, and for the L-layer in the elements at the end of the periodic system, starting from $Z \gtrsim 90$ [293]. A vacancy in light atoms, and in the intermediate and outer shells of heavy atoms, is filled mainly in a radiationless process. And, if a vacancy in

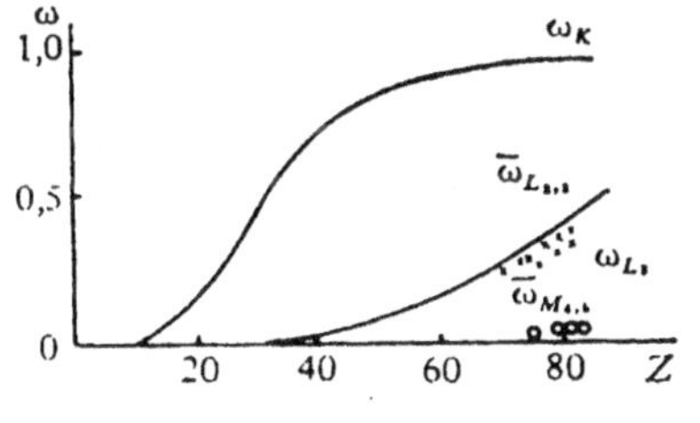

Fig. 9.1. Fluorescence yield of the K-layer, L_3-subshell, and the average fluorescence

$$\bar{\omega}_{nl} = \sum_j [j]\,\omega_{nlj}/(4l+2).$$

yields of the $L_{2,3}$- and $M_{4,5}$-shells, determined from the formula $\bar{\omega}_{nl} = \Sigma[j]\omega_{nlj}/(4l+2)$. The experimental data are from [259, 293].

a subvalent shell can be filled via a Coster–Kronig or super Coster–Kronig transition, the contribution made by radiative transitions to the total width of the level can, for practical purposes, be neglected and the total width considered equal to the Auger width of the level.

The effective nuclear charge increases when an atom is excited and ionized; therefore, the fluorescence yield in ions and, to a lesser extent, in excited atoms, tends to increase.

The radiative and Auger widths of the levels, as well as the fluorescence yield as summed quantities are more weakly dependent on relativistic and correlation effects than are the probabilities of the individual transitions; when added together the corrections are partially compensated. For this reason, calculating radiation widths by the nonrelativistic Hartree–Fock–Slater method for even K- and L-vacancies differs from the results obtained through relativistic calculations by as much as 10% [162].

We will examine the natural widths of the levels and the fluorescence yield as a function of the many-electron quantum numbers, and the approximate invariance of these quantities for some configurations.

The radiative or Auger width of a level is the sum of the partial widths

$$\Gamma_A(K\gamma J)=\sum_{K'l}\Gamma_A(K\gamma J\to K'\varepsilon l), \tag{9.54}$$

$$\Gamma_r(K\gamma J)=\sum_{K'}\Gamma_r(K\gamma J\to K'), \tag{9.55}$$

that correspond to transitions into all levels of a given final configuration. In the nonrelativistic theory we have

$$\Gamma_A(K\gamma J\to K'\varepsilon l)=\sum_{\gamma'J'j}\left|\langle K\gamma J|H^e|K'\gamma' J'\varepsilon ljJ\rangle\right|^2, \tag{9.56}$$

$$\Gamma_r(K\gamma J\to K')=$$
$$=\frac{4}{3\hbar^3c^3}\sum_{\gamma'J'}\frac{[E(K\gamma J)-E(K'\gamma'J')]^3}{2J+1}\,|\langle K\gamma J\|D^{(1)}\|K'\gamma'J'\rangle|^2. \tag{9.57}$$

If we assume that the radial integrals for the transitions are independent of the many-electron quantum numbers and say, in addition, that the energy of a radiative transition in Eq. (9.57) can be replaced by the average energy of the emission spectrum between given configurations, Eq. (7.31), the summation in Eqs. (9.56) and (9.57) can then be done for the general case [270] in a second quantization representation [40]. When this is done the sum is given in terms of the matrix element of the effective operator for the wave functions of the level being examined:

$$\sum_{\gamma'} \left| \langle K\gamma | T | K'\gamma' \rangle \right|^2 = \langle K\gamma | T^{ef} | K\gamma \rangle. \tag{9.58}$$

In the case of a radiative dipole transition,

$$K_0 n_1 l_1^{N_1} n_2 l_2^{N_2} \to K_0 n_1 l_1^{N_1+1} n_2 l_2^{N_2-1}$$

(K_0 are any passive shells, open or closed) the effective operator is found to be

$$T^{ef} = \left\{ -\sum_{xx'} (-1)^x [x, x'] \begin{Bmatrix} l_1 & l_1 & x \\ l_2 & l_2 & 1 \end{Bmatrix} (V_1^{(xx')} \cdot V_2^{(xx')}) + \frac{N_2}{2l_2+1} \right\} \times$$
$$\times e^2 \langle l_1 \| C^{(1)} \| l_2 \rangle^2 \langle n_1 l_1 | r | n_2 l_2 \rangle^2. \tag{9.59}$$

Here, $V_i^{(xx')}$ is the operator given by Eq. (1.172) consisting of unit tensors and acting on the electrons in the ith shell.

In the approximation being used, the radiative partial width will be invariant and independent of the quantum numbers of the level only when T^{ef} is a scalar. Both ranks x and x' attain only null values when one shell, $l_1^{N_1}$ or $l_2^{N_2}$, in the K configuration is either closed or contains no electrons. Thus, the partial radiative width does not depend (and, with due regard for the approximate nature of our model, is weakly dependent) on the quantum numbers of the level in the following circumstances:

$$K_0 n_1 l_1^{N_1} n_2 l_2^{4l_2+2} \to K_0 n_1 l_1^{N_1+1} n_2 l_2^{4l_2+1}, \tag{9.60}$$

$$K_0 n_1 l_1^0 n_2 l_2^{N_2} \to K_0 n_1 l_1^1 n_2 l_2^{N_2-1}. \tag{9.61}$$

The second of these two expressions is rather exotic; therefore, we will only give an expression for the partial radiative width for Eq. (9.60) [found by setting the ranks $x = x' = 0$ in Eq. (9.59) and using the formula for the submatrix element $V^{(00)}$, Eq. (1.79)] [270]:

$$\Gamma_r (K_0 n_1 l_1^{N_1} n_2 l_2^{4l_2+2} \gamma J \to K_0 n_1 l_1^{N_1+1} n_2 l_2^{4l_2+1}) = \frac{4l_1+2-N_1}{4l_1+2} B(n_2 l_2, n_1 l_1), \tag{9.62}$$

where

$$B(n_2 l_2,\ n_1 l_1)=\frac{8e^2}{3\hbar^3 c^3}\ E_e(K-K')^3\,\langle l_1 \| C^{(1)} \| l_2\rangle^2\,\langle n_1 l_1 | r | n_2 l_2\rangle^2. \tag{9.63}$$

It can be proved in an analogous fashion that the partial Auger widths are approximately invariant for the transitions

$$K_0\, n_1\, l_1^{N_1}\, n_2\, l_2^{4l_2+2}\, n_3\, l_3^{4l_3+2} \rightarrow K_0\, n_1\, l_1^{N_1+1}\, n_2\, l_2^{4l_2+1}\, n_3\, l_3^{4l_3+1}\, \varepsilon\, l, \tag{9.64}$$

$$K_0\, n_1\, l_1^{N_1}\, n_2\, l_2^{4l_2+2} \rightarrow K_0\, n_1\, l_1^{N_1+1}\, n_2\, l_2^{4l_2}\, \varepsilon\, l, \tag{9.65}$$

and an Auger width that is independent of the term is given by [270]

$$\Gamma_A(K_0\, n_1\, l_1^{N_1}\, n_2\, l_2^{4l_2+2}\, n_3\, l_3^{4l_3+2}\, \gamma J \rightarrow K_0\, n_1\, l_1^{N_1+1}\, n_2\, l_2^{4l_2+1}\, n_3\, l_3^{4l_3+1}\, \varepsilon\, l)=$$
$$=\frac{4l_1+2-N_1}{4l_1+2}\ s_A(n_2 l_2 n_3 l_3,\ n_1 l_1\, \varepsilon\, l), \tag{9.66}$$

$$\Gamma_A(K_0\, n_1\, l_1^{N_1}\, n_2\, l_2^{4l_2+2}\, \gamma J \rightarrow K_0\, n_1\, l_1^{N_1+1}\, n_2\, l_2^{4l_2}\, \varepsilon\, l)=$$
$$=\frac{4l_1+2-N_1}{4l_1+2}\ s_A(n_2 l_2^2,\ n_1 l_1\, \varepsilon\, l), \tag{9.67}$$

where s_A is the total line strength of the Auger transitions in the two-electron model given by Eq. (8.28), and the K_0 are any open or closed shells that do not take part in the transition.

The total radiative and Auger widths will be approximately invariant if all partial widths have this property. This occurs for the narrower class of configurations that contain only a single open shell:

$$\Gamma_r(K_0'\, n_1\, l_1^{N_1}\, \gamma J)=\frac{4l_1+2-N_1}{4l_1+2}\ \sum_{n_2 l_2} B(n_2 l_2,\ n_1 l_1)\quad (n_2 l_2 \in K_0'), \tag{9.68}$$

$$\Gamma_A(K_0'\, n_1\, l_1^{N_1}\, \gamma J)=$$
$$=\frac{4l_1+2-N_1}{4l_1+2}\ \sum_{n_2 l_2 n_3 l_3 l} s_A(n_2 l_2 n_3 l_3,\ n_1 l_1\, \varepsilon\, l)\quad (n_2 l_2,\ n_3 l_3 \in K_0'), \tag{9.69}$$

where the K_0' are closed shells. It is assumed in Eq. (9.68) that the $n_1 l_1^{N_1}$ is an inner shell and that transitions into it can only be made from closed shells. If the $n_1 l_1^{N_1}$ is an outer shell that has been excited above from closed shells, the radiative width is determined by the $l_1^{N_1} \rightarrow l_1^{N_1-1}$ transitions and the $4l_1 + 2 - N_1$ in Eq. (9.68) must be replaced by N_1.

The quantities s_A and B are weakly (only in terms of the radial wave functions) dependent on N; therefore,

$$\frac{\Gamma(K_0' nl^N \gamma J)}{\Gamma(K_0' nl^{N'} \gamma' J')} \approx \frac{4l+2-N}{4l+2-N'}. \tag{9.70}$$

which is satisfied for Auger, radiative, and total widths. Note that if a change in the number of electrons in an nl-shell leads to energy forbidding of some Auger transitions, Eq. (9.70) can only be used when the contribution of these widths can be excluded from the total or the Auger width for another configuration.

The fluorescence yield for a $K = K_0\ nl^N$ configuration is also weakly dependent on the term and even (because the same $4l + 2 - N$ factors are present in the equations for the Auger and radiative widths) on the number of electrons in an open shell.

Invariance in the level widths also takes place in relativistic theory. The respective formulas are obtained from Eqs. (9.62) and (9.66)-(9.70) if the $l_i^{N_i}$ shells are replaced by the subshells $l_i j_i^{N_i}$, $4l_i + 2$ is replaced by $2j_i + 1$, Eq. (8.31) is used for the s_A, and

$$B(n_2 l_2 j_2,\ n_1 l_1 j_1) = \sum_t \frac{2(2t+1)(t+1)}{[(2t+1)!!]^2\, t} \left(\frac{\bar{E}_e(K-K')}{\hbar c}\right)^{2t+1} \times$$
$$\times \{ |(n_1 l_1 j_1 \| {}_eO^{(t)} \| n_2 l_2 j_2)|^2 + |(n_1 l_1 j_1 \| {}_mO^{(t)} \| n_2 l_2 j_2)|^2 \}, \tag{9.71}$$

is used for B, where $\bar{E}_e(K - K')$ is the average energy of a radiative transition.

A special case of invariance that has been examined is the well-known weak dependence of the widths of the one-vacancy levels $\Gamma(nlj^{-1})$ on the quantum number j. The few exceptions are due to correlation effects or energy forbidding of Coster–Kronig transitions for a single value of j.

We can use the average partial width of a level given as the total line strength of transitions between configurations K and K' divided by the statistical weight of the initial configuration $g(K)$:

$$\bar{\Gamma}_A(K, K' \varepsilon l) = \frac{g(K,\ K')}{g(K)}\, s_A(n_2 l_2 n_3 l_3,\ n_1 l_1 \varepsilon l), \tag{9.72}$$

$$\bar{\Gamma}_r(K, K') = \frac{g(K,\ K')}{g(K)}\, B(n_2 l_2,\ n_1 l_1). \tag{9.73}$$

to estimate the width of a level, which depends on its quantum numbers. The statistical factor $g(K, K')$ is found from Eq. (8.27).

Chapter 10

BREMSSTRAHLUNG AND X-RAY SCATTERING

In this chapter we treat jointly x-ray scattering by free atoms and electron bremsstrahlung in the atoms. The spectra for both processes are continuous, although at certain energies resonance features can appear in them. The theoretical description of these processes combines the need to go beyond the limits of first-order perturbation theory and to use dispersion relations, the weak dependence of the cross sections on the many-electron quantum numbers and the applicability of the one-electron model that this is based upon, and the comparatively minor role played by correlation effects.

In the broadest sense of the concept, bremsstrahlung includes synchrotron radiation – an important source of x rays. However, it is not an atomic process; therefore, we will not discuss it here, especially since synchrotron radiation has been thoroughly examined in many monographs and review articles.

10.1. Electron Bremsstrahlung in the Field of an Atom

According to classical electrodynamics, bremsstrahlung intensity is directly proportional to the square of the acceleration of a charged particle, i.e., inversely proportional to the square of its mass. For this reason, proton or ion bremsstrahlung is much weaker than electron bremsstrahlung. In this section we will examine the fundamental approximations used to describe the bremsstrahlung that occurs when electrons are scattered by free atoms (for short, we will call it bremsstrahlung).

The continuous spectrum for bremsstrahlung extends from frequencies that are nearly zero (an electron transfers only a negligible fraction of its energy to a photon) to the high-energy edge $\omega_1 = \varepsilon_1/\hbar$, where ε_1 is the electron's initial kinetic energy (the photon acquires all of the electron's energy). Bremsstrahlung intensity increases as the velocity of the incident electron increases. Because of this process, electrons begin to lose a significant portion of their energy when $\varepsilon_1 \gtrsim 10$ keV and the main portion when ε_1 is greater than 10–100 MeV; i.e., when the bremsstrahlung embraces the x-ray frequencies.

The radiation that appears when an electron is decelerated in an atomic field comprises two components [295, 296]: the "direct" or electron radiation in the atomic static field, and the polarization or "atomic" radiation that is due to the atom being perturbed by the incident electron, which induces a dipole moment in the atom. The second component becomes important when the photon's energy is nearly equal to the shell's ionization energy [296]. We will begin by examining the more thoroughly studied and more common "direct" electron bremsstrahlung in an atomic field.

The basic quantity that describes bremsstrahlung is the photon's differential cross section with respect to frequency. If we multiply this quantity by the photon's energy and integrate over the entire spectrum we find the total amount of energy lost by an electron due to radiation (this is usually defined relative to its total energy E_1, and the electron flux is assumed to be unity)

$$\Phi_{rad} = \frac{1}{E_1} \int_0^{\varepsilon_1/\hbar} \hbar\omega \frac{d\sigma}{d\omega} d\omega \tag{10.1}$$

or the average loss in the electron's energy per unit path length is

$$-\frac{dE_1}{dx} = \mathcal{N} \int_0^{\varepsilon_1/\hbar} \hbar\omega \frac{d\sigma}{d\omega} d\omega = \mathcal{N} E_1 \Phi_{rad}, \tag{10.2}$$

where $\mathcal{N}$ is the number of atoms per unit volume, and ε_1 is the electron's initial kinetic energy.

A doubly differential cross section, $d^2\sigma/d\omega\, d\Omega_k$, is used to describe the photons' angular distribution (Ω_k is the photon's spatial escape angle), and when experiments are interpreted by coincidence, a triple differential cross section (supplemented by the angle at which the electron moves after radiation) and more complex cross sections are also used [297, 298].

In the asymptotic limit as $\hbar\omega/\varepsilon_1 \to 0$ the bremsstrahlung cross section is given in terms of the cross section of the electron's elastic scattering by the atom, and as $\hbar\omega/\varepsilon_1 \to 1$ it is associated with the cross section for an atom's radiation accompanying capture of the electron or its inverse process – the atomic photoeffect [297].

We can ignore relativistic effects [299] and examine the bremsstrahlung in a nonrelativistic approximation if the energy of the incident electron is no more than a few tens of keV.

In first-order perturbation theory, starting from Eqs. (1.240) and (1.243), the amplitude of a process corresponding to a free electron making a transition from a state having a wave vector $\mathbf{q}_1$ and spin projection μ_1 into a state $\mathbf{q}_2\mu_2$ by emitting a photon having a wave vector $\mathbf{k}$ and polarization $\hat{\varepsilon}_{\mathbf{k}\rho}$ is

$$\langle\gamma\,\mathbf{q}_2\,\mu_2\,n_{\mathbf{k}\rho}+1\,|\,H_1'\,|\,\gamma\,\mathbf{q}_1\,\mu_1\,n_{\mathbf{k}\rho}\rangle = = -\frac{e}{m}\left[\frac{2\pi\hbar\,(n_{\mathbf{k}\rho}+1)}{\omega V}\right]^{\frac{1}{2}}\langle\gamma\,\mathbf{q}_2\,\mu_2\,|\,\hat{\varepsilon}_{\mathbf{k}\rho}\cdot\mathbf{p}\,e^{-i\mathbf{k}\cdot\mathbf{r}}\,|\,\gamma\,\mathbf{q}_1\,\mu_1\rangle. \tag{10.3}$$

The summation over the coordinates of the electrons in the system has been omitted in the operator, since the matrix element goes to zero because the wave functions of a free electron are orthogonal if the operator does not act on its coordinates. In this regard the state γ of a "passive" atom must remain unchanged during scattering in single-electron approximation. The matrix element of the H_2' operator, Eq. (1.240), for a one-photon transition is zero.

The function $v_1^{-1/2}(2\pi)^{3/2}\psi_{q_1\mu_1}{}^{+}$, normalized to unit electron flux density, must be used as the wave function for the initial state of an electron $\mathbf{q}_1\mu_1$ (the bremsstrahlung cross section will then match the probability), and the function $\psi_{q_2\mu_2}$, Eq. (4.40), normalized to $\delta(\mathbf{q}_2 - \mathbf{q}_2\;)$, is used for the wave function $\mathbf{q}_2\mu_2$ of the final state. Because the final state contains a photon and an electron, the differential df in Eq. (1.244) must be taken as

$$df = d\,\mathbf{q}_2\,\frac{V d\mathbf{k}}{(2\pi)^3}\,. \tag{10.4}$$

Substituting Eqs. (10.3) and (10.4) into Eqs. (1.244) and (1.245), and assuming that $n_{\mathbf{k}\rho} = 0$, we find

$$d\sigma_e = \frac{e^2}{2\pi m^2\omega}\,|\,\langle\gamma\,\mathbf{q}_2\,\mu_2\,|\,\hat{\varepsilon}_{\mathbf{k}\rho}\cdot\mathbf{p}\,e^{-i\mathbf{k}\cdot\mathbf{r}}\,|\,\gamma\,\mathbf{q}_1\,\mu_1\rangle\,|^2\times$$

$$\times\,\delta\Big(\frac{q_1^2\hbar^2}{2m}-\frac{q_2^2\hbar^2}{2m}-\hbar\omega\Big)\,d\mathbf{q}_2\,d\mathbf{k}. \tag{10.5}$$

The differential cross section can be integrated with respect to dq_2, which is done with the relations

$$d\mathbf{k}=\frac{\omega^2}{c^3}\,d\omega\,d\Omega_k,\quad d\mathbf{q}_2=q_2^2dq_2\,d\Omega_q, \tag{10.6}$$

$$\delta\Big(-\hbar\omega+\frac{\hbar^2q_1^2}{2m}-\frac{\hbar^2q_2^2}{2m}\Big)=\frac{m}{\hbar^2q_2}\,\delta(q_2-\bar{q}_2),\ \bar{q}_2=\Big[-\frac{2m\omega}{\hbar}+q_1^2\Big]^{\frac{1}{2}}. \tag{10.7}$$

In the nonrelativistic approximation being used, the exponent in the operator is ordinarily replaced by unity. Transforming the operator from its p-form into its r-form and writing it in terms of the dipole moment $\mathbf{D}$, we finally obtain [39]

$$\frac{d^3\sigma_e}{d\Omega_2\,d\Omega_k\,d\omega}=\frac{m\omega^3q_2}{2\pi\hbar^2c^3}\,|\,\hat{\epsilon}_{\mathbf{k}\rho}\cdot\langle\gamma\mathbf{q}_2\mu_2\,|\,\mathbf{D}\,|\,\gamma\mathbf{q}_1\mu_1\rangle\,|^2. \tag{10.8}$$

Here, $\bar{q}_2$ is again replaced by q_2.

A bremsstrahlung spectrum that is integral with respect to the angles is obtained by integrating Eq. (10.8) with respect to $d\Omega_2$ and averaging over the possible mutual orientations of the $\mathbf{q}_1$ and $\mathbf{k}$ vectors [39]. When this is done, the expansions $\psi^{\mp}$, Eq. (4.40), and the orthonormality of spherical functions are used:

$$\frac{1}{4\pi}\int d\Omega_1\,d\Omega_2\,|\,\hat{\epsilon}_{\mathbf{k}\rho}\cdot\langle\psi^{-}_{\mathbf{q}_2\mu_2}|\,r\,|\,\psi^{+}_{\mathbf{q}_1\mu_1}\rangle\,|^2=$$
$$=\frac{1}{4\pi q_1^2q_2^2}\sum_{l_1m_1l_2m_2}|\,\langle q_2l_2m_2\mu_2\,|\,\hat{\epsilon}_{\mathbf{k}\rho}\cdot\mathbf{r}\,|\,q_1l_1m_1\mu_1\rangle\,|^2. \tag{10.9}$$

By recoupling the angular momenta of the atom and free electron into total angular momenta, averaging the cross section over the projections of the atom's initial state and the incident electron's spin, integrating with respect to $d\Omega_k$, and summing over the direction in which the photon is polarized and the projection μ_2, we arrive at the formula

$$\frac{d\sigma_e}{d\omega}=\frac{4\pi^2\hbar\omega^3}{3c^3q_1^2}\sum_{l}\sum_{l'=l\pm1}[J]^{-1}\sum_{J'J''j_1j_2}|\,\langle\gamma J\varepsilon_2l_2j_2J'\,\|\,D^{(1)}\,\|\,\gamma J\varepsilon_1l_1j_1J''\rangle\,|^2.$$

Here, the free electron's radial wave functions have been normalized to $\delta(\varepsilon - \varepsilon')$.

As the electron's kinetic energy increases, its radial wave function begins to oscillate periodically inside the atom, except near the nucleus which region makes the fundamental contribution to the cross section. For atoms having large Z and starting from incident electron energies of $\varepsilon_1 \gtrsim 100$ keV, and from $\varepsilon_1 \gtrsim 5$ keV for light atoms, x-ray bremsstrahlung is mainly determined by the nuclear field [299], and the simple one-electron model may be used; i.e., the internal structure of the atom can be ignored and the process thought of as an electron making a radiative transition between states in a continuous spectrum in the field of a point charge Ze.

The dipole transition integral for one-electron wave functions is given in terms of a hypergeometric function, whereby the bremsstrahlung cross section is described by Sommerfeld's equation:

$$\omega \frac{d\sigma_e}{d\omega} = \frac{16}{3} \pi^2 \alpha^3 a_0^2 \frac{\nu_1^2 \chi_0}{\left(\exp(2\pi\nu_1) - 1\right)\left(1 - \exp(-2\pi\nu_2)\right)} \times$$
$$\times \frac{d}{d\chi_0} \left| F(i\nu_1, i\nu_2, 1, \chi_0) \right|^2, \tag{10.11}$$

where

$$\nu_1 = \frac{Ze^2}{\hbar v_1}, \quad \nu_2 = \frac{Ze^2}{\hbar v_2}, \quad \chi_0 = -\frac{4\nu_1 \nu_2}{(\nu_2 - \nu_1)^2}. \tag{10.12}$$

Here, F is a hypergeometric function, and v_1 and v_2 are the initial and final velocities of the electron. Since $v_1 > v_2$, $\nu_2 > \nu_1$.

According to Eqs. (10.11) and (10.12), the bremsstrahlung cross section is proportional to the square of the nuclear charge, which follows from simple qualitative considerations: the acceleration of the electron and the strength with which it interacts with the nucleus are proportional to Z, and the radiation intensity, being given in terms of the square of the acceleration, is $\sim Z^2$.

At high electron velocities, where the condition that $\nu_1, \nu_2 << 1$ is satisfied, expanding the function $\chi_0(d/d\chi_0)F$ in powers of ν_1 and ν_2 and keeping only the first term we can, in an approximation, replace Eq. (10.11) by [300]

$$\omega \frac{d\sigma_e}{d\omega} = \frac{16}{3} \alpha^3 Z^2 q_1^{-2} \ln \frac{q_1 + q_2}{q_1 - q_2}. \tag{10.13}$$

If only $\nu_1 << 1$ and ν_2 has any value (including those near the high-frequency limit), then Eq. (10.11) is approximated to the form

$$\omega \frac{d\sigma_e}{d\omega} = \frac{16}{3} \alpha^3 a_0^2 \nu_1 \nu_2 \frac{1-\exp(-2\pi\nu_1)}{1-\exp(-2\pi\nu_2)} \ln \frac{q_1+q_2}{q_1-q_2}, \tag{10.14}$$

which is distinguished from Eq. (10.13) by Elvert's correction factor:

$$\frac{\nu_2}{\nu_1} \frac{1-\exp(-2\pi\nu_1)}{1-\exp(-2\pi\nu_2)}. \tag{10.15}$$

When the condition that $\hbar\omega/\varepsilon_1 >> \nu_1{}^{-1}$ is satisfied (in practice, it is satisfied for all x-ray wavelengths), we can use another approximation:

$$\omega \frac{d\sigma_e}{d\omega} = \frac{16\pi}{3\sqrt{3}} \alpha^3 a_0^2 \nu_1^2 \times$$
$$\times \left\{ 1 + \frac{\sqrt[3]{9/4}\,\Gamma^2\left(\frac{1}{3}\right)}{10\pi\sqrt{3}} \nu_1^{-\frac{2}{3}} \left(\frac{\varepsilon_1}{\hbar\omega}\right)^{\frac{3}{2}} \left(2 - \frac{\hbar\omega}{\varepsilon_1}\right) \right\}, \tag{10.16}$$

where $\Gamma(n)$ is the gamma function. If only the first term inside the brackets remains Eq. (10.16) reverts to the classical Kramer's formula, which is the special case of the general classical formula [301].

The Born approximation is widely used [289, 54, 300] when the bremsstrahlung of fast electrons (ν_1, ν_2 << 1) is examined. The matrix element of the operator for an interaction between an electron and an electromagnetic field, Eq. (10.3), goes to zero when plane waves are used if the electron's momentum is not conserved during the transition; therefore, second-order perturbation theory must be used. The H' in the transition amplitude defined by the general equation, Eq. (1.252), is a perturbation operator, in this case equal to the sum of the operators of the interaction between the electron being scattered and the electromagnetic field (H') and the Coulomb interaction between this electron and the atom (H^e). A transition from the initial state $\mathbf{q}_1$, $n_{\mathbf{k}\rho} = 0$ into the final state $\mathbf{q}_2$, $n_{\mathbf{k}\rho} = 1$ can be made through the following intermediate states:

$$1)\ \mathbf{q}' = \mathbf{q}_1 - \mathbf{k},\ n_{\mathbf{k}\rho} = 1; \quad 2)\ \mathbf{q}'' = \mathbf{q}_2 + \mathbf{k},\ n_{\mathbf{k}\rho} = 0. \tag{10.17}$$

When this happens, the "surplus" momentum

$$\hbar\mathbf{q} = \hbar(\mathbf{q}_1 - \mathbf{q}_2 - \mathbf{k}) \tag{10.18}$$

is transferred to the nucleus (the laws of conservation of energy and momentum when a free electron emits a photon can only be satisfied when a third body takes part).

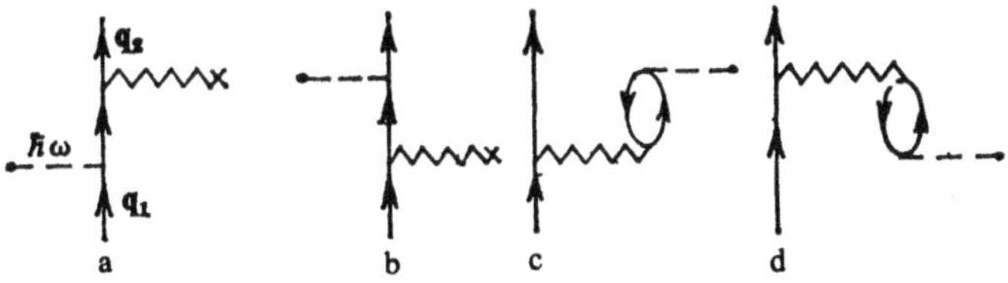

Fig. 10.1. Diagrams of the amplitudes of electron (a, b) and "atomic" (c, d) bremsstrahlung in lowest-order perturbation theory.

Only between states having the same number of photons does the matrix element for the H^e operator not go to zero; that of the H' operator does not go to zero if the electron's momentum is conserved. Further, the matrix element of H^e is given in much the same way as it was when the ionization of an atom by electrons was examined (Sec. 4.1), and the operator for H' is given through the use of Eq. (10.3).

The final expression for the cross section of electron bremsstrahlung scattering by a nucleus coincides with Eq. (10.13) in the Born approximation.

When screening of the nuclear field by the electron shells is taken into account the Born differential bremsstrahlung cross section $d^3\sigma/d\omega d\Omega_k d\Omega_2$ can be given in the form of the radiation cross section in the nuclear field multiplied by the correction factor

$$\left| 1 - \frac{1}{Z} F_e(q) \right|^2 . \qquad (10.19)$$

Here, $F_e(q)$ is the atomic form factor

$$F_e(q) = \langle \gamma | e^{i\mathbf{q}\cdot\mathbf{r}} | \gamma \rangle = \int \rho(\mathbf{r}) e^{i\mathbf{q}\cdot\mathbf{r}} d\mathbf{r}, \qquad (10.20)$$

where $\rho(\mathbf{r})$ is the probability density of finding the electrons in the atom.

The contribution from screening effects increases in the keV and MeV ranges as the velocity of the scattered electron decreases and the number of electrons in the atom increases [297, 300].

Dividing $\omega d\sigma_e/d\omega$ by v_1^2 yields a quantity that is a fairly weak function of the element's atomic number and the incident electron's energy [297]. As a function of $\hbar\omega/\varepsilon_1$ this quantity has rather flat form (when electron screening of the nucleus is not taken into consideration, it diverges logarithmically at the low-frequency limit $\hbar\omega/\varepsilon_1 \to 0$).

Two amplitude terms in the bremsstrahlung in a static approximation, which correspond to two intermediate states, Eq. (10.17), are shown by the diagrams of Fig. 10.1a,b. The small $\times$ denotes the atom's static field. The two other diagrams in this figure describe (in the lowest order, which is suitable for fast electrons, $\varepsilon_1 >> I_{nl}$) the "atomic" radiation. According to the diagram of Fig. 10.1d a virtual "vacancy−electron" pair is created by the interaction between a free electron and an atom; the electron then returns to the initial state and emits a photon.

By defining the contribution from the diagrams of Fig. 10.1c, d and using Eq. (4.26) in the matrix element of the h^e to perform an integration over the radius-vector of a free electron that can be described by a plane wave function, we obtain the following expression in the dipole approximation for the amplitude of "atomic" bremsstrahlung [296]

$$M_{at}(\omega, q) = -\frac{4\pi\omega}{q^2} \sum_{a \neq \varkappa} \frac{\langle a | e^{-i\mathbf{q}\cdot\mathbf{r}} | \varkappa \rangle \langle \varkappa | \hat{e} \cdot \mathbf{p} | a \rangle}{(\varepsilon_\varkappa - \varepsilon_a)^2 - \omega^2} \quad (\mathbf{q} = \mathbf{q}_1 - \mathbf{q}_2) \qquad (10.21)^*$$

When the energy of the emitted photon approaches the ionization energy of one of the atom's shells, $I_a \approx \varepsilon_\kappa - \varepsilon_a$, the amplitude increases resonantly, especially for small q. For these q, Eq. (10.21) can be simplified and the amplitude given, with the help of Eqs. (5.57) and (5.60), in terms of the atom's dynamic polarizability $\alpha(\omega)$ [296]:

$$M_{at}(\omega, q) = -\frac{4\pi\omega}{q^2} \alpha(\omega)\, \hat{e} \cdot \mathbf{q}. \qquad (10.22)^*$$

Estimates of M_{at}/M_{el} done in [295, 296] show that when $\hbar\omega \approx I_a$ this ratio may be much greater than unity. If $\hbar\omega >> I_a$, the amplitude of M_{at} for fast electrons is compensated by a term that describes screening, and the bremsstrahlung can then be examined in a static approximation, taking only the Coulomb field of the nucleus into account [296].

Consequently, when only a small amount of momentum is transferred, a structure caused by "atomic" radiation is imposed on the monotonically changing spectrum of electron bremsstrahlung near frequencies $\hbar\omega = I_a$.

When substituting both quantities into the process cross section and integrating over dq the interference term may be neglected, since different ranges of integration make the fundamental contribution to electron and "atomic" radiation [296].

The expression

$$\frac{d\sigma_a}{d\omega} = \frac{16}{3} \left(\frac{\omega}{c}\right)^3 \frac{|\alpha(\omega)|^2}{v_1^2} \ln \frac{v_1}{\omega \bar{r}}, \qquad (10.23)^*$$

is obtained for the "atomic" bremsstrahlung cross section for fast electrons [296], where $\bar{r}$ is the average radius of the subshell that makes the fundamental contribution to polarizability, and v_1 is the velocity of the incident electron.

Because the dynamic polarizability function is analytical, its real part, according to Cauchy's theorem, is given in terms of its imaginary part through the dispersion relation

$$\mathrm{Re}\,\alpha(\omega) = \frac{2}{\pi} P \int_0^\infty \frac{\mathrm{Im}\,\alpha(\omega')\,\omega'\,d\omega'}{(\omega')^2 - \omega^2} \tag{10.24}$$

(P denotes the principal value of the integral), and the imaginary part of $\alpha(\omega)$ is given by the optical theorem, Eq. (5.60), in terms of the photoabsorption cross section:

$$\mathrm{Im}\,\alpha(\omega) = \frac{c\,\sigma^{ph}(\omega)}{4\pi\omega}. \tag{10.25}$$

The relation between the "atomic" bremsstrahlung cross section and the photoabsorption cross section leads to the existence of similar structures in both spectra for the same values of photon energies. For example, the giant maxima in the $4d$-photoabsorption spectra of La and Xe caused by the positive potential barrier in the effective potential for a photoelectron correspond to quasiresonances of a similar shape in the bremsstrahlung spectra for these same elements [302, 303].

A resonance caused by the electron component of radiation may appear at the high-frequency limit of a bremsstrahlung spectrum [297, 304]. If an electron transfers almost all of its energy to a photon, then, when there is an effective potential having a positive barrier, a slow electron can form a metastable complex with a target atom (in other words, a quasidiscrete level having positive energy in an inner well may cause the free electron's wave function to collapse). The appearance of narrow maxima in the bremsstrahlung spectra of metallic La and Ce were explained in this manner [305]. Computation of the spectrum for a free La atom showed that a similar structure, caused by resonance in the scattered electron's f-partial wave, does indeed appear in the spectrum [304].

When the kinetic energy of the electron impinging on an atom is high, multipole and relativistic effects become significant in the bremsstrahlung spectrum [299, 297]. Comparing the results from relativistic calculations of $d\sigma/d\omega$ in the static approximation with Dirac–Fock–Slater wave functions of atoms $2 \le Z \le 92$ at electron kinetic energies of $1\ \mathrm{keV} \le \varepsilon_1 \le 2000\ \mathrm{keV}$ with results from nonrelativistic calculations according to Sommerfeld's formula shows [306] that the contribution from relativistic effects to the differential cross section at an energy of 50 keV is not more than 20%, whereas it reaches 100% at the high-frequency limit when the energy is 500 keV.

The expression for the electron bremsstrahlung cross section in relativistic theory is obtained the same way as in the nonrelativistic case. In doing this, the relativistic operator for an interaction between the electron and the electromagnetic field, Eq. (1.238), must be used and the density of states of the scattered electron is found, starting from the expression for its total energy (E_2). Then,

$$d\mathbf{q}_2 = \frac{q_2 E_2}{\hbar^3 c^3} dE_2 d\Omega_2, \tag{10.26}$$

$$df = \frac{V}{(2\pi)^3} \frac{q_2 E_2}{\hbar^3 c^3} k^2 dk\, d\Omega_k dE_2 d\Omega_2. \tag{10.27}$$

The wave function for a free electron in the initial state is normalized to unit electron flux density

$$\psi_i = \frac{1}{\sqrt{v_1}} (2\pi)^{\frac{3}{2}} \psi^+_{\mathbf{q}_1}(\mathbf{r}, \sigma) = \sqrt{\frac{(2\pi)^3 E_1}{\hbar q_1 c^2}} \psi^+_{\mathbf{q}_1}(\mathbf{r}, \sigma), \tag{10.28}$$

and the function in the final state is $\psi_{\mathbf{q}_1}{}^-$, normalized to $\delta(\mathbf{q}_1 - \mathbf{q}_1')$. The relativistic functions $\psi^\pm$ in an atom's spherical field have the form of expansions of Eq. (5.24) in terms of the generalized spherical harmonics of Eq. (1.177). Considering what has been said, the electron bremsstrahlung cross section is obtained in first-order perturbation theory as the following:

$$\frac{d^3\sigma_e}{d\omega\, d\Omega_k\, d\Omega_2} = \frac{(2\pi)^3 e^2}{c^4 \hbar^3} \frac{kE_1 E_2 q_2}{q_1} |\langle \psi^-_{\mathbf{q}_2} | \boldsymbol{\alpha} \cdot \hat{\mathbf{e}}_{\mathbf{k}p} e^{-i\mathbf{k}\cdot\mathbf{r}} | \psi^+_{\mathbf{q}_1} \rangle|^2. \tag{10.29}$$

Integration over the electron scattering angles is done the same way as in the nonrelativistic approximation – by substituting the expansions for $\psi^\pm$, Eq. (5.24), into Eq. (10.29) and using the orthonormality of the generalized spherical harmonics. The one-electron matrix elements of the relativistic operators for multipole transitions between the states of a free electron having specific angular momenta have the same spin-angular part as in a transition between discrete spectrum states, Eqs. (1.280) and (1.281). If unpolarized electrons are used in an experiment and the bremsstrahlung polarization is not measured, the cross section must be averaged over the values of the electron spin projection in the initial state and summed over two values of the spin projection in the final state and over both photon polarization directions. Finally, the differential cross section with respect to photon frequency is given by

$$\omega \frac{d\sigma_e}{d\omega} = \frac{2\pi^2 k^3 E_1 E_2}{q_1^3 q_2 c^3 \hbar^3} \sum_{j_1 l_1 j_2 l_2} \sum_t k^{2t} \frac{(2t+1)(t+1)}{[(2t+1)!!]^2 t} \times$$

$$\times \{ | (q_2 j_2 l_2 \| {}_eO^{(t)} \| q_1 j_1 l_1) |^2 + | (q_2 j_2 l_2 \|_m O^{(t)} \| q_1 j_1 l_1) |^2 \}, \tag{10.30}$$

where ${}_eO^{(t)}$ and ${}_mO^{(t)}$ are the one-electron operators for electric, Eq. (1.277), and magnetic, Eq. (1.278), multipole transitions; the wave functions have been normalized over the scale of the wave number.

Another equivalent expression for the differential cross section with respect to frequency was presented in [307] along with a relativistic expression for the differential cross section with respect to frequency and escape direction of the photon; however, they still contain a summation over the projection of the total angular momentum of the scattered electron.

The bremsstrahlung of a relativistic electron was examined with due regard for the dynamic polarizability of the complex atom that scatters it and rather simple expressions were obtained for the cross section at photon frequencies $\hbar\omega \gtrsim I_{nlj}$ and for the ultrarelativistic case of $\hbar\omega >> I_{nlj}$ [308].

The angular distribution of electron bremsstrahlung when $\varepsilon_1 < mc^2$ is similar to the distribution of electric dipole radiation having an intensity maximum directed perpendicular to the plane of electron motion ($\mathbf{q}_1$, $\mathbf{q}_2$). The radiation from an ultrarelativistic electron $\varepsilon_1 >> mc^2$ is concentrated along $\mathbf{q}_1$ in a cone whose angular divergence is equal to mc^2/ε_1.

10.2. X-Ray Scattering by Atoms

In practice, only the electron shells of an atom take part in x-ray scattering because the operator for an interaction between the atom and the electromagnetic field contains the reciprocal of the scattering particle's mass; therefore, the cross section for this process is inversely proportional to the square of its mass. Scattering from the nucleus only begins to appear for gamma rays at photon energies of 1 MeV and above, and can be ignored in the x-ray region [309, 310].

X-ray scattering can be subdivided into elastic (coherent or Rayleigh) scattering and inelastic (noncoherent) scattering. In elastic scattering the photon's energy is conserved and thus its frequency, but the direction of the photon's wave vector and/or its polarization may be altered. Inelastic scattering can be further subdivided into:

– Compton scattering, in which part of the quantum's energy is transferred to one of the atomic electrons, which then escapes from the atom;

– Raman scattering, in which inelastic scattering of a photon leads to the excitation of the atom to a discrete level (the photon's energy is altered by an amount equal to the transition energy in the atom).

In both cases the photon loses some of its energy; therefore, its frequency decreases.

Elastic scattering of x rays by atoms arranged in a certain order produces diffraction effects. X-ray diffraction by crystals and liquids is a familiar phenomenon. Even in a gas there are some correlations in the mutual arrangement of the atoms or molecules (they cannot come any closer to one another than some distance); therefore, weak interatom interference is possible even in a high-density gas [1, 311]. A gas may be thought of as a disordered conglomeration of atoms if the average distance between atoms is much greater than the wavelength of the scattered rays. In this case the total intensity of the scattered radiation is determined by a monatomic quantity – the scattering cross section σ_s. Similar to what we did for the absorption of x rays, we can introduce the linear scattering coefficient

$$\mu_s = \sigma_s N_0, \tag{10.31}$$

where N_0 is the number of atoms per unit volume of the gas. Instead of μ_s, the mass scattering coefficient [1]

$$\mu_m = \mu_s/\rho = \sigma_s/M, \tag{10.32}$$

is more often used, where M is the mass of the atom.

X-ray scattering in a given direction is described by the differential scattering cross section (the z axis is aligned with the direction of the primary beam of x rays):

$$d\sigma_s = \frac{dn(\omega_2;\ \theta,\ \varphi)}{j_1(\omega_1)}. \tag{10.33}$$

Here, $j_1(\omega_1)$ is the flux density of photons having a frequency of ω_2, $\omega_2 + d\omega_2$ impinging per unit time into the spatial angle Ω, $\Omega + d\Omega$;

$$dn(\omega_2;\ \theta,\ \varphi) = j_2(\omega_2;\ \theta,\ \varphi)\, d\omega_2\, dS = j_2(\omega_2;\ \theta,\ \varphi)\, R^2\, d\omega_2\, d\Omega, \tag{10.34}$$

where j_2 is the flux density of the scattered photons, R is the distance from the atom to the point at which they are measured, and dS and $d\Omega$ are elements of the area and the spatial angle, respectively.

By substituting Eq. (10.34) into (10.33) and using the relationship between flux density and intensity

$$I_1(\omega_1) = \hbar\omega_1 j_1(\omega_1),\quad I_2(\omega_2;\ \theta,\ \varphi) = \hbar\omega_2 j_2(\omega_2;\ \theta,\ \varphi), \tag{10.35}$$

we find

$$\frac{d^2\sigma_s}{d\Omega\, d\omega_2} = \frac{I_2(\omega_2;\ \theta,\ \varphi)\, R^2}{I_1(\omega_1)}\ \frac{\omega_1}{\omega_2}\quad \left(\int \frac{d^2\sigma_s}{d\Omega d\omega_2}\, d\Omega d\omega_2 = \sigma_s\right). \tag{10.36}$$

For elastic scattering, instead of Eq. (10.36), we have

$$\frac{d\sigma_{es}}{d\Omega} = \frac{I_s(\theta,\ \varphi)\, R^2}{I_1}. \tag{10.36a}$$

We will derive an expression for the differential scattering cross section in a nonrelativistic approximation. Photon scattering is a two-photon process: one quantum of x rays having frequency ω_1, wave vector $\mathbf{k}_1$, and polarization ρ_1 disappears and another photon $\omega_2\mathbf{k}_2\rho_2$ appears. The basic term in the operator for an interaction between an atom and electromagnetic field H_1', Eq. (1.240), is a one-photon operator; therefore, in first-order perturbation theory its matrix element makes no contribution to the scattering cross section. It is defined by the matrix element of the operator

$$H_2' = \frac{e^2}{2mc^2} \sum_j \mathbf{A}(\mathbf{r}_j)^2, \tag{10.37}$$

and by the contribution from the operator

$$H_1' = -\frac{e}{mc} \sum_j \mathbf{p}_j \cdot \mathbf{A}(\mathbf{r}_j) \tag{10.38}$$

in second-order perturbation theory [the summation in Eqs. (10.37) and (10.38) is carried out over the coordinates of all of the atom's electrons]. Both parts in the scattering amplitude are quadratic with respect to the vector potential and have the same order of magnitude.

Only the following term in the expansion of this operator for the initial state with one photon $\mathbf{k}_1\rho_1$ and the final state with one photon $\mathbf{k}_2\rho_2$ makes a contribution to the matrix element for this operator:

$$2\hat{e}_{\mathbf{k}_1\rho_1} \cdot \hat{e}_{\mathbf{k}_2\rho_2} a_{\mathbf{k}_1\rho_1} a^*_{\mathbf{k}_2\rho_2} e^{i(\mathbf{k}_1-\mathbf{k}_2)\cdot\mathbf{r}}. \tag{10.39}$$

Using the expressions for the matrix elements for the operators a, Eq. (1.242), and a^*, Eq. (1.243), and introducing the notation

$$\hat{e}_1 \equiv \hat{e}_{\mathbf{k}_1\rho_1},\quad \hat{e}_2 \equiv \hat{e}_{\mathbf{k}_2\rho_2},\quad \mathbf{k} = \mathbf{k}_1 - \mathbf{k}_2, \tag{10.40}$$

we can write the matrix element for the operator H_2' as

$$M'_{fi} = \langle f | H'_2 | i \rangle = \frac{2\pi\hbar e^2}{mV\sqrt{\omega_1\omega_2}} \hat{e}_1 \cdot \hat{e}_2 \langle \gamma_2 | \sum_j e^{i\mathbf{k}\cdot\mathbf{r}_j} | \gamma_1 \rangle, \tag{10.41}$$

where γ_1 and γ_2 are the atom's initial and final states, and $\hbar\mathbf{k}$ is the momentum transferred to the atom by the photon.

A transition such as $i \equiv \gamma_1\mathbf{k}_1\rho_1 \to f \equiv \gamma_2\mathbf{k}_2\rho_2$ under the action of the H_1' can only be made through intermediate states:

n, in which there are no photons and the atom is in an excited state γ;

n', in which there is at least one $\mathbf{k}_1\rho_1$ and $\mathbf{k}_2\rho_2$ photon and the atom is in a state γ'.

The corresponding matrix elements are

$$\langle n | H'_1 | i \rangle = -\frac{e}{m} \left[\frac{2\pi\hbar}{V\omega_1}\right]^{\frac{1}{2}} \langle \gamma | \sum_j \hat{e}_1 \cdot \mathbf{p}_j e^{i\mathbf{k}_1\cdot\mathbf{r}_j} | \gamma_1 \rangle, \tag{10.42}$$

$$\langle f | H'_1 | n \rangle = -\frac{e}{m} \left[\frac{2\pi\hbar}{V\omega_2}\right]^{\frac{1}{2}} \langle \gamma_2 | \sum_j \hat{e}_2 \cdot \mathbf{p}_j e^{-i\mathbf{k}_2\cdot\mathbf{r}_j} | \gamma \rangle, \tag{10.43}$$

$$\langle n' | H'_1 | i \rangle = -\frac{e}{m} \left[\frac{2\pi\hbar}{V\omega_2}\right]^{\frac{1}{2}} \langle \gamma' | \sum_j \hat{e}_2 \cdot \mathbf{p}_j e^{-i\mathbf{k}_2\cdot\mathbf{r}_j} | \gamma_1 \rangle, \tag{10.44}$$

$$\langle f | H'_1 | n' \rangle = -\frac{e}{m} \left[\frac{2\pi\hbar}{V\omega_1}\right]^{\frac{1}{2}} \langle \gamma_2 | \sum_j \hat{e}_1 \cdot \mathbf{p}_j e^{i\mathbf{k}_1\cdot\mathbf{r}_j} | \gamma' \rangle. \tag{10.45}$$

The process amplitude defined by the operator H_1' in second-order perturbation theory is, according to Eq. (1.254), equal to

$$M''_{if} = \sum_n \frac{\langle f | H'_1 | n \rangle \langle n | H'_1 | i \rangle}{E_i - E_n} + \sum_{n'} \frac{\langle f | H'_1 | n' \rangle \langle n' | H'_1 | i \rangle}{E_i - E_{n'}}, \tag{10.46}$$

where

$$E_i = E(\gamma_1) + \hbar\omega_1, \quad E_n = E(\gamma), \quad E_{n'} = E(\gamma') + \hbar\omega_1 + \hbar\omega_2. \tag{10.47}$$

The total amplitude when Eqs. (10.42)–(10.45) are substituted into (10.47) is obtained as

$$M_{fi}=\frac{2\pi e^2}{mV}\frac{\hbar}{\sqrt{\omega_1\omega_2}}\left\{\hat{e}_1\cdot\hat{e}_2\langle\gamma_2|\sum_j e^{i\mathbf{k}\cdot\mathbf{r}_j}|\gamma_1\rangle+\right.$$

$$+\frac{1}{m}\sum_{\gamma}\frac{\langle\gamma_2|\sum_j \hat{e}_2\cdot\mathbf{p}_j e^{-i\mathbf{k}_2\cdot\mathbf{r}_j}|\gamma\rangle\langle\gamma|\sum_j \hat{e}_1\cdot\mathbf{p}_j e^{i\mathbf{k}_1\cdot\mathbf{r}_j}|\gamma_1\rangle}{E(\gamma_1)-E(\gamma)+\hbar\omega_1}+$$

$$\left.+\frac{1}{m}\sum_{\gamma'}\frac{\langle\gamma_2|\sum_j \hat{e}_1\cdot\mathbf{p}_j e^{i\mathbf{k}_1\cdot\mathbf{r}_j}|\gamma'\rangle\langle\gamma'|\sum_j \hat{e}_2\cdot\mathbf{p}_j e^{-i\mathbf{k}_2\cdot\mathbf{r}_j}|\gamma_1\rangle}{E(\gamma_1)-E(\gamma')-\hbar\omega_2}\right\}. \quad (10.48)$$

In order that the expression remain valid for the photon energy at which the denominator goes to zero, the energy of the intermediate level must be replaced by the complex quantity [289, 309, 310]

$$E(\gamma)\to E(\gamma)-\frac{i\Gamma(\gamma)}{2}, \quad (10.49)$$

where $\Gamma(\gamma)$ is the natural width of this level [cf. Eq. (2.87)].

Let γ_2 be a discrete spectrum state. Then, when $\gamma_1 = \gamma_2$ and $\omega_1 = \omega_2$, Eq. (10.48) describes elastic scattering, and when $\gamma_1 \neq \gamma_2$ and $\omega_1 \neq \omega_2$ it describes Raman scattering. Equations for the cross sections of these processes are obtained by substituting the amplitude, Eq. (10.48), into the formula for the transition probability, Eq. (1.244), and using the relationship between the cross section and the probability, Eq. (1.245). There is only one free particle in the final state – a photon; therefore, we must take

$$df=\frac{V}{(2\pi)^3}k_2^2 d\Omega\, dk_2. \quad (10.50)$$

Because there is a δ-function of energy, integration over dk_2 is reduced to the auxiliary factor $(\hbar c)^{-1}$. Finally,

$$\frac{d\sigma_s(\gamma_1\to\gamma_2)}{d\Omega}=\left(\frac{e^2}{mc^2}\right)^2\frac{\omega_2}{\omega_1}\left|\hat{e}_1\cdot\hat{e}_2\langle\gamma_2|\sum_j e^{i\mathbf{k}\cdot\mathbf{r}_j}|\gamma_1\rangle+\right.$$

$$+\frac{1}{m}\sum_{\gamma}\frac{\langle\gamma_2|\sum_j \hat{e}_2\cdot\mathbf{p}_j e^{-i\mathbf{k}_2\cdot\mathbf{r}_j}|\gamma\rangle\langle\gamma|\sum_j \hat{e}_1\cdot\mathbf{p}_j e^{i\mathbf{k}_1\cdot\mathbf{r}_j}|\gamma_1\rangle}{E(\gamma_1)-E(\gamma)+\hbar\omega_1}+$$

$$\left.+\frac{1}{m}\sum_{\gamma'}\frac{\langle\gamma_2|\sum_j \hat{e}_1\cdot\mathbf{p}_j e^{i\mathbf{k}_1\cdot\mathbf{r}_j}|\gamma'\rangle\langle\gamma'|\sum_j \hat{e}_2\cdot\mathbf{p}_j e^{-i\mathbf{k}_2\cdot\mathbf{r}_j}|\gamma_1\rangle}{E(\gamma_1)-E(\gamma')-\hbar\omega_2}\right|^2, \quad (10.51)$$

where e^2/mc^2 is equal to the classical electron radius r_0. The first term in the scattering amplitude is called normal, and the terms in the denominator that contain energy are called dispersion, or anomalous terms.

For soft x rays whose wavelengths exceed the size of the atom the exponential terms in Eq. (10.51) are practically constant inside the atom. Then,

$$\frac{d\sigma_s(\gamma_1 \to \gamma_2)}{d\Omega} = r_0^2 \frac{\omega_2}{\omega_1} \Bigg| \hat{e}_1 \cdot \hat{e}_2 \delta(\gamma_1, \gamma_2) + \tag{10.52}$$

$$+ \frac{1}{m} \sum_{\gamma} \frac{\langle \gamma_2 | \sum_j \hat{e}_2 \cdot \mathbf{p}_j | \gamma \rangle \langle \gamma | \sum_j \hat{e}_1 \cdot \mathbf{p}_j | \gamma_1 \rangle}{E(\gamma_1) - E(\gamma) + \hbar\omega_1} +$$

$$+ \frac{1}{m} \sum_{\gamma'} \frac{\langle \gamma_2 | \sum_j \hat{e}_1 \cdot \mathbf{p}_j | \gamma' \rangle \langle \gamma' | \sum_j \hat{e}_2 \cdot \mathbf{p}_j | \gamma_1 \rangle}{E(\gamma_1) - E(\gamma') - \hbar\omega_2} \Bigg|^2 .$$

Because the wave functions are orthogonal, a factor $\delta(\gamma_1, \gamma_2)$ appears; consequently, in this approximation Raman scattering is described only by dispersion terms.

As the x-ray wavelengths λ decrease, the exponents in Eq. (10.51) cease to be constant quantities inside the atom. The shorter the λ, the faster the function $e^{i\mathbf{k}\cdot\mathbf{r}}$ oscillates, which reduces the elastic and Raman scattering cross sections [289].

The role of the dispersion terms increases in a resonant manner as the incident photon's energy approaches the atom's excitation energy. This phenomenon is called anomalous or resonance scattering. As we move away from typical atomic frequencies x-ray scattering is determined mainly by the normal term and can be approximated as follows:

$$\frac{d\sigma_s(\gamma_1 \to \gamma_2)}{d\Omega} = r_0^2 \frac{\omega_2}{\omega_1} (\hat{e}_1 \cdot \hat{e}_2)^2 \Big| \langle \gamma_2 | \sum_j e^{i\mathbf{k}\cdot\mathbf{r}_j} | \gamma_1 \rangle \Big|^2 . \tag{10.53}$$

If $\omega_1 = \omega_2$, the multiplier on the square of the absolute value of the amplitude coincides with the cross section of x-ray scattering by a free classical electron (this scattering can only be elastic)

$$\left(\frac{d\sigma_s}{d\Omega} \right)_0 = r_0^2 (\hat{e}_1 \cdot \hat{e}_2)^2 . \tag{10.54}$$

Equation (10.54) is called Thomson's formula.

The differential cross section given by Eq. (10.53), just like Eq. (10.54), reaches a maximum when the incident and scattered photons are polarized in the same or opposite directions.

We will expand the vector $\hat{\epsilon}_2$ into two independent components: $\hat{\epsilon}_{2\perp}$, which is perpendicular to $\hat{\epsilon}_1$, and $\hat{\epsilon}_{2\parallel}$, which is in the plane $(\mathbf{k}_2, \hat{\epsilon}_1)$ $(\epsilon_{2\parallel} \perp \epsilon_{2\perp})$. We designate φ as the angle between the planes $(\mathbf{k}_1, \mathbf{k}_2)$ and $(\mathbf{k}_2, \epsilon_1)$ and θ as the scattering angle (the angle between the directions of incident and scattered photons). This allows us to write

$$(\hat{\epsilon}_1 \cdot \hat{\epsilon}_2)^2 = 1 - \sin^2\theta \cos^2\varphi . \tag{10.55}$$

If unpolarized x rays are scattered, Eq. (10.55) must be averaged over the direction of the first photon, i.e., over the angle φ. Considering that

$$\overline{\cos^2\varphi} = 1/2 \tag{10.56}$$

the dependence of the angular scattering cross section for unpolarized x rays on the scattering angle is described by the factor $(1 + \cos^2\theta)$:

$$\frac{d\sigma_s}{d\Omega} = \frac{r_0^2}{2} (1 + \cos^2\theta) \left| \langle \gamma | \sum_j e^{i\mathbf{k}\cdot\mathbf{r}_j} | \gamma \rangle \right|^2 . \tag{10.57}$$

This type of angular distribution is typical of the scattering by a classical electron:

$$\left(\frac{d\sigma_s}{d\Omega}\right)_0 = \frac{r_0^2}{2} (1 + \cos^2\theta) . \tag{10.58}$$

Integrating Eq. (10.58) with respect to the angles we obtain an expression for total Thomson scattering:

$$\sigma_0 = \frac{8\pi}{3} r_0^2 \approx 6.65 \cdot 10^{-27}\ \mathrm{m}^2 . \tag{10.59}$$

According to Eqs. (10.57) and (10.58), the cross section for elastic scattering by an atom can be given in terms of the cross section for scattering by a free classical electron

$$\frac{d\sigma_{el}}{d\Omega} = \left(\frac{d\sigma_s}{d\Omega}\right)_0 |f|^2 , \tag{10.60}$$

where f is the atomic form factor that contains the entire dependence of the scattering on the structure of the atom

$$f=\sum_{nl} N_{nl}\langle nl|j_0|nl\rangle. \tag{10.61}$$

The operator $e^{i\mathbf{k}\cdot\mathbf{r}}$ is expanded in a series over the spherical function operators $C^{(t)}$ of different rank t, Eq. (4.31). If the scattered atom has spherical symmetry, the atom's total angular momentum J must be conserved during the scattering process and the amplitude must not be a function of its projection M [78]. For this reason, only the scalar portion – the single term with $t = 0$ – must be considered in the expansion of the operator. The matrix element for the scalar operator, according to Sec. 1.3, is equal to the sum of the one-electron matrix elements, each of which has been multiplied by the number of electrons in the respective shell:

$$f=\langle\gamma|\sum_{j} e^{i\mathbf{k}\cdot\mathbf{r}_j}|\gamma\rangle. \tag{10.62}$$

Here,

$$\langle nl|j_0|nl\rangle=\int_0^\infty P_{nl}^2(r)j_0(kr)\,dr=\int_0^\infty P_{nl}^2(r)\frac{\sin kr}{kr}\,dr, \tag{10.63}$$

$$k=|\mathbf{k}_1-\mathbf{k}_2|=2k_1\sin\frac{\theta}{2}=\frac{4\pi}{\lambda_1}\sin\frac{\theta}{2}, \tag{10.64}$$

where j_0 is a spherical Bessel function of order zero, and θ is the angle between the wave vectors of the incident and scattered photons.

The field of an atom having open shells is not completely spherically symmetric; however, because the model is approximate this distortion can be neglected and Eq. (10.62) can also be used.

When $\theta = 0$ (forward scattering), we have

$$\langle nl|j_0|nl\rangle=1,\quad f=\sum_{nl} N_{nl}=N, \tag{10.65}$$

i.e., in the model being used the atom scatters elastically forward as N free electrons.

This result, like the physical sense of the form factor, becomes clearer if the form factor of Eq. (10.62) is given as a spherically averaged probability density of finding electrons in the atom $\rho(r)$, Eq. (1.158):

$$f = 4\pi \int \rho(r) \frac{\sin kr}{kr} r^2 dr. \tag{10.66}$$

This formula was first obtained classically as a factor that corrects the scattering cross section of the atom's electron charge distribution. Forward scattering of electromagnetic waves by the various points of the atom does not produce a phase shift; therefore, the distributed charge scatters as N point electrons in this direction. When observed at a different angle the waves scattered by the various points of the atom travel unequal paths and are added with a phase shift; therefore,

$$f(\theta) \leqslant f(0) = N. \tag{10.67}$$

At small scattering angles each shell i makes a contribution to the form factor $f_i(k)$, which is approximately equal to the number of electrons in the shell. As follows from the results of systematic calculations of the elastic scattering cross section [312], $f_i(k)$ is weakly dependent on the transferred momentum, while $\hbar k$ is less than the average momentum of an electron in an orbit having a principal quantum number n:

$$k < k_n \approx \frac{1}{\hbar a_n} = \frac{Z\alpha mc}{n^2 \hbar^2}, \tag{10.68}$$

where a_n is the radius of the nth Bohr orbit.

When k exceeds k_n, $f_i(k)$ begins to decrease rapidly. Because k_n is inversely proportional to the radius of the respective orbit, the contribution to scattering from the outer, and then the intermediate, shells initially decreases as k increases and when the amount of transferred momentum is large, x rays are scattered mainly by the deepest K-shell.

The approximation of the form factor loses its validity as the role of dispersion terms increases, which happens at photon energies near the values of the electron binding energies in the atom and, for large values of transferred momentum and large angles θ, when the normal term is small.

For forward elastic scattering, when the polarization is unchanged, dispersion terms can be considered as a correction to the form factor. This correction has real and imaginary parts in the denominator of Eq. (10.51):

$$f(\omega) = N + f'(\omega) + if''(\omega). \tag{10.69}$$

The imaginary part f'' is given in the dipole approximation as the atom's dynamic polarizability and, via the optical theorem, Eq. (5.60), in terms of the atom's photoabsorption cross section $\sigma^{ph}(\omega)$ [309, 313]:

$$f''(\omega) = \frac{\omega}{4\pi c r_0} \sigma^{ph}(\omega). \tag{10.70}$$

Because the scattering amplitude is analytic, the real part $f'(\omega)$ is associated with $f''(\omega)$ by a dispersion relation and can therefore be given in terms of $\sigma^{ph}(\omega)$:

$$f'(\omega) = \frac{2}{\pi} P \int_0^\infty \frac{\omega_1 f''(\omega_1)}{\omega_1^2 - \omega^2} d\omega_1 = \frac{1}{2\pi^2 c r_0} P \int_0^\infty \frac{\omega_1^2 \sigma^{ph}(\omega)}{\omega_1^2 - \omega^2} d\omega_1. \tag{10.71}$$

The atomic form factor for the forward elastic scattering of x rays makes it possible for us to find the index of refraction for these rays [309],

$$n(\omega) = 1 - 2\pi \left(\frac{c}{\omega}\right)^2 r_0 \sum_a N_a f_a(\omega), \tag{10.72}$$

where N_a is the number of a atoms per unit volume of the gas, and $f_a(\omega)$ is the corresponding form factor for forward elastic scattering.

Equation (10.69) also approximately describes scattering at small angles [$f'(\omega)$ and $f''(\omega)$ are weakly dependent on the scattering angle θ]. Sometimes Eq. (10.69) is extended to larger angles for soft x rays [313], although this is done without sufficient theoretical foundation [314, 315].

Almost all calculations for normal and anomalous x-ray scattering are done in the one-electron approximation; good agreement between these results and experimental data attest to the fact that the individual electrons scatter x rays almost independently of one another [312]. Notice that in such an approximation, dispersion terms may yield nonphysical resonances corresponding to transitions that are forbidden by the Pauli principle into closed shells; however, they go to zero when summed over all of the atom's electrons [312, 315]. Correlation effects make a noticeable contribution to the total cross section and to the angular distribution, mainly when the energies of the photons being scattered are low – 100 eV and below [316, 315]. Their relative influence decreases for heavy atoms having a large number of inner shells.

On the other hand, relativistic and multipole effects become important as the energy of the x rays increases (from roughly 20 keV for $_{13}$Al and 150 keV for $_{82}$Pb [312, 309]).

Expressions for the x-ray amplitudes and scattering cross sections in relativistic theory are obtained in the same way as in a nonrelativistic approximation. The relativistic operator for an interaction between an electromagnetic field and an atom, Eq. (1.238), contains only a single term, linear in **A**. Thus, the unknown equation can easily be obtained from Eq. (10.51), by omitting the normal term, replacing **p** with $\boldsymbol{\alpha}$ in the dispersion terms, and multiplying the cross section by an auxiliary factor $(mc)^4$ that considers the difference between the constants that appear in the operators H', Eq. (1.238), and H_1', Eq. (10.38). Also, the wave functions, the energy, and the mass of an electron are all changed by relativistic quantities:

$$\frac{d\sigma_s(\gamma_1 \to \gamma_2)}{d\Omega} = e^4 \frac{\omega_2}{\omega_1} \left| \sum_{\gamma} \frac{(\gamma_2 | \sum_j \hat{e}_2 \cdot \boldsymbol{\alpha}_j e^{-i\mathbf{k}_2 \cdot \mathbf{r}_j} | \gamma)(\gamma | \sum_j \hat{e}_1 \cdot \boldsymbol{\alpha}_j e^{i\mathbf{k}_1 \cdot \mathbf{r}_j} | \gamma_1)}{E(\gamma_1) - E(\gamma) + \hbar\omega_1} + \right.$$
$$\left. + \sum_{\gamma'} \frac{(\gamma_2 | \sum_j \hat{e}_1 \cdot \boldsymbol{\alpha}_j e^{i\mathbf{k}_1 \cdot \mathbf{r}_j} | \gamma')(\gamma' | \sum_j \hat{e}_2 \cdot \boldsymbol{\alpha}_j e^{-i\mathbf{k}_2 \cdot \mathbf{r}_j} | \gamma_1)}{E(\gamma_1) - E(\gamma') - \hbar\omega_2} \right|^2. \quad (10.73)$$

The summation in Eq. (10.73) must be done for electron states having positive and negative energy. A normal or form factor term appears in the nonrelativistic limit from sums of the first type, and dispersion terms appear from sums of the second type [78].

The approximately relativistic corrections to the normal term when $\gamma_1 = \gamma_2$ can be accounted for by introducing a modified form factor, the contribution to which from an *nlj* electron is [315]

$$f_{nlj}(k) = 4\pi \int_0^\infty \rho_{nlj}(r) \frac{\sin kr}{kr} \frac{mc^2}{mc^2 - I_{nlj} - V_{nlj}(r)} r^2 \, dr. \quad (10.74)$$

Here, I_{nlj} is the binding energy of an *nlj* electron, $V_{nlj}(r)$ is the potential energy of an *nlj* electron, and $\rho_{nlj}(r)$ is the relativistic probability density of finding the electron in the atom:

$$\rho_{nlj}(r) = \frac{1}{4\pi r^2} [P^2_{nlj}(r) + Q^2_{nlj}(r)]. \quad (10.75)$$

Taking relativistic effects into consideration attracts the wave functions of the inner electrons to the nucleus and, thus, increases the form factor and the elastic scattering cross section [315].

In comparison with elastic scattering, the Raman scattering of x rays is a weak and as yet little studied effect. In Raman scattering the quantum energy for all scattering angles changes by a discrete amount – the excitation energy of the atom:

$$\hbar\omega_k = \hbar\omega - (E_n - E_0) \tag{10.76}$$

(E_0 is the energy of the atom's ground state, and E_n is the energy in the excited state); therefore, lines having a shifted frequency may appear when monochromatic rays act on the atom. These lines are distinguished from emission lines by the fact that they are functions of not only the structure of the atom's energy levels, but of the frequency of the photons being scattered as well. A weak Raman line is ordinarily seen at the tail of a Compton scattering maximum [10].

Even more intense resonance Raman lines can appear when the energy of the rays being scattered is approximately equal to the binding energy of an electron in the atom's inner shell [317, 310]. The resonance Raman effect is due to the term in Eq. (10.51) containing the denominator $E(\gamma_1) - E(\gamma) + \hbar\omega_1$. Resonance scattering occurs at an energy of $\hbar\omega_1 \approx E(\gamma) - E(\gamma_1)$ through an intermediate virtual state $\gamma \equiv n_1 l_1^{-1} nl$ having a vacancy in an inner shell and an excited electron nl, into the final state of the system $n_2 l_2^{-1} nl + \hbar\omega_2$. (For example, a resonance Raman scattering maximum appears when x rays having $\hbar\omega_1 \approx I_{1s} - I_{3p}$ are scattered by neon, an ion in the $2p^{-1}3p$ state and a photon $\hbar\omega_2 = I_{1s} - I_{2p}$ are created; the intermediate state is $1s^{-1}3p$.) An intermediate state can also decay in a radiationless process (the final state contains an electron in the continuous spectrum) – the corresponding process is called a resonance Raman Auger effect [318].

Inelastic Compton scattering from the electrons in a shell becomes possible if the photon energy is greater than the shell's ionization energy. Because a free electron appears in the scattering process, the shifted line has a broad intensity distribution [289].

The Compton scattering amplitude in a nonrelativistic approximation is described by the same general formula, Eq. (10.48), except that γ_2 corresponds to a state in the ion + free electron system.

The role played by Compton scattering increases in comparison with elastic Rayleigh scattering as the photon frequency increases. This is because the oscillating term $e^{i\mathbf{k}\cdot\mathbf{r}}$ in the Compton scattering amplitude is compensated by the complex conjugate wave function of the free electron $e^{-i\mathbf{q}_2\cdot\mathbf{r}}$ (if its momentum is much greater than the momentum of an electron in the initial bound state, $\mathbf{q}_2 \approx \mathbf{k}_1 - \mathbf{k}_2 = \mathbf{k}$).

If the energy that a photon transfers to an atom's electron is much greater than the electron's binding energy, this electron will scatter x rays in much the same way that a free electron scatters x rays – the contribution from dispersion terms to the scattering amplitude will be insignificant. Since it is mainly when the transferred momenta and photon scattering energies are high that the Compton effect becomes the dominant process in x-ray scattering, only the normal term in the amplitude is ordinarily considered when describing this effect. Omitting dispersion terms from the matrix element in Eq. (10.48), substituting it into Eq. (1.244), and putting it into concrete form for the final state of a system containing a photon and a free electron, we find

$$df=\frac{V}{(2\pi)^3}\,k_2^2\,dk_2\,d\Omega\,d\mathbf{q}_2, \tag{10.77}$$

(for brevity's sake, we will no longer consider the electron's spin):

$$\frac{d^3\sigma_C}{d\omega_2\,d\Omega\,d\mathbf{q}_2}=\hbar r_0^2\,\frac{\omega_2}{\omega_1}\,(\hat{\epsilon}_1\cdot\hat{\epsilon}_2)^2\sum_{\gamma_2}\left|\langle\gamma_2\,\mathbf{q}_2|\sum_j e^{i\mathbf{k}\cdot\mathbf{r}_j}|\gamma_1\rangle\right|^2\times$$

$$\times\,\delta\left(E(\gamma_1)+\hbar\omega_1-E(\gamma_2)-\hbar\omega_2-\frac{\hbar^2q_2^2}{2m}\right), \tag{10.78}$$

where $\mathbf{q}_2$ is the wave vector of the free electron, γ_1 and γ_2 are the atom's initial and final states, ω_1 and ω_2 are the frequencies of the incident and scattered photons, $\hat{\epsilon}_1$ and $\hat{\epsilon}_2$ are its unit polarization vectors, $d\Omega$ is an element of the scattered photon's spatial escape angle, and $\mathbf{k}=\mathbf{k}_1-\mathbf{k}_2$.

The differential cross section can be written in terms of the scattering cross section of a classical electron, Eq. (10.54):

$$\frac{d^2\sigma_C}{d\omega_2\,d\Omega}=\left(\frac{d\sigma_s}{d\Omega}\right)_0\hbar\,\frac{\omega_2}{\omega_1}\sum_{\gamma_2}\int\left|\langle\gamma_2\,\mathbf{q}_2|\sum_j e^{i\mathbf{k}\cdot\mathbf{r}_j}|\gamma_1\rangle\right|^2\times$$

$$\times\,\delta\left(E(\gamma_1)-E(\gamma_2)+\hbar(\omega_1-\omega_2)-\frac{\hbar^2q_2^2}{2m}\right)d\mathbf{q}_2. \tag{10.79}$$

The impulse approximation [14, 319, 320] is widely used in calculating the Compton scattering cross section. This approximation follows from Eq. (10.79) when a number of assumptions are made: 1) an atom's electrons scatter x rays independently of one another; 2) the wave function of a Compton electron corresponds to the plane wave

$$\psi_{\mathbf{q}_2}(\mathbf{r})=(2\pi)^{-\frac{3}{2}}\,e^{i\mathbf{q}_2\cdot\mathbf{r}}; \tag{10.80}$$

3) we can omit the binding energy of an electron $E(\gamma_2)-E(\gamma_1)$ in the δ-function, which is much less than the transferred energy.

The one-electron matrix element of the $e^{i\mathbf{k}\cdot\mathbf{r}}$ operator for Eq. (10.80) is

$$\langle \mathbf{q}_2 | e^{i\mathbf{k}\cdot\mathbf{r}} | a\rangle = (2\pi)^{-\frac{3}{2}} \int \psi_a(\mathbf{r})\, e^{i(\mathbf{k}-\mathbf{q}_2)\cdot\mathbf{r}}\, d\mathbf{r}. \tag{10.81}$$

According to the law of conservation of momentum, $\hbar(\mathbf{q}_2 - \mathbf{k})$ is equal to the electron's momentum in the initial state $\hbar\mathbf{q}_1$, and the right-hand side of Eq. (10.81) is (to within $\hbar^{-3/2}$) equal to the wave function of an electron a in the momentum representation:

$$u_a(\mathbf{p}) = \hbar^{-3/2} u_a(\mathbf{q}), \quad u_a(\mathbf{q}) = (2\pi)^{-\frac{3}{2}} \int \psi_a(\mathbf{r})\, e^{-i\mathbf{q}\cdot\mathbf{r}}\, d\mathbf{r}. \tag{10.82}$$

Consequently,

$$\frac{d^2\sigma_C}{d\omega_2 d\Omega} = \left(\frac{d\sigma_s}{d\Omega}\right)_0 \hbar \frac{\omega_2}{\omega_1} \sum_a \int |u_a(\mathbf{q}_1)|^2\, \delta\!\left(\hbar\omega_1 - \hbar\omega_2 - \frac{\hbar^2 q_1^2}{2m}\right) d\mathbf{q}_2. \tag{10.83}$$

Using the laws of conservation of energy and momentum and the smallness of $\mathbf{q}_1$ in comparison with $\mathbf{k}$, we obtain

$$\omega_1 - \omega_2 \approx \frac{\hbar q_2^2}{2m} = \frac{\hbar(\mathbf{q}_1 + \mathbf{k})^2}{2m} \simeq \frac{\hbar}{2m}(k^2 + 2\mathbf{q}_1\cdot\mathbf{k}) \tag{10.84}$$

or, by aligning the z axis with the wave vector $\mathbf{k}$, we have

$$\omega_1 - \omega_2 = \frac{\hbar}{2m}(k^2 + 2kq_{1z}). \tag{10.85}$$

For wavelengths which satisfy the condition that $\lambda_1, \lambda_2 >> \lambda_2 - \lambda_1$ this relation, with due regard for Eq. (10.64), becomes

$$\Delta\lambda = \lambda_2 - \lambda_1 = \frac{2h}{mc}\sin^2\frac{\theta}{2} + \frac{2\hbar\lambda_1}{mc}\sin\frac{\theta}{2}\, q_{1z}, \tag{10.86}$$

where θ is the photon scattering angle, and λ_1 and λ_2 are the wavelengths before and after scattering, respectively.

The first term in Eq. (10.86) coincides with the familiar formula for Compton scattering by a free, motionless electron. This shift is independent of the photon's frequency and, when $\theta = \pi/2$, is equal to a universal constant – the Compton wavelength

$$\lambda_C = h/mc = 0.00243 \text{ nm}. \tag{10.87}$$

The second term in Eq. (10.86) describes the additional shift in the wavelength due to a momentum component of the atom's electron in the direction of the vector $\mathbf{k}$. Distribution of electrons in an atom according to their momenta leads to broadening in the shifted line and determines its shape (the Compton profile).

That $d\mathbf{q}_2 = d\mathbf{q}_1$ for given $\mathbf{k}_1$ and $\mathbf{k}_2$ follows from the relationship $\mathbf{q}_2 = \mathbf{q}_1 + \mathbf{k}$. The δ-function in Eq. (10.83) limits integration over dq_1 in the momentum space by a plane that is perpendicular to the transferred momentum. The Compton scattering cross section in the impulse representation becomes [320]

$$\frac{d^2\sigma_C}{d\omega_2 d\Omega}=\left(\frac{d\sigma_s}{d\Omega}\right)_0 \frac{\omega_2}{\omega_1}\frac{m}{\hbar k} J(Q), \tag{10.88}$$

where Q is the projection of the wave vector of an atom's electron in the direction of the photon's transferred momentum, which is defined according to Eq. (10.85) by

$$Q=\frac{\mathbf{q}\cdot\mathbf{k}}{k}=\frac{m}{\hbar k}(\omega_1-\omega_2)-\frac{k}{2}. \tag{10.89}$$

$J(Q)$ is the distribution function for the electron wave vectors in the atom

$$J(Q)=\sum_a J_a(Q), \tag{10.90}$$

which is the sum of the distribution functions of the individual electrons

$$J_a(Q)\equiv J_a(q_z)=\int\int |u_a(\mathbf{q})|^2\, dq_x\, dq_y. \tag{10.91}$$

The equation for $u_a(\mathbf{q})$ can be simplified by using factorization of the coordinate wave function, Eq. (1.6), and expanding $e^{-i\mathbf{q}\cdot\mathbf{r}}$ over the spherical functions, Eq. (4.31):

$$u_{nlm}(\mathbf{q})=\sqrt{\frac{2}{\pi}}\, i^{-l}\, Y_{lm}(\hat{\mathbf{q}})\int_0^\infty P_{nl}(r)\, j_l(qr)\, r\, dr. \tag{10.92}$$

For an atom with closed shells we can write $J(Q)$ in terms of the one-dimensional density $|u_{nl}(q)|^2$, which is a function of only the absolute value of q:

$$J(Q)=\sum_{nl} N_{nl} J_{nl}(Q),\quad J_{nl}(Q)=\frac{1}{2}\int_{|Q|}^\infty |u_{nl}(q)|^2\, q\, dq, \tag{10.93}$$

$$u_{nl}(q)=\sqrt{\frac{2}{\pi}}\int_0^\infty P_{nl}(r)j_l(qr)\,r\,dr. \tag{10.94}$$

This is done by summing the $|u_{nlm}(\mathbf{q})|^2$ over the m and using Eq. (5.20) and, when open shells are present, by averaging the $|u_{nlm}(\mathbf{q})|^2$ over the angles.

The impulse approximation was first obtained, starting from a simple semiclassical model: an atom's electrons scatter x rays like free electrons, and their binding manifests itself only through the distribution of the momenta of the atom's electrons.

Relativistic effects for heavy atoms can be approximately accounted for by calculating the Compton profile $J(Q)$ using the relativistic wave functions

$$J_{nlj}(Q)=\frac{1}{2}\int_{|Q|}^\infty |u_{nlj}(q)|^2\,q\,dq=\frac{1}{2}\int_{|Q|}^\infty [|u^P_{nlj}(q)|^2+|u^Q_{nlj}(q)|^2]\,q\,dq, \tag{10.95}$$

where u^P and u^Q are the large and the small components of a one-electron wave function in momentum space:

$$u^P_{nlj}(q)=\sqrt{\frac{2}{\pi}}\int_0^\infty P_{nlj}(r)j_l(qr)\,r^2\,dr, \tag{10.96}$$

$$u^Q_{nlj}(q)=\begin{cases}\sqrt{\frac{2}{\pi}}\int_0^\infty Q_{n\bar{l}j}(r)j_{l+1}(qr)\,r^2\,dr & \left(j=l+\frac{1}{2}\right),\\ \sqrt{\frac{2}{\pi}}\int_0^\infty Q_{n\bar{l}j}(r)j_{l-1}(qr)\,r^2\,dr & \left(j=l-\frac{1}{2}\right).\end{cases} \tag{10.97}$$

A subsequent examination of Compton scattering in relativistic theory leads to an improvement in not only $J(Q)$, but in the coefficients on it in the cross section equation [321]. For the case $\theta = 2\pi$ the differential cross section becomes, to within $\hbar^2 q_1^2/m^2c^2$,

$$\frac{d^2\sigma_C}{d\omega_2\,d\Omega}=\frac{r_0^2\,mc\omega_2\left(1+\frac{\hbar\omega_1}{mc^2}\right)B}{\hbar A_1(\omega_1+\omega_2)^2\left(1+\frac{\hbar\omega}{mc^2}\right)}J(Q), \tag{10.98}$$

where

$$A_1 = 1 + \frac{Q}{mc}, \quad A_2 = 1 - \frac{Q}{mc}, \tag{10.99}$$

$$B = \frac{\omega_1 A_1}{\omega_2 A_2} + \frac{\omega_2 A_2}{\omega_1 A_1} - \frac{4}{A_1 A_2} + \frac{4}{A_1^2 A_2^2}. \tag{10.100}$$

The Compton profiles $J_{nlj}(Q)$, calculated by means of the Hartree–Fock wave functions for free atoms of the elements $1 \leq Z \leq 36$ and by using relativistic wave functions for the elements $37 \leq Z \leq 102$, have been tabulated in [320]. Relativistic effects lead to flattening of the Compton profile.

A general equation for the differential Compton scattering cross section at free, spin-polarized electrons was derived in [322]. We will present only its special case – the familiar Klein–Nishina–Tamm formula that describes the cross section for Compton scattering of a photon by a free unpolarized electron, averaged over the initial and summed over the final polarizations of the photon and the electron:

$$\frac{d\sigma_C}{d\Omega} = \frac{r_0^2}{2}\left(\frac{\omega_2}{\omega_1}\right)^2 \left\{ \frac{\omega_1}{\omega_2} + \frac{\omega_2}{\omega_1} - \sin^2\theta \right\}, \tag{10.101}$$

where θ is the photon scattering angle, and ω_2 is related to ω_1 by Eq. (10.85). When $\omega_1 = \omega_2$ Eq. (10.101) becomes the Thomson formula, Eq. (10.58).

When the energy of the x rays is about the same as the electron binding energy in the atom, second-order dispersion terms in the process amplitude begin to play an important role in Compton scattering. These terms cause resonances in the inelastic scattering spectrum at the ionization thresholds of the atom's inner shells [323].

Appendix 1

ELECTRON BINDING ENERGIES IN FREE ATOMS

Experimental data from [324, 9, 325, 326]; when these are not available for free atoms, calculated values for solids [324, 9] are given. The error is on the order of 0 to ±3 in the least significant digit.

<table>
<tr><th rowspan="2" colspan="2">Element</th><th colspan="10">Subshell</th></tr>
<tr><th>K</th><th>L_1</th><th>L_2</th><th>L_3</th><th>M_1</th><th>M_2</th><th>M_3</th><th>M_4</th><th>M_5</th><th>N_1</th></tr>
<tr><td>1</td><td>H</td><td>13,60</td><td></td><td></td><td></td><td></td><td></td><td></td><td></td><td></td><td></td></tr>
<tr><td>2</td><td>He</td><td>24,59</td><td></td><td></td><td></td><td></td><td></td><td></td><td></td><td></td><td></td></tr>
<tr><td>3</td><td>Li</td><td>64,39</td><td>5,39</td><td></td><td></td><td></td><td></td><td></td><td></td><td></td><td></td></tr>
<tr><td>4</td><td>Be</td><td>123,6</td><td>9,32</td><td></td><td></td><td></td><td></td><td></td><td></td><td></td><td></td></tr>
<tr><td>5</td><td>B</td><td>200,8</td><td>12,93</td><td>8,30</td><td></td><td></td><td></td><td></td><td></td><td></td><td></td></tr>
<tr><td>6</td><td>C</td><td>296,1</td><td>16,59</td><td colspan="2">11,26</td><td></td><td></td><td></td><td></td><td></td><td></td></tr>
<tr><td>7</td><td>N</td><td>403</td><td>20,33</td><td colspan="2">14,53</td><td></td><td></td><td></td><td></td><td></td><td></td></tr>
<tr><td>8</td><td>O</td><td>538</td><td>28,48</td><td colspan="2">13,62</td><td></td><td></td><td></td><td></td><td></td><td></td></tr>
<tr><td>9</td><td>F</td><td>694</td><td>37,86</td><td colspan="2">17,42</td><td></td><td></td><td></td><td></td><td></td><td></td></tr>
<tr><td>10</td><td>Ne</td><td>870,21</td><td>48,47</td><td>21,66</td><td>21,56</td><td></td><td></td><td></td><td></td><td></td><td></td></tr>
<tr><td>11</td><td>Na</td><td>1079,1</td><td>70,9</td><td>38,46</td><td>38,02</td><td>5,14</td><td></td><td></td><td></td><td></td><td></td></tr>
<tr><td>12</td><td>Mg</td><td>1311,3</td><td>96,5</td><td>57,8</td><td>57,6</td><td>7,65</td><td></td><td></td><td></td><td></td><td></td></tr>
<tr><td>13</td><td>Al</td><td>1567</td><td>126</td><td>81</td><td>80</td><td>10,62</td><td>5,99</td><td></td><td></td><td></td><td></td></tr>
<tr><td>14</td><td>Si</td><td>1844</td><td>154</td><td>104</td><td>104</td><td>13,46</td><td colspan="2">8,15</td><td></td><td></td><td></td></tr>
<tr><td>15</td><td>P</td><td>2148</td><td>191</td><td>135</td><td>134</td><td>16,15</td><td colspan="2">10,49</td><td></td><td></td><td></td></tr>
<tr><td>16</td><td>S</td><td>2476</td><td>232</td><td>170</td><td>168</td><td>20,20</td><td colspan="2">10,36</td><td></td><td></td><td></td></tr>
<tr><td>17</td><td>Cl</td><td>2829</td><td>277</td><td>208</td><td>206</td><td>24,59</td><td colspan="2">12,97</td><td></td><td></td><td></td></tr>
<tr><td>18</td><td>Ar</td><td>3206,3</td><td>326,3</td><td>250,8</td><td>248,6</td><td>29,24</td><td>15,94</td><td>15,76</td><td></td><td></td><td></td></tr>
<tr><td>19</td><td>K</td><td>3614,3</td><td>384</td><td>303,3</td><td>300,6</td><td>37</td><td>24,82</td><td>24,49</td><td></td><td></td><td>4,34</td></tr>
<tr><td>20</td><td>Ca</td><td>4048</td><td>441</td><td>360</td><td>357</td><td>48</td><td>34,7</td><td>34,3</td><td></td><td></td><td>6,11</td></tr>
<tr><td>21</td><td>Sc</td><td>4494</td><td>503</td><td>408</td><td>404</td><td>56</td><td>33</td><td>33</td><td>8,0</td><td></td><td>6,56</td></tr>
<tr><td>22</td><td>Ti</td><td>4970</td><td>567</td><td>465</td><td>459</td><td>64</td><td>39</td><td>38</td><td colspan="2">8</td><td>6,82</td></tr>
<tr><td>23</td><td>V</td><td>5470</td><td>633</td><td>525</td><td>518</td><td>72</td><td>44</td><td>43</td><td colspan="2">8</td><td>6,74</td></tr>
<tr><td>24</td><td>Cr</td><td>5995</td><td>702</td><td>589</td><td>580</td><td>80</td><td>49</td><td>48</td><td colspan="2">8,2</td><td>6,77</td></tr>
</table>

Element		Subshell K	L_1	L_2	L_3	M_1	M_2	M_3	M_4	M_5	N_1
25	Mn	6544	755	656	645	89	55	53	9		7,43
26	Fe	7117	851	726	713	98	61	59	9		7,90
27	Co	7715	931	800	785	107	68	66	9		7,86
28	Ni	8338	1015	877	860	117	75	73	10		7,64
29	Cu	8986	1103	958	938	127	82	80	11	10,4	7,73
30	Zn	9663	1198	1052	1029,1	141	98,7	96,1	12	11,2	9,39
		$1s_{1/2}$	$2s_{1/2}$	$2p_{1/2}$	$2p_{3/2}$	$3s_{1/2}$	$3p_{1/2}$	$3p_{3/2}$	$3d_{3/2}$	$3d_{5/2}$	$4s_{1/2}$

Element		Subshell K	L_1	L_2	L_3	M_1	M_2	M_3
31	Ga	10371	1302	1146	1119	162	111	107
32	Ge	11107	1413	1251	1220	184	130	125
33	As	11871	1531	1362	1327	208	151	145
34	Se	12662	1656	1479	1439	234	173	166
35	Br	13481	1787	1602	1556	262	197	189
36	Kr	14327	1923	1731	1678	292,8	221,8	214,2
37	Rb	15203	2068	1867	1807	325	254,3	245,4
38	Sr	16108	2219	2010	1943	361	288	278
39	Y	17041	2375	2158	2083	397	315	304
40	Zr	18002	2536	2311	2227	434	348	335
41	Nb	18990	2702	2469	2375	472	382	367
42	Mo	20006	2872	2632	2527	511	416	399
43	Tc	21050	3048	2800	2683	551	451	432
44	Ru	22123	3230	2973	2844	592	488	466
45	Rh	23225	3418	3152	3010	634	526	501
46	Pd	24357	3611	3337	3180	677	565	537
47	Ag	25520	3812	3530	3357	724	608	577
48	Cd	26715	4022	3732	3542	775	659	625
		$1s_{1/2}$	$2s_{1/2}$	$2p_{1/2}$	$2p_{3/2}$	$3s_{1/2}$	$3p_{1/2}$	$3p_{3/2}$

Element		Subshell M_4	M_5	N_1	N_2	N_3	N_4	N_5	O_1
31	Ga	21	20	11,87	6,00				
32	Ge	33	32	14,28		7,90			
33	As	46	45	18,96		9,79			
34	Se	61	60	22,19		9,75			
35	Br	77	76	23,80		11,81			

Element		Subshell M_4	M_5	N_1	N_2	N_3	N_4	N_5	O_1
36	Kr	95,0	93,80	27,51	14,65	14,00			
37	Rb	117,4	116	32	21,77	20,71			4,18
38	Sr	144	142,3	40	29,2	28,2			5,69
39	Y	163	161	48	30	29	6,38		6,22
40	Zr	187	185	56	35	33		8,61	6,84
41	Nb	212	209	62	40	38		7,17	6,88
42	Mo	237	234	68	45	42		8,56	7,10
43	Tc	263	259	74	49	45		8,6	7,28
44	Ru	290	286	81	53	49		8,50	7,37
45	Rh	318	313	87	58	53		9,56	7,46
46	Pd	347	342	93	63	57	8,78	8,34	
47	Ag	379	375,6	101	69	63	11	10	7,58
48	Cd	419	412	112	78	71	18,28	17,58	8,99
		$3d_{3/2}$	$3d_{5/2}$	$4s_{1/2}$	$4p_{1/2}$	$4p_{3/2}$	$4d_{3/2}$	$4d_{5/2}$	$5s_{1/2}$

Element		Subshell K	L_1	L_2	L_3	M_1	M_2	M_3	M_4	M_5
49	In	27944	4242	3943	3735	830	707	669	455	447
50	Sn	29204	4469	4160	3933	888	761	719	497	489
51	Sb	30496	4703	4385	4137	949	817	771	542	533
52	Te	31820	4945	4618	4347	1012	876	825	589	578
53	I	33176	5195	4858	4563	1078	937	881	638	626
54	Xe	34561	5453	5107	4787	1149	1002	941	689	676,7
55	Cs	35987	5717	5362	5014	1220	1068	1005	746	731,6
56	Ba	37442	5991	5626	5249	1293	1138	1063	797	788
57	La	38928	6269	5894	5486	1365	1207	1124	851	834
58	Ce	40446	6552	6167	5726	1437	1275	1184	903	885
59	Pr	41995	6839	6444	5968	1509	1342	1244	954	934
60	Nd	43575	7132	6727	6213	1580	1408	1303	1005	983
61	Pm	45188	7432	7017	6464	1653	1476	1362	1057	1032
62	Sm	46837	7740	7315	6720	1728	1546	1422	1110	1083
		$1s_{1/2}$	$2s_{1/2}$	$2p_{1/2}$	$2p_{3/2}$	$3s_{1/2}$	$3p_{1/2}$	$3p_{3/2}$	$3d_{3/2}$	$3d_{5/2}$

Element		Subshell N_1	N_2	N_3	N_4	N_5	$N_{6,7}$	O_1	O_2	O_3	O_4	P_1
49	In	126	90	82	21	20		11,03	5,79			
50	Sn	141	102	93	29	28		13,10		7,34		

Element		Subshell										
		N_1	N_2	N_3	N_4	N_5	$N_{6,7}$	O_1	O_2	O_3	O_4	P_1
51	Sb	157	114	104	38	37		16,86		8,61		
52	Te	174	127	117	48	46		17,84		9,01		
53	I	193	141	131	58	56		20,61		10,45		
54	Xe	213,3	157	145,5	69,5	67,5		23,40	13,43	12,13		
55	Cs	233	174	164	81	79		25	19,07	17,21		3,89
56	Ba	254	193	181	101	98,3		31	24,8	22,8		5,21
57	La	273	210	196	105	103		36	22	19	5,75	5,58
58	Ce	291	225	209	114	111	6	39	25	22		5,54
59	Pr	307	238	220	121	117	6	41	27	24		5,47
60	Nd	321	250	230	126	122	6	42	28	25		5,52
61	Pm	335	261	240	131	127	6	43	28	25		5,58
62	Sm	349	273	251	137	132	6	44	29	25		5,64
		$4s_{1/2}$	$4p_{1/2}$	$4p_{3/2}$	$4d_{3/2}$	$4d_{5/2}$	$4f_{5/2,\,7/2}$	$5s_{1/2}$	$5p_{1/2}$	$5p_{3/2}$	$5d_{3/2}$	$6s_{1/2}$

Элемент		Subshell							
		K	L_1	L_2	L_3	M_1	M_2	M_3	M_4
77	Ir	76115	13422	12828	11219	3175	2912	2554	2119
78	Pt	78399	13883	13277	11567	3300	3030	2649	2206
79	Au	80729	14356	13738	11923	3430	3153	2748	2295
80	Hg	83108	14845	14214	12288	3567	3283	2852	2390
81	Tl	85536	15350	14704	12662	3710	3420	2961	2490
82	Pb	88011	15867	15206	13041	3857	3560	3072	2592
83	Bi	90534	16396	15719	13426	4007	3704	3185	2696
84	Po	93105	16933	16242	13815	4160	3850	3300	2800
85	At	95730	17485	16780	14213	4315	4000	3415	2905
86	Rn	98400	18053	17335	14615	4480	4155	3535	3015
87	Fr	101135	18640	17905	15030	4650	4320	3660	3135
88	Ra	103920	19240	18490	15450	4825	4490	3790	3253
89	Ac	106760	19850	19090	15875	5005	4660	3920	3375
90	Th	109655	20475	19700	16305	5187	4835	4050	3495
		$1s_{1/2}$	$2s_{1/2}$	$2p_{1/2}$	$2p_{3/2}$	$3s_{1/2}$	$3p_{1/2}$	$3p_{3/2}$	$3d_{3/2}$

Element	N_1	N_2	N_3	N_4	N_5	N_6	N_7	O_1	O_2	O_3	O_4	O_5	P_1
63 Eu	364	286	262	143	137	6		45	30	26			5,67
64 Gd	380	300	273	150	143	6		46	31	27	6		6,15
65 Tb	398	315	285	157	150	6		48	32	28			5,86
66 Dy	416	331	297	164	157	6		50	33	28			5,94
67 Ho	434	348	310	172	164	6		52	34	29			6,02
68 Er	452	365	323	181	172	6		54	35	30			6,11
69 Tm	471	382	336	190	181	7		56	36	30			6,18
70 Yb	490	399	349	200	190	8	7	58	37	31			6,25
71 Lu	514	420	366	213	202	13	12	62	39	32	5,43		7,0
72 Hf	542	444	386	229	217	21	20	68	43	35	6,8		7,5
73 Ta	570	469	407	245	232	30	28	74	47	38	8,3		7,89
74 W	599	495	428	261	248	38	36	80	51	41	9,0		7,98
75 Re	629	522	450	278	264	47	45	86	56	45	9,6		7,88
76 Os	660	551	473	295	280	56	54	92	61	49	9,6		8,73
	$4s_{1/2}$	$4p_{1/2}$	$4p_{3/2}$	$4d_{3/2}$	$4d_{5/2}$	$4f_{5/2}$	$4f_{7/2}$	$5s_{1/2}$	$5p_{1/2}$	$5p_{3/2}$	$5d_{3/2}$	$5d_{5/2}$	$6s_{1/2}$

Element	O_1	O_2	O_3	O_4	O_5	P_1	P_2	P_3	$P_{4,5}$	Q_1
77 Ir	99	66	53	9,6		9,05				
78 Pt	106	71	57	9,6		8,96				
79 Au	114	76	61	12,5	11,1	9,23				
80 Hg	134	90	72	14	12	10,44				
81 Tl	139	98	79	21	19	8	6,11			
82 Pb	153	111	90	27	25	10	7,42			
83 Bi	167	125	101	34	32	12	7,29			
84 Po	183	137	108	39	35	15	8,42			
85 At	198	150	118	45	42	19	9,0			
86 Rn	215	165	129	53	48	24	14	10,8		
87 Fr	235	184	144	65	60	33	19	14		4,0
88 Ra	260	205	158	75	70	40	25	19		5,3
89 Ac	280	222	173	90	82	45	29	22	5,2	6,3
90 Th	295	238	185	100	92	50	33	25	6	6
	$5s_{1/2}$	$5p_{1/2}$	$5p_{3/2}$	$5d_{3/2}$	$5d_{5/2}$	$6s_{1/2}$	$6p_{1/2}$	$6p_{3/2}$	$6d_{3/2,5/2}$	$7s_{1/2}$

Element		Subshell M_5	N_1	N_2	N_3	N_4	N_5	N_6	N_7
77	Ir	2044	693	581	497	314	298	67	64
78	Pt	2126	727	612	522	335	318	78	75
79	Au	2210	764	645	548	357	339	91	87
80	Hg	2300	809	686	584	385	366	107	103
81	Tl	2394	852	726	615	411	391	127	123
82	Pb	2490	899	769	651	441	419	148	144
83	Bi	2588	946	813	687	472	448	170	165
84	Po	2685	992	855	720	500	475	190	184
85	At	2785	1043	902	760	533	504	212	205
86	Rn	2890	1095	950	795	565	535	235	230
87	Fr	3000	1155	1005	840	604	572	265	255
88	Ra	3110	1213	1062	885	640	608	293	285
89	Ac	3225	1275	1118	930	680	645	320	312
90	Th	3337	1335	1173	972	718	682	350	340
		$3d_{5/2}$	$4s_{1/2}$	$4p_{1/2}$	$4p_{3/2}$	$4d_{3/2}$	$4d_{5/2}$	$4f_{5/2}$	$4f_{7/2}$

Element		Subshell K	L_1	L_2	L_3	M_1	M_2	M_3	M_4	M_5
63	Eu	48522	8056	7621	6981	1805	1618	1484	1164	1135
64	Gd	50243	8380	7935	7247	1884	1692	1547	1220	1189
65	Tb	51999	8711	8256	7518	1965	1768	1612	1277	1243
66	Dy	53792	9050	8585	7794	2048	1846	1678	1335	1298
67	Ho	55622	9398	8922	8075	2133	1926	1746	1395	1354
68	Er	57489	9754	9267	8361	2220	2008	1815	1456	1412
69	Tm	59393	10118	9620	8651	2309	2092	1885	1518	1471
70	Yb	61335	10490	9981	8946	2401	2178	1956	1580	1531
71	Lu	63320	10876	10355	9250	2499	2270	2032	1647	1596
72	Hf	65350	11275	10742	9564	2604	2369	2113	1720	1665
73	Ta	67419	11684	11139	9884	2712	2472	2197	1796	1737
74	W	69529	12103	11546	10209	2823	2577	2283	1874	1811
75	Re	71681	12532	11963	10540	2937	2686	2371	1953	1887
76	Os	73876	12972	12390	10876	3054	2797	2461	2035	1964
		$1s_{1/2}$	$2s_{1/2}$	$2p_{1/2}$	$2p_{3/2}$	$3s_{1/2}$	$3p_{1/2}$	$3p_{3/2}$	$3d_{3/2}$	$3d_{5/2}$

Appendix 2

BASIC FORMULAS FOR SUMMING THE 6-j AND 9-j SYMBOLS

$$\sum_x [x]\{x j_1 j_2\} = [j_1, j_2]. \tag{A2.1}$$

$$\sum_x (-1)^x [x]\{x j j\} = (-1)^{2j}[j]. \tag{A2.2}$$

$$\sum_x [x]\begin{Bmatrix} j_1 & j_2 & j_3 \\ j_1 & j_2 & x \end{Bmatrix} = (-1)^{2j_3}\{j_1 j_2 j_3\}. \tag{A2.3}$$

$$\sum_x (-1)^x [x]\begin{Bmatrix} j_1 & j_1 & j_3 \\ l_1 & l_1 & x \end{Bmatrix} = \delta(j_3, 0)\,(-1)^{-j_1-l_1}[j_1, l_1]^{\frac{1}{2}}. \tag{A2.4}$$

$$\sum_x [x]\begin{Bmatrix} j_1 & j_2 & j_3 \\ l_1 & l_2 & l_3 \\ j_3 & l_3 & x \end{Bmatrix} = \delta(l_1, j_2)\,[j_2]^{-1}\{j_1 j_2 j_3\}\{j_2 l_2 l_3\}. \tag{A2.5}$$

$$\sum_x (-1)^x [x]\begin{Bmatrix} j_1 & j_2 & j_3 \\ l_1 & l_2 & l_3 \\ l_3 & j_3 & x \end{Bmatrix} = \frac{\delta(j_1, l_2)}{2j_1+1}(-1)^{j_1+j_2+l_1+l_2}\{j_1 l_1 l_3\}\{j_2 l_2 j_3\}. \tag{A2.6}$$

(A2.7)

$$\sum_x [x]\begin{Bmatrix} j_1 & j_2 & j_3 \\ l_1 & l_2 & x \end{Bmatrix}\begin{Bmatrix} j_1 & j_2 & j_3' \\ l_1 & l_2 & x \end{Bmatrix} = \frac{\delta(j_3, j_3')}{2j_3+1}\{j_1 j_2 j_3\}\{l_1 l_2 j_3\}. \tag{A2.8}$$

$$\sum_x (-1)^x [x]\begin{Bmatrix} j_1 & j_2 & j_3 \\ l_1 & l_2 & x \end{Bmatrix}\begin{Bmatrix} j_1 & l_1 & j_3' \\ j_2 & l_2 & x \end{Bmatrix} = (-1)^{-j_3-j_3'}\begin{Bmatrix} j_1 & j_2 & j_3 \\ l_2 & l_1 & j_3' \end{Bmatrix}. \tag{A2.9}$$

$$\sum_x [x]\begin{Bmatrix} j_1 & j_2 & j_3 \\ l_1 & l_2 & l_3 \\ k_1 & k_2 & x \end{Bmatrix}\begin{Bmatrix} j_3 & k_2 & k_3 \\ k_1 & l_3 & x \end{Bmatrix} = (-1)^{2k_3}\begin{Bmatrix} j_1 & j_2 & j_3 \\ k_2 & k_3 & l_2 \end{Bmatrix}\begin{Bmatrix} l_1 & l_2 & l_3 \\ k_3 & k_1 & j_1 \end{Bmatrix}.$$

$$\sum_x (-1)^x [x]\begin{Bmatrix} j_1 & j_2 & j_3 \\ l_1 & l_2 & l_3 \\ k_1 & k_2 & x \end{Bmatrix}\begin{Bmatrix} j_3 & k_1 & k_3 \\ k_2 & l_3 & x \end{Bmatrix} = \tag{A2.10}$$

$$= (-1)^{j_1+j_2+j_3+l_1+l_2+l_3-k_1-k_2+2k_3}\begin{Bmatrix} j_1 & j_2 & j_3 \\ k_3 & k_1 & l_1 \end{Bmatrix}\begin{Bmatrix} l_1 & l_2 & l_3 \\ k_2 & k_3 & j_2 \end{Bmatrix}.$$

$$\sum_{x_1 x_2} [x_1, x_2]\begin{Bmatrix} j_1 & j_2 & j_3 \\ l_1 & l_2 & x_1 \\ k_1 & k_2 & x_2 \end{Bmatrix}\begin{Bmatrix} j_1' & j_2' & j_3 \\ l_1 & l_2 & x_1 \\ k_1 & k_2 & x_2 \end{Bmatrix} = \frac{\delta(j_1, j_1')\,\delta(j_2, j_2')}{(2j_1+1)(2j_2+1)} \tag{A2.11}$$

$$\times \{j_1 j_2 j_3\}\{j_1 l_1 k_1\}\{j_2 l_2 k_2\}.$$

$$\sum_{x_1 x_2} (-1)^{x_1} [x_1, x_2] \begin{Bmatrix} j_1 & j_2 & j_3 \\ l_1 & l_2 & x_1 \\ k_1 & k_2 & x_2 \end{Bmatrix} \begin{Bmatrix} j_1' & j_2' & j_3 \\ l_2 & l_1 & x_1 \\ k_1 & k_2 & x_2 \end{Bmatrix} = \tag{A2.12}$$

$$= (-1)^{l_1+k_1+l_2-k_2+j_2+j_2'-j_3} \begin{Bmatrix} j_1 & j_2 & j_3 \\ l_1 & k_2 & j_2' \\ k_1 & l_2 & j_1' \end{Bmatrix}.$$

$$\sum_{x} (-1)^{2x} [x] \begin{Bmatrix} j_1 & j_2 & j_3 \\ l_3 & k_3 & x \end{Bmatrix} \begin{Bmatrix} l_1 & l_2 & l_3 \\ j_2 & x & k_2 \end{Bmatrix} \begin{Bmatrix} k_1 & k_2 & k_3 \\ x & j_1 & l_1 \end{Bmatrix} = \begin{Bmatrix} j_1 & j_2 & j_3 \\ l_1 & l_2 & l_3 \\ k_1 & k_2 & k_3 \end{Bmatrix}. \tag{A2.13}$$

$$\sum_{x} (-1)^{x} [x] \begin{Bmatrix} j_1 & x & l_1 \\ l_2 & k_3 & j_2 \end{Bmatrix} \begin{Bmatrix} j_2 & x & l_2 \\ l_3 & k_1 & j_3 \end{Bmatrix} \begin{Bmatrix} j_3 & x & l_3 \\ l_1 & k_2 & j_1 \end{Bmatrix} =$$

$$= (-1)^{j_1+j_2+j_3+k_1+k_2+k_3+l_1+l_2+l_3} \begin{Bmatrix} k_1 & k_2 & k_3 \\ j_1 & j_2 & j_3 \end{Bmatrix} \begin{Bmatrix} k_1 & k_2 & k_3 \\ l_1 & l_2 & l_3 \end{Bmatrix}. \tag{A2.14}$$

$$\sum_{x_1 x_2} [x_1, x_2] \begin{Bmatrix} j_1 & j_2 & x_1 \\ j_3 & j_4 & l_1 \end{Bmatrix} \begin{Bmatrix} j_3 & j_4 & x_1 \\ j_1 & j_2 & x_2 \end{Bmatrix} \begin{Bmatrix} j_1 & j_2 & l_2 \\ j_3 & j_4 & x_2 \end{Bmatrix} = \begin{Bmatrix} j_1 & j_2 & l_2 \\ j_3 & j_4 & l_1 \end{Bmatrix}. \tag{A2.15}$$

$$\sum_{x_1 x_2} (-1)^{x_1+x_2} [x_1, x_2] \begin{Bmatrix} j_1 & j_2 & x_1 \\ j_3 & j_4 & l_1 \end{Bmatrix} \begin{Bmatrix} j_1 & j_3 & x_2 \\ j_4 & j_2 & x_1 \end{Bmatrix} \begin{Bmatrix} j_1 & j_4 & l_2 \\ j_2 & j_3 & x_2 \end{Bmatrix} = \tag{A2.16}$$

$$= (-1)^{l_1} \delta(l_1, l_2) [l_1]^{-1} \{j_1 j_4 l_1\} \{j_2 j_3 l_1\}.$$

Appendix 3

TRANSFORMATION MATRICES FOR THREE AND FOUR ANGULAR MOMENTA

Intermediate angular momenta are enclosed in parentheses.

$$\langle j_1 j_2 (j_{12}) j_3 j \,|\, j_1 j_3 (j_{13}) j_2 j \rangle = (-1)^{j_2+j_3+j_{12}+j_{13}} [j_{12}, j_{13}]^{\frac{1}{2}} \begin{Bmatrix} j_1 & j_2 & j_{12} \\ j & j_3 & j_{13} \end{Bmatrix}. \tag{A3.1}$$

$$\langle j_1 j_2 (j_{12}) j_3 j \,|\, j_3 j_2 (j_{32}) j_1 j \rangle = (-1)^{2j} [j_{12}, j_{32}]^{1/2} \begin{Bmatrix} j_1 & j_2 & j_{12} \\ j_3 & j & j_{32} \end{Bmatrix}. \tag{A3.2}$$

$$\langle j_1, j_2 j_3 (j_{23}) j \,|\, j_1 j_2 (j_{12}) j_3 j \rangle = (-1)^{j_1+j_2+j_3+j} [j_{12}, j_{23}]^{\frac{1}{2}} \begin{Bmatrix} j_1 & j_2 & j_{12} \\ j_3 & j & j_{23} \end{Bmatrix}. \tag{A3.3}$$

$$\langle j_1, j_2 j_3 (j_{23}) j \,|\, j_1 j_3 (j_{13}) j_2 j \rangle = (-1)^{j_1+j_{23}+j} [j_{13}, j_{23}]^{\frac{1}{2}} \begin{Bmatrix} j_1 & j_3 & j_{13} \\ j_2 & j & j_{23} \end{Bmatrix}. \tag{A3.4}$$

$$\langle j_1 j_2 (j_{12}), j_3 j_4 (j_{34}) j \,|\, j_1 j_3 (j_{13}), j_2 j_4 (j_{24}) j \rangle = [j_{12}, j_{34}, j_{13}, j_{24}]^{\frac{1}{2}} \begin{Bmatrix} j_1 & j_2 & j_{12} \\ j_3 & j_4 & j_{34} \\ j_{13} & j_{24} & j \end{Bmatrix}. \tag{A3.5}$$

$$\langle j_1 j_2 (j_{12}) j_3 (j_{123}) j_4 j \,|\, j_1 j_4 (j_{14}) j_3 (j_{143}) j_2 j \rangle =$$
$$= (-1)^{j_{12}-j_{14}-j_{123}+j_{143}} [j_{12}, j_{123}, j_{14}, j_{143}]^{\frac{1}{2}} \begin{Bmatrix} j_1 & j_{12} & j_2 \\ j_{14} & j_3 & j_{143} \\ j_4 & j_{123} & j \end{Bmatrix}. \tag{A3.6}$$

$$\langle j_1 j_2 (j_{12}) j_3 (j_{123}) j_4 j \,|\, j_1 j_3 (j_{13}) j_4 (j_{134}) j_2 j \rangle = (-1)^{j_2+j_4+j_{12}-j_{13}-j_{123}-j_{134}]} [j_{12}, j_{13}, j_{123}, j_{134}]^{\frac{1}{2}} \times$$
$$\times \begin{Bmatrix} j_1 & j_2 & j_{12} \\ j_{123} & j_3 & j_{13} \end{Bmatrix} \begin{Bmatrix} j_{13} & j_2 & j_{123} \\ j & j_4 & j_{134} \end{Bmatrix}. \tag{A3.7}$$

$$\langle j_1 j_2 (j_{12}) j_3 (j_{123}) j_4 j \,|\, j_3 j_4 (j_{34}) j_1 (j_{341}) j_2 j \rangle = (-1)^{j_1+j_2+j_3+j_4-j} [j_{12}, j_{123}, j_{34}, j_{341}]^{\frac{1}{2}} \times$$
$$\times \begin{Bmatrix} j_1 & j_2 & j_{12} \\ j & j_{34} & j_{341} \end{Bmatrix} \begin{Bmatrix} j_{12} & j_3 & j_{123} \\ j_4 & j & j_{34} \end{Bmatrix}. \tag{A3.8}$$

$$\langle j_1 j_2 (j_{12}), j_3 j_4 (j_{34}) j \,|\, j_1 j_3 (j_{13}) j_2 (j_{132}) j_4 j \rangle = (-1)^{j_3 - j_4 + j_{13} - j} [j_{12}, j_{34}, j_{13}, j_{132}]^{\frac{1}{2}} \times$$
$$\times \begin{Bmatrix} j_1 & j_2 & j_{12} \\ j_{132} & j_3 & j_{13} \end{Bmatrix} \begin{Bmatrix} j_3 & j_4 & j_{34} \\ j & j_{12} & j_{132} \end{Bmatrix}. \qquad \text{(A3.9)}$$

$$\langle j_1 j_2 (j_{12}), j_3 j_4 (j_{34}) j \,|\, j_1 j_3 (j_{13}) j_4 (j_{134}) j_2 j \rangle = (-1)^{j_1 - j_2 + j_3 + j_4 - j_{12} - j_{34}} [j_{12}, j_{13}, j_{34}, j_{134}]^{\frac{1}{2}} \times$$
$$\times \begin{Bmatrix} j_1 & j_2 & j_{12} \\ j & j_{34} & j_{134} \end{Bmatrix} \begin{Bmatrix} j_4 & j_3 & j_{34} \\ j_1 & j_{134} & j_{13} \end{Bmatrix}. \qquad \text{(A3.10)}$$

$$\langle j_1 j_2 (j_{12}), j_3 j_4 (j_{34}) j \,|\, j_1 j_4 (j_{14}) j_3 (j_{143}) j_2 j \rangle = (-1)^{j_1 - j_3 - j_{12}} [j_{12}, j_{24}, j_{14}, j_{143}]^{\frac{1}{2}} \times$$
$$\times \begin{Bmatrix} j_1 & j_2 & j_{12} \\ j & j_{34} & j_{143} \end{Bmatrix} \begin{Bmatrix} j_3 & j_4 & j_{34} \\ j_1 & j_{143} & j_{14} \end{Bmatrix}. \qquad \text{(A3.11)}$$

$$\langle j_1 j_2 (j_{12}), j_3 j_4 (j_{34}) j \,|\, j_3 j_2 (j_{32}) j_1 (j_{321}) j_4 j \rangle = (-1)^{j_3 - j_4 + j_{12} - j} [j_{12}, j_{34}, j_{32}, j_{321}]^{\frac{1}{2}} \times$$
$$\times \begin{Bmatrix} j_1 & j_2 & j_{12} \\ j_3 & j_{321} & j_{32} \end{Bmatrix} \begin{Bmatrix} j_3 & j_4 & j_{34} \\ j & j_{12} & j_{321} \end{Bmatrix}. \qquad \text{(A3.12)}$$

$$\langle j_1 j_2 (j_{12}), j_3 j_4 (j_{34}) j \,|\, j_3 j_2 (j_{32}) j_4 (j_{324}) j_1 j \rangle = (-1)^{2j_1 - j_2 + j_4 + j_{32} - j_{34}} [j_{12}, j_{34}, j_{32}, j_{324}]^{\frac{1}{2}} \times$$
$$\times \begin{Bmatrix} j_1 & j_2 & j_{12} \\ j_{34} & j & j_{324} \end{Bmatrix} \begin{Bmatrix} j_3 & j_4 & j_{34} \\ j_{324} & j_2 & j_{32} \end{Bmatrix}. \qquad \text{(A3.13)}$$

REFERENCES

1. M. A. Blokhin, *The Physics of X Rays* [in Russian], Gostekhizdat, Moscow (1982).
2. M. A. Blokhin, *Methods of X-Ray Spectroscopic Research*, Pergamon, Oxford (1965).
3. I. B. Borovskii, *The Physical Foundations of X-Ray Spectroscopic Research* [in Russian], Izd. MGU, Moscow (1956).
4. É. S. Parilis, *The Auger Effect* [in Russian], Fan, Tashkent (1969).
5. K. Siegbahn et al., *ESCA. Atomic, Molecular, and Solid-State Structure Studied by Means of Electron Spectroscopy*, Uppsala (1967).
6. T. M. Zimkina and V. A. Fomichev, *Ultrasoft X-Ray Spectroscopy* [in Russian], Izd. LGU, Leningrad (1971).
7. V. V. Nemoshkalenko and V. G. Aleshin, *The Theoretical Foundations for X-Ray Emission Spectroscopy* [in Russian], Naukova Dumka, Kiev (1974).
8. D. Chattarji, *Theory of Auger Transitions*, Academic Press, New York (1976).
9. T. A. Carlson, *Photoelectron and Auger Spectroscopy*, Plenum Press, New York (1975).
10. B. K. Agarwal, *X-Ray Spectroscopy*, Springer, Berlin (1979).
11. E. E. Koch, R. Haensel, and Ch. Kunz (eds.), *Vacuum UV Radiation Physics*, Pergamon Press, Oxford (1974).
12. B. Crasemann (ed.), *Atomic Inner Shell Processes*, Vol. I. *Ionization and Transition Probabilities*, Academic Press, New York (1975).
13. F. J. Wuilleumier (ed.), *Photoionization and Other Probes of Many-Electron Interactions*, Plenum Press, New York (1976).
14. B. Williams (ed.), *Compton Scattering. The Investigation of Electron Momentum Distribution*, McGraw-Hill, New York (1977).
15. C. R. Brundle and A. D. Baker (eds.), *Electron Spectroscopy*, Vol. II. *Theory, Techniques, and Applications*, Academic Press, New York (1978).
16. C. A. Nicolaides and D. R. Beck (eds.), *Excited States in Quantum Chemistry*. NATO ASI Series; Ser. C, Vol. 46, Reidel, Dordrecht (1978).
17. D. J. Fabian, H. Kleinpoppen, and L. M. Watson (eds.), *Inner Shell and X-Ray Physics of Atoms and Solids*, Plenum Press, New York (1981).
18. C. Bonnelle and C. Mandé (eds.), *Advances in X-Ray Spectroscopy*, Pergamon, Oxford (1982).

19. S. Flügge (ed.), "Corpuscles and radiation in matter," *Handbuch der Physik. Encyclopedia of Physics,* Vol. 31, Springer, Berlin (1982).

20. H. O. Lutz, J. S. Briggs, and H. Kleinpoppen (eds.), *Fundamental Processes in Energetic Atomic Collisions,* NATO ASI Series; Ser. B, Vol. 103, Plenum Press, New York (1983).

21. B. Crasemann (ed.), "X-Ray and Atomic Inner-Shell Physics," *Proceedings of the AIP Conference, No. 94,* American Institute of Physics, New York (1982)

22. I. Lindgren, A. Rosen, and S. Svanberg (eds.), *Atomic Physics,* Vol. 8, Plenum Press, New York (1983).

23. *New Trends in Atomic Physics,* Part 2, North-Holland, Amsterdam (1984).

24. A. Meisel and J. Finster (eds.), *X-Ray and Inner-Shell Processes in Atoms, Molecules and Solids,* Karl Marx University, Leipzig (1984).

25. B. Crasemann (ed.), *Atomic Inner-Shell Physics,* Plenum Press, New York (1985).

26. U. Fano and J. W. Cooper, "Spectral distribution of atomic oscillator strengths," *Rev. Mod. Phys.,* **40,** No. 3, 441–507 (1968).

27. M. Ya. Amusia, "Collective effects in photoionization of atoms," in: *Advances in Atomic and Molecular Physics,* Vol. 17, Academic Press, New York (1981), pp. 2–54.

28. M. Ya. Amusia and N. A. Cherepkov, "Many-electron correlations in scattering processes," in: *Case Studies in Atomic Physics,* Vol. 5, North- Holland, Amsterdam (1975).

29. J. P. Connerade, "Non-Rydberg spectroscopy of atoms," *Contemp. Phys.,* **19,** No. 5, 415–448 (1978).

30. H. Aksela, "Diagram Auger spectra of free atoms: a critical comparison with theory," *J. Electron Spectrosc. Relat. Phenom.,* **19,** No. 4, 371–391 (1980).

31. M. O. Krause, "Electron spectrometry experiments with the use of synchrotron radiation," in: *Atomic Physics,* Vol. 9, World Scientific, Washington (1984).

32. G. Racah, "Theory of complex spectra II," *Phys. Rev.,* **62,** 438–462 (1942).

33. G. Racah, "Theory of complex spectra III," *Phys. Rev.,* **63,** 367–382 (1943).

34. U. Fano and G. Racah, *Irreducible Tensorial Sets,* Academic Press, New York (1959).

35. J. C. Slater, *Quantum Theory of Atomic Structure,* McGraw-Hill, New York (1960), Vols. 1 and 2.

36. A. P. Yutsis, I. B. Levinson, and V. V. Vanagas, *Mathematical Apparatus of the Theory of Angular Momentum,* National Science Foundation, Washington (1962).

37. D. R. Hartree, *The Calculation of Atomic Structures,* Wiley, New York (1957).

38. I. B. Levinson and A. A. Nikitin, *Handbook for Theoretical Computation of Line Intensities in Atomic Spectra,* Israel Program for Scientific Translations, Jerusalem (1965); D. Davey and Co., New York.

39. I. I. Sobel'man, *Introduction to the Theory of Atomic Spectra,* Pergamon, Oxford (1972).

40. B. R. Judd, *Second Quantization and Atomic Spectroscopy,* Johns Hopkins University Press, Baltimore (1967).

41. B. R. Judd, *Topics in Atomic Theory,* University of Canterbury, Christchurch (1968); B. G. Wybourne, *Symmetry Principles and Atomic Spectroscopy,* Wiley Interscience, New York (1970).

42. A. P. Yutsis and A. A. Bandzaitis, *The Theory of Angular Momentum in Quantum Mechanics,* 2nd edn. [in Russian], Mokslas, Vilnius (1977).

43. A. P. Yutsis and A. Yu. Savukinas, *The Mathematical Foundations of Atomic Theory* [in Russian], Mintis, Vilnius (1973).

44. D. A. Varshalovich, A. N. Moskalev, and V. K. Khersonskii, *A Quantum Theory of Angular Momentum* [in Russian], Nauka, Leningrad (1975).

45. C. Froese-Fischer, *The Hartree–Fock Method for Atoms,* Wiley, New York (1977).

46. R. D. Cowan, *The Theory of Atomic Structure and Spectra,* University of California, Berkeley (1981).

47. I. Lindgren and J. Morrison, "Atomic many-body theory," in: *Springer Series in Chemical Physics,* Vol. 13, Springer, Berlin (1982).

48. A. A. Nikitin and Z. B. Rudzikas, *Fundamentals of the Theory of Atom and Ion Spectra* [in Russian], Nauka, Moscow (1983).

49. Z. B. Rudzikas and Yu. M. Kanyauskas, *Quasispin and Isospin in Atomic Theory* [in Russian], Mokslas, Vilnius (1984).

50. A. Messiah, *Quantum Mechanics*, North-Holland, Amsterdam (1961).

51. C. W. Nielson and G. F. Koster, *Spectroscopic Coefficients for the* p^n, d^n, *and* f^n *Configurations*, Mass. Inst. Technol., Cambridge (1963).

52. R. I. Karaziya, Ya. I. Vizbaraite, Z. B. Rudzikas, and A. P. Yutsis, *Tables for the Calculation of Matrix Elements of Atomic Quantities*, ANL Transl. 563, National Technical Information Service, Springfield (1968).

53. Yu. M. Kanyauskas, B. Ch. Shimonis, and Z. B. Rudzikas, "The hole representation and the system of wave function phases and coefficients of fractional parentage for l^N configurations," *Lit. Fiz. Sb.*, **25**, No. 2, 19-30 (1985).

54. H. A. Bethe and E. E. Salpeter, *Quantum Mechanics of One- and Two-Electron Atoms*, Springer, Berlin–Gotingen–Heidelberg (1957).

55. M. A. El'yashevich, *Atomic and Molecular Spectroscopy* [in Russian], Fizmatgiz, Moscow (1962).

56. L. D. Landau and E. M. Lifshits, *Quantum Mechanics. Nonrelativistic Theory*, 2nd edn., Pergamon, London (1965).

57. A. Hibbert, "Model potentials in atomic structure," in: *Advances in Atomic and Molecular Physics*, Vol. 18, Academic Press, New York (1982), pp. 309–340.

58. P. O. Bogdanovich, I. I. Boruta, and A. Yu. Savukinas, "Obtaining Hartree–Fock functions for mixing terms," *Lit. Fiz. Sb.*, **17**, No. 6, 740-748 (1977).

59. C. C. J. Roothan and P. S. Bagus, "Atomic self-consistent field calculations by the expansion method," in: *Methods in Computational Physics*, Vol. 2, Academic Press, New York (1963), pp. 47-94.

60. P. O. Bogdanovich, R. I. Karaziya, and I. I. Boruta, "The orthogonality of wave functions to the functions of energetically lower configurations and Brillouin's theorem for the $n_1l^{N_1}n_2l^{N_2}$ configuration," *Lit. Fiz. Sb.*, **20**, No. 2, 15–24 (1980).

61. V. Ch. Shimonis, Yu. M. Kanyauskas, and Z. B. Rudzikas, "An isospin basis for the $n_1l^{N_1}n_2l^{N_2}$ configurations," *Lit. Fiz. Sb.*, **22**, No. 4, 3–15 (1982).

62. Ch. Froese-Fischer, "Hartree–Fock calculations for atoms with inner-shell vacancies," *Phys. Rev. Lett.*, **38**, No. 19, 1075–1077 (1977).

63. Ch. Froese, "The orthogonality assumption in the Hartree–Fock approximation," *Can. J. Phys.*, **45**, No. 1, 7–12 (1967).

64. V. V. Murakhtanov, T. I. Guzhavina, and L. N. Mazalov, "The variational principle for calculating excited states," *Zh. Strukt. Khim.*, **20**, No. 6, 1106–1110 (1979).

65. P. S. Bagus, "Self-consistent-field wave functions for hole states of some Ne-like and Ar-like ions," *Phys. Rev.*, **139**, No. 3A, 619–634 (1965).

66. L. A. Vainshtein, I. I. Sobel'man, and E. A. Yukov, *Excitation of Atoms and Spectral Line Broadening* [in Russian], Nauka, Moscow (1979).

67. I. P. Grant, "Relativistic calculation of atomic structures," *Adv. Phys.*, **19**, No. 82, 747–811 (1970).

68. L. Armstrong, Jr., "Relativistic effects in many-body systems," in: *Atomic Physics*, Vol. 8, Plenum Press, New York (1983), pp. 129-147.

69. J. B. Mann and W. R. Johnson, "Breit interaction in multielectron atoms," *Phys. Rev.*, **A4**, No. 1, 41–50 (1971).

70. J. P. Desclaus, C. M. Moser, and G. Werhaegen, "Relativistic energies of excited states of atoms and ions of the second period," *J. Phys. B*, **4**, No. 3, 296–310 (1971).

71. I. P. Grant and N. C. Pyper, "Breit interaction in multiconfiguration relativistic atomic calculations," *J. Phys. B,* **9**, No. 5, 761–774 (1976).

72. J. Sucher, "Foundations of the relativistic theory of many-electron bound states," *Int. J. Quant. Chem.,* **25**, No. 1, 3–21 (1984).

73. J. P. Desclaux, "Relativity in electronic structure: how and why?" in: *Atomic Physics of Highly Ionized Atoms, Ser. B,* Vol. 96, NATO ASI, R. Marrus (ed.), Plenum Press, New York (1983), pp. 75–111.

74. L. Armstrong, Jr., "Relativistic effects in atomic fine structure, Part I," *J. Math. Phys.,* **7**, No. 11, 1891–1899 (1966); Part II, *ibid.,* **9**, No. 7, 1083–1086 (1968).

75. L. Armstrong and S. Feneuille, "Relativistic effects in the many-electron atom," in: *Advances in Atomic and Molecular Physics,* Vol. 10, Academic Press, New York (1974), pp. 1–52.

76. J. Bauche, C. Bauche-Arnoult, E. Luc-Koenig, and M. Klapish, "Nonrelativistic energies from relativistic radial integrals in atoms and ions," *J. Phys. B,* **15**, No. 15, 2325–2338 (1982).

77. F. P. Larkins, "Relativistic *LS* multiplet energies for atoms and ions," *J. Phys. B,* **9**, No. 1, 37–46 (1976).

78. A. I. Akhiezer and V. B. Berestetskii, *Quantum Electrodynamics,* Wiley-Interscience, New York (1965).

79. S. A. Kuchas, A. V. Karosene, and R. I. Karaziya, "On the applicability of Hartree–Fock–Pauli approximation for studying the energy characteristics of the inner electrons," *Lit. Fiz. Sb.,* **18**, No. 5, 593–602 (1978).

80. S. J. Rose, I. P. Grant, and N. C. Pyper, "The direct and indirect effects in the relativistic modification of atomic valence orbitals," *J. Phys. B,* **11**, No. 7, 1171–1176 (1978).

81. H. W. Kugel and D. E. Murnik, "The Lamb shift in hydrogenic solids," *Rep. Prog. Phys.,* **40**, No. 3, 297–343 (1977).

82. G. W. F. Drake, "Quantum electrodynamic effects in few-electron atomic systems," in: *Advances in Atomic and Molecular Physics,* Vol. 18, Academic Press, New York (1982), pp. 399–460.

83. P. J. Mohr, "Lamb shift in a strong Coulomb potential," *Phys. Rev. Lett.,* **34**, No. 16, 1050–1052 (1975).

84. K. N. Huang, M. Aoyagi, M. H. Chen, et al., "Neutral-atom electron binding energies from relaxed-orbital relativistic Hartree–Fock calculations," *At. Data Nucl. Data Tables,* **18**, No. 3, 243–291 (1976).

85. J. P. Desclaux, "The status of relativistic calculations for atoms and molecules," *Phys. Scr.,* **21**, 436–442 (1980).

86. M. H. Chen, B. Crasemann, M. Aoyagi, et al., "Theoretical atomic inner-shell energy levels, $70 \leq Z \leq 106$," *At. Data Nucl. Data Tables,* **26**, No. 6, 561–574 (1981).

87. I. P. Grant and B. J. McKenzie, "The transverse electron–electron interaction in atomic structure calculations," *J. Phys. B,* **13**, No. 14, 2671–2681 (1980).

88. M. E. Rose, *Multipole Fields,* Wiley, New York (1955).

89. L. Scofield, "Radiative decay rates of vacancies in the *K* and *L* shells," *Phys. Rev.,* **179**, No. 1, 9–16 (1969).

90. O. Goscinski, G. Howat, and T. Aberg, "On transition energies and probabilities by a transition operator method," *J. Phys. B,* **8**, No. 1, 11–19 (1975).

91. J. H. Scofield, "Exchange corrections of *K* x-ray emission rates," *Phys. Rev. A,* **9**, No. 3, 1041–1049 (1974).

92. S. T. Manson, "Systematics of zeroes in dipole matrix elements for photoionizing transitions: nonrelativistic calculations," *Phys. Rev. A,* **31**, No. 6, 3698–3703 (1985).

93. Yu. M. Kanyauskas, I. S. Kychkin, and Z. B. Rudzikas, "A relativistic examination of electron transitions in multielectron atoms," *Lit. Fiz. Sb.,* **14**, No. 3, 463–475 (1974).

94. Z. Rudzikas, M. Szulkin, and I. Martinson, "Oscillator strengths and the accuracy of atomic wave functions," *Phys. Scr.,* **28**, No. 2, 141–144 (1984).

95. O. Sinanoglu, "A many-electron theory of atoms, molecules, and their interactions," *Adv. Chem. Phys.*, **6**, 315–412 (1964).

96. M. A. Braun, A. D. Gurchumeliya, and U. I. Safronova, *The Relativistic Theory of an Atom* [in Russian], Nauka, Moscow (1984).

97. P. O. Löwdin, "Correlation problem in many-electron quantum mechanics, Part I. Review of different approaches and discussion of some current ideas," *Adv. Chem. Phys.*, **2**, 207–322 (1959).

98. J. C. Slater, "Exchange in spin-polarized energy bands," *Phys. Rev.*, **165**, No. 2, 658–669 (1968).

99. B. R. Judd, "Atomic shell theory recast," *Phys. Rev.*, **162**, No. 1, 28–37 (1967).

100. S. A. Kuchas and R. I. Karaziya, "The spin-polarized Hartree-Fock method and the localization of electrons in an inner shell," *Lit. Fiz. Sb.*, **26**, No. 6, 666–675 (1986).

101. V. Hinze, "The unrestricted Hartree-Fock method," *Czech. J. Phys.*, **13**, No. 9, 619–630 (1963).

102. A. P. Jucys, "On the extended method of calculations of atomic structures," *Int. J. Quant. Chem.*, **1**, No. 4, 311–319 (1967).

103. A. P. Yutsis and V. M. Lazauskas, "Hartree–Fock equations for nonorthogonal radial orbitals," in: *Problems in Theoretical Physics, Part I, Quantum Mechanics* [in Russian], Izd. LGU, Leningrad (1974), pp. 108–117.

104. M. A. Klapisch, "Program for atomic wave function computations by the parametric potential method," *Comp. Phys. Commun.*, **2**, No. 5, 239–260 (1971).

105. K. Dietz, O. Lechtenfeld, and G. Weymans, "Optimized mean fields for atoms, Part I. Mean field methods for the description of *N*-fermion systems," *J. Phys. B*, **15**, No. 23, 4301–4314 (1982).

106. J. P. Connerade, K. Dietz, M. Ohno, and G. Weymans, "*g*-Hartree mean-field analysis of inner-shell many-body effects for small- and large-*Z* atoms," *J. Phys. B*, **18**, No. 12, L351–L356 (1985).

107. W. Kohn and P. Vashishta, "General density functional theory," in: *Theory of the Inhomogeneous Electron Gas*, S. Lundqvist and N. M. March (eds.), Plenum Press, New York (1983), pp. 79–147.

108. B. R. Judd, "Complex atomic spectra," *Rep. Prog. Phys.*, **48**, No. 7, 907–954 (1985).

109. K. Rajnak and B. G. Wybourne, "Configuration interaction effects in l^N configurations," *Phys. Rev.*, **132**, No. 1, 280–290 (1963).

110. P. O. Bogdanovich and G. L. Zhukauskas, "An approximate calculation of the superposition of configurations in atomic spectra," *Lit. Fiz. Sb.*, **23**, No. 3, 18–33 (1983).

111. B. G. Wybourne, *Spectroscopic Properties of Rare Earths*, Interscience, New York (1965).

112. V. F. Demekhin, T. M. Poltinikova, Yu. I. Bairachnyi, T. V. Shelkovich, and V. L. Sukhorukov, "Multiplet structure of x-ray and electron spectra of the rare-earth elements," *Izv. Akad. Nauk SSSR, Ser. Fiz.*, **38**, No. 3, 593–598 (1974).

113. V. F. Demekhin, V. L. Sukhorukov, V. A. Yavna, S. A. Kulagina, S. A. Prosandeev, and Yu. I. Bairachnyi, "The effect of configuration interaction on the structure of x-ray spectra," *Izv. Akad. Nauk SSSR, Ser. Fiz.*, **40**, No. 2, 255–262 (1976).

114. A. F. Starace, "The quantum defect theory approach," *op. cit.* [13], pp. 395–406.

115. U. Fano, "Evolution of quantum defect methods," *Commun. At. Mol. Phys.*, **13**, No. 4, 157–169 (1983).

116. J. J. Chang, "Relativistic quantum-defect theory. General formulation," *Phys. Rev.*, **28**, No. 2, 592–603 (1965).

117. U. Fano, "Interaction between configurations with several open shells," *Phys. Rev.*, **140**, No. 1A, 67–75 (1965).

118. C. Froese-Fischer, "Brillouin's theorem for excited $nl^q nl^{q'}$ configurations," *J. Phys. B*, **6**, No. 10, 1933–1941 (1973).

119. R. I. Karaziya and L. S. Rudzikaite, "Peculiarities in the mixing of electron configurations that differ by the quantum number of a single electron," *Lit. Fiz. Sb.*, **27**, No. 2, 144–155 (1987).

120. S. A. Kuchas, R. I. Karaziya, and V. I. Tutlis, "Some regularities governing mixing of $(s + d)^{N+1}$-configurations," *Lit. Fiz. Sb.*, **24**, No. 416–28 (1984).

121. F. Sasaki and M. Yoshimine, "Configuration-interaction study of atoms, Part I. Correlation energies of B, C, N, O, F, and Ne," *Phys. Rev. A*, **9**, No. 1, 17–25 (1974).

122. A. P. Yutsis, "The Fock equations in a multiconfiguration approximation," *Zh. Eksp. Teor. Fiz.*, **23**, No. 2, 129–139 (1952).

123. A. P. Yutsis, Ya. I. Vizbaraite, I. V. Batarunas, and V. I. Kavetskis, "On the multiconfiguration approximation and the future of its development," *Tr. Akad. Nauk Lit. SSR, Ser. B.*, **2(14)**, No. 1, 3–16 (1958).

124. L. Armstrong, Jr., "An open-shell random phase approximation," *J. Phys. B*, **7**, No. 17, 2320–2331 (1974).

125. N. A. Cherepkov and L. V. Chernysheva, "Random phase approximations with exchange for open-shell atoms: photoionization of Cl," *Phys. Lett. A*, **60**, No. 2, 103–105 (1977).

126. G. V. Merkelis, Yu. M. Kanyauskas, and Z. B. Rudzikas, "Formal methods of stationary perturbation theory in atoms," *Lit. Fiz. Sb.*, **25**, No. 5, 21–30 (1985).

127. G. V. Merkelis, G. A. Gaigalas, and Z. B. Rudzikas, "The irreducible tensorial form and a graphical representation of the effective Hamiltonian of an atom in first- and second-order stationary perturbation theory," *Lit. Fiz. Sb.*, **25**, No. 5, 14–31 (1985).

128. H. P. Kelly, "Application of many-body diagram techniques in atomic physics," *Adv. Chem. Phys.*, **14**, 129–190 (1969).

129. H. P. Kelly, "Many-body calculations of photoionization," *op. cit.* [22], pp. 305–337.

130. U. Fano, "Effects of configuration interaction on intensities and phase shifts," *Phys. Rev.*, **124**, No. 6, 1866–1878 (1961).

131. F. Comber Farnoux, "Multichannel scattering theory of the resonant Auger effect in photoelectron spectroscopy," *Phys. Rev. A*, **25**, No. 1, 287–303 (1982).

132. V. V. Balashov, S. S. Lipovetskii, and V. S. Senashenko, "On a unified description of the profile of resonance lines in the energy spectra of scattered and emitted electrons," *Zh. Eksp. Teor. Fiz.*, **63**, No. 5, 1622–1627 (1972).

133. S. S. Lipovetsky and V. S. Senashenko, "On the shape of the $(2s^2)^1S$ resonance in the spectra of electrons ejected from He atoms," *J. Phys. B*, **7**, No. 6, 693–703 (1974).

134. F. H. Mies, "Configuration theory. Effects of overlapping resonances," *Phys. Rev.*, **175**, No. 1, 164–175 (1968).

135. T. Åberg and G. Howat, "Theory of the Auger effect," *op. cit.* [19], pp. 469–619.

136. L. Armstrong, Jr., C. E. Theodosiou, and M. J. Wall, "Interference between radiative emission and autoionization in the decay of excited states of atoms," *Phys. Rev. A*, **18**, No. 6, 2538–2549 (1978).

137. T. Åberg, "A scattering approach to the decay of metastable states," *Phys. Scr.*, **21**, No. 3/4, 495–502 (1980).

138. G. Wendin, "Application of many-body problems to atomic physics," *op. cit.* [23], pp. 555–642.

139. G. Wendin and M. Ohno, "Strong dynamical effects of many-electron interactions in photoelectron spectra from 4*s* and 4*p* core levels," *Phys. Scr.*, **14**, No. 4, 148–161 (1976).

140. A. I. Udris and R. I. Karaziya, "The change in atomic quantities when vacancies are created in an atom's inner electron shell," *Lit. Fiz. Sb.*, **17**, No. 2, 159–170 (1977).

141. R. I. Karaziya, A. I. Udris, and D. V. Grabauskas, "Using the chemical shifts in electron levels to study the effective charge distribution of atoms in compounds," *Zh. Strukt. Khim.*, **18**, No. 4, 653–660 (1977).

142. S. Swensson, N. Mårtensson, E. Basilier, et al., "Lifetime broadening and Cl-resonances observed in ESCA," *Phys. Scr.*, **14**, No. 4, 141–147 (1976).

143. L. S. Cederbaum, J. Schirmer, W. Domcke, and W. von Nessen, "Complete breakdown of the quasiparticle picture for inner valence electrons," *J. Phys. B*, **10**, No. 15, L549–L553 (1977).

144. A. B. Migdal, *Finite Fermi Systems Theory and the Properties of Atomic Nuclei*, 2nd edn. [in Russian], Nauka, Moscow (1983).

145. A. S. Kheifets, M. Ya. Amusia, and V. G. Yarzemsky, "On the validity of the quasiparticle approximation in photoelectron spectroscopy," *J. Phys. B*, **18**, No. 11, L343–L350 (1985).

146. C. Bauche-Arnoult, J. Bauche, and M. Klapisch, "Variance of the distributions of energy levels and of the transition arrays in atomic spectra," *Phys. Rev. A*, **20**, No. 6, 2424–2439 (1979).

147. R. I. Karaziya, "The collapse of an excited electron's orbit and the characteristics of atomic spectra," *Usp. Fiz. Nauk*, **135**, No. 1, 79–115 (1981).

148. J. P. Connerade, "The physics of non-Rydberg states," *op. cit.*, [23], pp. 643–696.

149. D. C. Griffin and K. L. Andrew, "Theoretical calculations of the *d*-, *f*-, and *g*-electron transition series," *Phys. Rev.*, **177**, No. 1, 62–71 (1969).

150. S. A. Kuchas, A. V. Karosene, and R. I. Karaziya, "The localization of a 4*f*-electron as a function of the term in the $4d^9 4f$ configuration for Xe, Cs, Ba, and La," *Izv. Akad. Nauk SSSR, Ser. Fiz.*, **40**, No. 2, 270–278 (1976).

151. S. A. Kuchas, A. V. Karosene, and R. I. Karaziya, "The effects of a potential barrier in the 4*d*-photoabsorption spectrum of double ion of barium," *Lit. Fiz. Sb.*, **23**, No. 3, 34–40 (1983).

152. I. M. Band, V. I. Fomichev, and M. B. Trzhaskovskaja, " Dirac-Fock atoms in the rare-earth region and the 4*f* wave function collapse phenomenon," *J. Phys. B*, **14**, No. 7, 1103–1118 (1981).

153. A. V. Karosene and A. A. Kantseryavichyus, "Toward interpreting gigantic resonances of absorption by 4*d*-electrons," *Izv. Akad. Nauk SSSR, Ser. Fiz.*, **49**, No. 8, 1501–1504 (1985).

154. T. Åberg, K. A. Jamisson, and P. Richard, "Theory of two-electron rearrangement *K* x-ray transitions," *Phys. Rev. A*, **15**, No. 1, 172–179 (1977).

155. G. Racah and Y. Shadmi, "The configurations $(3d + 4s)^N$ in the second spectra of the iron group," *Bull. Res. Council Israel*, **8F**, No. 1, 15–23 (1959).

156. A. Yu. Kantseryavichyus, "On the mixing of $3d^n 4s^k$ configurations in atoms and ions of the transition elements," *Lit. Fiz. Sb.*, **17**, No. 3, 295–303 (1977).

157. D. R. Beck and C. A. Nicolaides, "Theory of one-electron binding energies including correlation, relativistic and radiative effects: application to free atoms and metals," *op. cit.* [16], pp. 329–359.

158. D. R. Beck and C. A. Nicolaides, "Specific correlation effects in inner-electron photoelectron spectroscopy," *Phys. Rev. A*, **26**, No. 2, 857–862 (1982).

159. W. Mehlhorn, "Structure and dynamics of atoms probed by inner-shell ionization," *op. cit.* [22], pp. 213–241.

160. M. H. Chen, B. Crasemann, N. Mårtensson, and B. Johansson, "Residual limitations of theoretical atomic-electron binding energies," *Phys. Rev. A*, **31**, No. 2, 556–563 (1985).

161. V. F. Demekhin, V. L. Sukhorukov, T. V. Shelkovich, V. A. Yavna, and Yu. I. Bairachnyi, "The multiconfiguration approximation for interpreting the x-ray and electron spectra of the transition elements," *Zh. Strukt. Khim.*, **20**, No. 1, 38–48 (1979).

162. M. H. Chen, "Relativistic calculation of atomic transition probabilities," *op. cit.* [25], pp. 31–95.

163. E. W. McDaniel, M. R. Flannery, E. W. Thomas, and S. T. Manson, "Selected bibliography on atomic collisions: data collections, bibliographies, review articles, books, and papers of particular tutorial value," *At. Data Nucl. Data Tables*, **33**, No. 1, 1–148 (1985).

164. H. S. W. Massey and E. H. S. Burhop, *Electronic and Ionic Impact Phenomena*, Vol. I, Oxford University, London (1969).

165. Y. K. Kim, "Theory of electron-atom collisions," in: *Physics of Ion—Ion and Electron—Ion Collisions*, F. Brouillard and J. W. McGowan (eds.), Plenum Press, New York (1983), pp. 101–165.

166. G. F. Drukarev, *Collisions of Electrons with Atoms and Molecules*, Plenum Press, New York (1987).

167. P. G. Burke, "R-matrix theory of atomic and molecular processes," in: *Atomic Physics*, Vol. 5, R. Marrus, M. Prior, and H. Shugart (eds.), Plenum Press, New York (1977), pp. 293–312.

168. C. Bottcher, D. C. Griffin, and M. S. Pindzola, "Excitation-autoionization of Ti^{3+}, Zr^{3+}, and Hf^{3+} in the distorted-wave approximation with exchange," *J. Phys. B*, **16**, No. 3, L65–L70 (1983).

169. W. Lotz, "Electron impact ionization cross sections and ionization rate coefficients for atoms and ions," *Z. Phys.*, **216**, No. 3, 241–247 (1968); **220**, No. 5, 466–472 (1969).

170. L. A. Vainshtein, I. I. Sobel'man, and E. A. Yukov, *The Cross Sections of Atom and Ion Excitation by Electrons* [in Russian], Nauka, Moscow (1972).

171. U. Fano, "Differential inelastic scattering of relativistic charged particles," *Phys. Rev.*, **102**, No. 2, 385–387 (1956).

172. J. H. Scofield, *K*- and *L*-shell ionization of atoms by relativistic electrons," *Phys. Rev. A*, **18**, No. 3, 963–970 (1978).

173. H. Kolbenstvedt, "Asymptotic expression for *K*-shell ionization cross section with electrons," *J. Appl. Phys.*, **46**, No. 6, 2771–2773 (1975).

174. P. Eschwey and P. Manakos, "Structure functions and ionization of isolated atoms by charged projectiles," *Z. Phys. A*, **308**, No. 3, 199–207 (1982).

175. V. F. Demekhin, Yu. I. Bairachnyi, V. L. Sukhorukov, and S. A. Prosandeev, "On interpreting the spectra of the *L*-series of elements from the iron group," *Izv. Akad. Nauk SSSR, Ser. Fiz.*, **40**, No. 2, 263–266 (1976).

176. H. Madison and R. Merzbacher, "Theory of charged-particle excitation," *op. cit.* [12], pp. 1–72.

177. P. Richard, "Ion-atom collisions," *op. cit.* [12], pp. 73–158.

178. M. Barrat, "The quasimolecular model in heavy particle collisions," *op. cit.* [13], pp. 229–253.

179. R. Hippler, "Impact ionization by fast projectiles," in: *Progress in Atomic Spectroscopy, Part C*, H. J. Beyer and H. Kleinpoppen (eds.), Plenum Press, New York (1984), pp. 511–575.

180. L. Kocbach, J. M. Hansteen, and R. Gundersen, "Review of the SCA method for inner shell ionization — results, limitations, and possibilities," *Nucl. Instrum. Methods*, **169**, No. 3, 281–291 (1980).

181. H. Bethe and R. Jackiw, *Intermediate Quantum Mechanics*, 2nd edn., Benjamin, New York (1968).

182. M. Gryzinski, "Classical theory of atomic collisions. Part I. Theory of inelastic collisions," *Phys. Rev.*, **138**, No. 2A, 336–358 (1965).

183. J. S. Briggs, "Molecular treatment of atomic collisions (inner shells)," *op. cit.* [20], pp. 421–445.

184. J. D. Garcia, R. J. Fortner, and T. M. Kavanagh, "Inner-shell vacancy production in ion—atom collisions," *Rev. Mod. Phys.*, **45**, No. 2, 111–177 (1973).

185. J. H. McGuire and L. Weaver, "Independent electron approximation for atomic scattering by heavy particles," *Phys. Rev. A*, **16**, No. 1, 41–47 (1977).

186. V. I. Matveev and É. S. Parilis, "Shakeup during electron transitions in atoms," *Usp. Fiz. Nauk*, **138**, No. 4, 573–602 (1982).

187. V. P. Sachenko and V. F. Demekhin, "Satellites of x-ray spectra," *Zh. Eksp. Teor. Fiz.*, **49**, No. 3, 765–769 (1965).

188. T. Åberg, "Theory of x-ray satellites," *Phys. Rev.*, **156**, No. 1, 35–41 (1967).

189. T. A. Carlson and C. W. Nestor, Jr., "Calculation of electron shake-off probabilities as the result of x-ray photoionization of the rare gases," *Phys. Rev. A*, **8**, No. 6, 2887–2894 (1973).

190. A. F. Starace, "Theory of atomic photoionization," *op. cit.* [19], pp. 1–121.

191. J. W. Cooper, "Photoionization of inner-shell electrons," *op. cit.* [12], pp. 159–199.

192. R. H. Pratt, A. Ron, and H. K. Tseng, "Atomic photoelectric effect above 10 keV," *Rev. Mod. Phys.*, **45**, No. 2, 273–325 (1973).

193. H. P. Kelly, S. C. Carter, and B. E. Norum, "Calculations of photoionization of the 4*d* subshells of Ba and Ba^{2+} sequence," *Phys. Rev. A*, **25**, No. 4, 2052–2055 (1982).

194. T. B. Lucatorto, T. J. McIlrath, J. Sugar, and S. M. Younger, "Radical redistribution of the 4*d* oscillator strength observed in the photoabsorption of the Ba, Ba^{+}, and Ba^{2+} sequence," *Phys. Rev. Lett.*, **47**, No. 16, 1124–1128 (1981).

195. R. I. Karaziya and S. A. Kuchas, "Exchange interaction and the characteristics of x-ray spectra," *Lit. Fiz. Sb.*, **25**, No. 6, 32–42 (1985).

196. Yu. M. Kanyauskas and R. I. Karaziya, "Algebraic expressions for the energy of terms for maximal multiplicity and the terms associated with them, as well as for the ground and highest levels," *Lit. Fiz. Sb.*, **25**, No. 2, 31–41 (1985).

197. R. I. Karaziya and I. I. Grudzinskas, "Expressions for the average energy of emission and photoexcitation spectra," *Lit. Fiz. Sb.*, **25**, No. 5, 31-42 (1985).

198. I. I. Grudzinskas, "Variance of the photoabsorption spectra for $n_1l_1^{4l_1+2}n_2l_2^{N_2} \rightarrow n_1l_1^{4l_1+1}n_2l_2^{N_2+1}$," *Lit. Fiz. Sb.*, **26**, No. 5, 536–547 (1986).

199. L. Armstrong, Jr. and W. R. Fielder, Jr., "Photoionization cross sections using the multiconfiguration Hartree–Fock and its extensions," *Phys. Scr.*, **21**, No. 3, 457–462 (1980).

200. R. J. Henry and L. Lipsky, "Multichannel photoionization of atomic systems," *Phys. Rev.*, **153**, No. 1, 51–56 (1967).

201. F. Combet Farnoux, "The close-coupling methods applied to photoionization of neutral atoms and positive ions," *op. cit.* [13], pp. 407–417.

202. M. Ya. Amus'ya, V. K. Ivanov, S. A. Sheinerman, and S. I. Sheftel', "The manifestation of electron shell relaxation during ionization processes in atoms," *Zh. Eksp. Teor. Fiz.*, **78**, No. 3, 910–923 (1980).

203. V. L. Sukhorukov, I. D. Petrov, V. F. Demekhin, and S. V. Lavrent'ev, "X-ray processes with the participation of subvalent electrons in Ar, Xe, and HCl," *Izv. Akad. Nauk SSSR, Ser. Fiz.*, **49**, No. 8, 1463–1470 (1985).

204. M. Ya. Amus'ya, V. K. Dolmatov, and V. K. Ivanov, "Photionization of atoms with semifilled shells," *Zh. Eksp. Teor. Fiz.*, **85**, No. 1, 115–123 (1983).

205. W. R. Johnson, "Relativistic many-body calculations," *op. cit.* [22], pp. 149–170.

206. M. Ya. Amusia, V. K. Ivanov, and V. A. Kupchenko, "Photoionization of inner shells," *J. Phys. B*, **14**, No. 21, L667–L671 (1981).

207. C. S. Fadley, "Basic concepts of x-ray photoelectron spectroscopy," *op. cit.* [15], pp. 1–156.

208. V. L. Jacobs, "Theory of atomic photoionization measurements," *J. Phys. B*, **5**, No. 12, 2257–2271 (1972).

209. D. Dill and U. Fano, "Parity unfavoredness and the distribution of photofragments," *Phys. Rev. Lett.*, **29**, No. 18, 1203–1205 (1972).

210. S. T. Manson and D. Dill, "The photoionization of atoms: cross sections and photoelectron angular distributions," *op. cit.* [15], pp. 157–195.

211. J. W. Cooper and S. T. Manson, "Photoionization in the soft x-ray range: angular distributions of photoelectrons and interpretation in terms of subshell structure," *Phys. Rev.*, **177**, No. 1, 157–163 (1969).

212. J. A. R. Samson, "Angular distribution of photoelectrons," *J. Opt. Soc. Am.*, **59**, No. 3, 356–357 (1969).

213. V. Schmidt, "Angular distribution of photoelectrons after photoionization by elliptically polarized light," *Phys. Rev. Lett.*, **45**, No. 1, 63–64 (1973).

214. T. E. H. Walker and J. T. Waber, "The relativistic theory of the angular distribution of photoelectrons in the *jj* coupling," *J. Phys. B*, **6**, No. 7, 1165–1175 (1973).

215. N. A. Cherepkov, "Angular distribution of photoelectrons having a defined spin orientation," *Zh. Eksp. Teor. Fiz.*, **65**, No. 3, 933–946 (1973).

216. C. M. Lee, "Spin polarization and angular distribution of photoelectrons," *Phys. Rev. A*, **10**, No. 5, 1598–1604 (1974).

217. U. Heinzmann, "Experimental determination of the phase differences of continuum wave functions describing the photoionization process of xenon atoms," *J. Phys. B*, **13**, No. 22, 4353–4381 (1980).

218. G. K. Wertheim, A. Rosencwaig, R. L. Cohen, and H. J. Guggenheim, "Exchange splitting in the 4*f* photoelectron spectra of the rare earths," *Phys. Rev. Lett.*, **27**, No. 8, 505–507 (1971).

219. E. I. Zabolotskii, Yu. P. Irkhin, and L. D. Finkel'shtein, "Calculating the spectra of 4*f* photoelectrons of the rare earths in crystals," *Fiz. Tverd. Tela*, **16**, No. 4, 1142–1150 (1974).

220. D. A. Shirley, S. T. Lee, T. Süzer, et al., "Electron correlation in atoms from photoelectron spectroscopy," in: *Atomic Physics*, Vol. 5, R. Marrus, M. Prior, and H. Shugart (eds.), Plenum Press, New York (1977), pp. 313–324.

221. J. E. Hansen, "Correlation in the $ns^2\ ^1S$ ground states of Ca, Sr, Ba, Zn, Cd, and Hg as determined by multiconfiguration Hartree–Fock calculations and photoelectron spectroscopy," *Phys. Rev. A*, **15**, No. 2, 810–813 (1977).

222. K. G. Dyall and F. P. Larkins, "Satellite structure in atomic spectra," *J. Phys. B*, **15**, No. 2, 203–231 (1982).

223. M. O. Krause, T. A. Carlson, and A. Fahlman, "Photoelectron spectrometry of manganese vapor between 12 and 110 eV," *Phys. Rev. A*, **30**, No. 3, 1316–1324 (1984).

224. J. M. Dyke, V. W. J. Gravenor, R. A. Lewis, and A. Morris, "Gas-phase high-temperature photoelectron spectra: an investigation of the transition metals iron, cobalt, and nickel," *J. Phys. B*, **15**, No. 24, 4523–4534 (1982).

225. R. L. Martin and D. A. Shirley, "Theory of the neon 1*s* correlation – peak intensities," *Phys. Rev. A*, **13**, No. 4, 1475–1483 (1976).

226. K. G. Dyall, F. P. Larkins, K. D. Bomben, and T. D. Thomas, "Satellite structure in the argon 2*p* x-ray photoelectron spectrum," *J. Phys. B*, **14**, No. 15, 2551–2558 (1981).

227. I. Lindau, "Resonance effects in photoelectron spectra," *op. cit.* [24], pp. 251–262.

228. R. A. Rosenberg, M. G. White, G. Thornton, and D. A. Shirley, "Selective resonant enhancement of electron correlation satellites in atomic barium," *Phys. Rev. Lett.*, **43**, No. 19, 1384–1387 (1979).

229. A. Fahlman, M. O. Krause, T. A. Carlson, and A. Svensson, "Xe 5*s*, 5*p* correlation satellites in the region of strong interchannel interactions," *Phys. Rev. A*, **31**, No. 2, 812–819 (1984).

230. M. Ohno and G. Wendin, "A many-body calculation of 3*p* XPS and Auger spectra for Zn," *J. Phys. B*, **12**, No. 8, 1305–1328 (1979).

231. R. I. Karaziya, D. V. Grabauskas, and A. A. Kiselev, "Shifts in hard x-ray lines when several vacancies are formed in electron shells," *Lit. Fiz. Sb.*, **24**, No. 2, 249–261 (1974).

232. V. M. Mikhailov and M. A. Khanonkind, "Ionization of an atom and its effect on the energy of x-ray transitions," *Izv. Akad. Nauk SSSR, Ser. Fiz.*, **35**, No. 1, 86–93 (1971).

233. P. G. Hansen, B. Jonson, G. J. Borchert, and O. W. B. Schult, "Mechanisms for energy shifts of atomic *K* x-rays," *op. cit.* [25], pp. 237–280.

234. M. A. Blokhin and I. G. Shveitser, *Handbook of X-Ray Spectra* [in Russian], Nauka, Moscow (1982).

235. M. Gavrila and J. E. Hansen, "Calculation of transition probabilities for two-electron one-photon and hypersatellite transitions for ions with two vacancies in the *K* shell," *J. Phys. B*, **11**, No. 8, 1353–1381 (1978).

236. J. P. Briand, "Correlation effects in x-ray emission spectroscopy," *op. cit.* [13], pp. 287–308.

237. V. L. Ginzburg, "On the nature of spontaneous emission," *Usp. Fiz. Nauk*, **140**, No. 4, 687–698 (1983).

238. E. G. Berezhko and N. M. Kabachnik, "Theoretical study of inner-shell alignment of atoms in electron impact ionization: angular distribution and polarization of x rays and Auger electrons," *J. Phys. B*, **10**, No. 12, 2467–2477 (1977).

239. E. T. Verkhovceva and P. S. Pogrebnjak, "Manifestation of the dipole relaxation process of super-Coster-Kronig type in the soft x-ray emission spectra of Kr and Xe," *J. Phys. B*, **13**, No. 18, 3535–3543 (1980).

240. M. Ohno, "Breakdown of the one-electron picture in $3d^{-1}$—$4p^{-1}$ x-ray emission spectrum of Xe," *J. Phys. B*, **15**, No. 14, 513–520 (1982).

241. K. Tsutsumi, "The x-ray non-diagram lines $K_{\beta'}$ of some compounds of the iron group," *J. Phys. Soc. Jpn.*, **14**, No. 12, 1696–1706 (1959).

242. V. I. Nefedov, "The multiplet structure of $K_{\alpha_{1,2}}$ lines of the transition elements," *Izv. Akad. Nauk SSSR, Ser. Fiz.*, **28**, No. 5, 816–822 (1964).

243. B. Ekstig, E. Källne, E. Noreland, and R. Manne, "Electron interaction in transition metal x-ray emission spectra," *Phys. Scr.*, **2**, No. 1/2, 38–44 (1970).

244. S. I. Salem and P. L. Lee, "The effect of partially filled shells on x-ray emission lines," in: 4th Int. Conf. on Atom. Phys., Heidelberg (1974), pp. 682–685.

245. B. Crasemann, "Relativity: x-ray and Auger transitions," *op. cit.* [17], pp. 97–107.

246. R. D. Deslattes, L. Jacobs, E. G. Kessler, and W. Schwitz, "Comparison of relativistic atomic SCF calculations with improved experimental data," *op. cit.* [18], pp. 144–152.

247. B. Crasemann, "Atomic inner-shell transitions. Theory and the need for experiments," *IEEE Trans. Nucl. Sci.*, **30**, No. 2, 887–890 (1983).

248. R. I. Karaziya and S. A. Kuchas, "Population of the Ar II and Ar III configurations during electron collision and subsequent processes and the structure of the emission spectrum $L_{2,3}$," *Lit. Fiz. Sb.*, **19**, No. 4, 495–504 (1979).

249. K. G. Dyall and I. P. Grant, "The K_β x-ray emission spectrum of argon: a theoretical account," *J. Phys. B*, **17**, No. 7, 1281–1300 (1984).

250. T. Åberg, "Two-photon emission, the radiative Auger effect, and the double Auger process," *op. cit.* [12], pp. 353–375.

251. V. V. Afrosimov and A. P. Shergin, "Correlated transitions in atoms with two inner shell vacancies," in: *Atomic Physics 6*, R. J. Damburg (ed.), Plenum Press, New York (1979), pp. 289–307.

252. G. B. Baptista, "Two-electron one-photon decay rates in doubly ionized atoms," *J. Phys. B*, **17**, No. 11, 2177–2188 (1984); **19**, No. 2, 159–169 (1986).

253. V. F. Demekhin and V. P. Sachenko, "The spectral location of K_α-satellites. Chemical bonding and the relative intensities of K_α-satellites," *Izv. Akad. Nauk SSSR, Ser. Fiz.*, **31**, No. 6, 900–911 (1967).

254. M. Ya. Amusia, "Single-photon and single-electron decay of double vacancy states in atoms," *Commun. At. Mol. Phys.*, **9**, No. 1, 23–34 (1979).

255. W. L. Luken, J. S. Greenberg, and P. Vincent, "X-ray transitions between correlated many-electron quantum states," *Phys. Rev. A*, **15**, No. 6, 2305–2311 (1977).

256. P. V. Vedrinskii and V. V. Kolesnikov, "Auger transitions and the shape of the x-ray spectrum," *Izv. Akad. Nauk SSSR, Ser. Fiz.*, **31**, No. 6, 891–897 (1967).

257. B. Crasemann, "Probing atomic inner shells," *op. cit.* [24], pp. 51 – 60.

258. M. H. Chen, B. Crasemann, M. Aoyagi, and H. Mark, "Interpretation of the silver *L* x-ray spectrum," *Phys. Rev. A,* **15**, No. 6, 2312 – 2323 (1977).

259. W. Melhorn, "Auger-electron spectroscopy of core levels of atoms," *op. cit.* [25], pp. 119 – 180.

260. F. P. Larkins, "Transition energies," *op. cit.* [12], pp. 377 – 409.

261. W. A. Coghlan and R. E. Clausing, "Auger catalog. Calculated transition energies listed by energy and element," *At. Data,* **5**, No. 4, 317 – 469 (1973).

262. F. P. Larkins, "Semi-empirical Auger electron energies, Part I. General method and *K-LL* line energies," *J. Phys. B,* **9**, No. 1, 47 – 58 (1976).

263. L. M. Kishinevskii, V. I. Matveev, and É. S. Parilis, "The Auger half effect," *Pis'ma Zh. Tekh. Fiz.,* **2**, No. 15, 710 – 714 (1976).

264. M. N. Mirachmedov and E. S. Parilis, "Auger and x-ray cascades following inner shell ionization," *op. cit.* [24], pp. 177 – 191.

265. E. J. McGuire, "Auger and Coster-Kronig transitions," *op. cit.* [12], pp. 293 – 330.

266. R. I. Karaziya, A. I. Udris, and I. I. Grudzinskas, "Equations for the probabilities of Auger transitions for configurations having one open shell," *Lit. Fiz. Sb.,* **15**, No. 4, 527 – 537 (1975).

267. I. I. Grudzinskas, R. I. Karaziya, and S. A. Kuchas, "Auger transitions during the decays of vacancies in the subvalent shells of Kr and Xe," *Lit. Fiz. Sb.,* **23**, No. 4, 23 – 33 (1983).

268. V. O. Kostroun, M. H. Chen, and B. Crasemann, "Atomic radiative transition probabilities to the $1s$ state and theoretical *K*-shell fluorescence yields," *Phys. Rev. A,* **3**, No. 2, 533 – 545 (1971).

269. J. N. Ginocchio, "Operator averages in a shell-model basis," *Phys. Rev. C,* **8**, No. 1, 135 – 145 (1973).

270. R. I. Karaziya, "The approximate invariance of Auger and radiative widths of levels and the fluorescence yield," *Lit. Fiz. Sb.,* **23**, No. 1, 6 – 16 (1983).

271. K. N. Huang, "Relativistic radiationless transitions in atoms," *J. Phys. B,* **11**, No. 5, 787 – 795 (1978).

272. H. Aksela, S. Aksela, G. M. Bancroft, et al., "$N_{4,5}OO$ resonance Auger spectra of Xe studied with selective excitation by synchrotron radiation," *Phys. Rev. A,* **33**, No. 6, 3867 – 3875 (1986).

273. B. Cleff and W. Mehlhorn, "On the angular distribution of Auger electrons following impact ionization," *J. Phys. B,* **7**, No. 5, 593 – 604 (1974).

274. N. M. Kabachnik and I. P. Sazhina, "Angular distribution and spin polarization of Auger electrons," *J. Phys. B,* **17**, No. 7, 1335 – 1342 (1984).

275. H. Klar, "Spin polarization of Auger electrons," *J. Phys. B,* **13**, No. 24, 4741 – 4749 (1980).

276. D. R. Beck and C. A. Nicolaides, "Theory of Auger energies in free atoms: application to the alkaline earths," *Phys. Rev.,* **33**, No. 6, 3885 – 3890 (1986).

277. K. G. Dyall and F. P. Larkins, "Satellite structure in atomic spectra, Part IV. The $L_{2,3}MM$ Auger spectrum for Argon," *J. Phys. B,* **15**, No. 17, 2793 – 2806 (1982).

278. W. Mehlhorn, W. Schmitz, and D. Stalherm, "Korrelationseffekte in den Auger-Spektren vol Edelgasen," *Z. Phys.,* **252**, No. 5, 399 – 411 (1972).

279. M. I. Bogdanovichene and R. I. Karaziya, "A joint examination of the Kr$M_{4,5}NN$ and Xe$N_{4,5}OO$ Auger spectra and the energy levels of Kr III and Xe III," *Lit. Fiz. Sb.,* **21**, No. 2, 39 – 52 (1981).

280. H. Aksela, S. Aksela, and H. Patana, "Auger energies of free atoms. Comparison between experiment and relativistic theory," *Phys. Rev. A,* **30**, No. 2, 858 – 864 (1984).

281. M. H. Chen, "Effects of relativity and correlation on *L-MM* Auger spectra," *Phys. Rev. A,* **31**, No. 1, 177 – 186 (1985).

282. H. P. K. Kelly, "Auger rates calculated for Ne^+," *Phys. Rev. A,* **11**, No. 2, 556 – 565 (1975).

283. M. Ya. Amus'ya, V. A. Kilin, A. N. Kolesnikova, and I. S. Li, "The deepening of vacancies in correlation decays of the two-hole states of atoms," *Pis'ma Zh. Eksp. Teor. Fiz.*, **10**, No. 17, 1029–1033 (1984).

284. M. Ohno and G. Wendin, "Many-electron theory of x-ray photoelectron spectra: *N*-shell linewidths in the $_{46}$Pd to $_{92}$U range," *Phys. Rev. A*, **31**, No. 4, 2318–2330 (1985).

285. M. Ya. Amusia, M. Kuchiev, S. A. Sheinerman, and S. J. Sheftel, "Inner-shell correlation in the formation of singly charged ions near the Ar *L*-shell ionization threshold," *J. Phys. B*, **10**, No. 14, L535–L539 (1977).

286. A. Niehaus, "Analysis of post-collision interactions in Auger processes following near-threshold inner-shell photoionization," *J. Phys. B*, **10**, No. 10, 1845–1857 (1977).

287. V. Schmidt, "Post-collision interaction in inner shell ionization," *op. cit.* [21], pp. 544–558.

288. T. C. Chang, D. Eastman, F. J. Himpsel, et al., "Observation of the transition from uncollapsed to collapsed excited *f*-wave functions in I^-, Xe, and Cs^+ via the giant post-collision interaction Auger effect," *Phys. Rev. Lett.*, **45**, No. 23, 1846–1849 (1980).

289. V. Heitler, *The Quantum Theory of Radiation*, 3rd edn., Oxford University Press, London–New York (1954).

290. Yu. S. Gordeev and G. N. Ogurtsov, "On interpreting the energy spectra of the electrons formed during atomic collisions," *Zh. Eksp. Teor. Fiz.*, **60**, No. 6, 2051–2059 (1971).

291. G. C. Nelson, W. John, and B. G. Saunders, "Widths of the $K_{\alpha 1}$ and $K_{\alpha 2}$ x-ray lines for $Z > 50$," *Phys. Rev.*, **187**, No. 1, 1–5 (1969).

292. S. I. Salem and P. L. Lee, "Experimental widths of *K* and *L* x-ray lines," *At. Data Nucl. Data Tables*, **18**, No. 3, 233–241 (1976).

293. W. Bambynek, B. Crasemann, R. W. Fink, et al., "X-ray fluorescence yields, Auger, and Coster-Kronig transition probabilities," *Rev. Mod. Phys.*, **44**, No. 4, 716–813 (1972).

294. M. H. Chen and B. Crasemann, "Multiplet effects on the $L_{2,3}$ fluorescence yield of multiply ionized Ar," *Phys. Rev. A*, **10**, No. 6, 2232–2239 (1974).

295. V. A. Zonn, "On the bremsstrahlung effect during electron-atom collisions," *Zh. Eksp. Teor. Fiz.*, **73**, No. 1, 128–133 (1977).

296. M. Ya. Amusia, "'Atomic' bremsstrahlung spectrum," *Commun. At. Mol. Phys.*, **11**, No. 3–5, 123–137 (1982).

297. R. H. Pratt and I. J. Feng, "Electron-atom bremsstrahlung," *op. cit.* [25], pp. 533–580.

298. E. Haug and M. Keppler, "Bremsstrahlung in the field of bound electrons," *J. Phys. B*, **17**, No. 10, 2075–2084 (1984).

299. R. H. Pratt, "Electron bremsstrahlung," *op. cit.* [18], pp. 411–422.

300. H. W. Koch and J. W. Motz, "Bremsstrahlung cross section formulas and related data," *Rev. Mod. Phys.*, **31**, No. 4, 920–955 (1959).

301. L. D. Landau and E. M. Lifshits, *The Classical Theory of Fields*, 3rd edn., Pergamon, New York (1971).

302. M. Ya. Amus'ya, T. M. Zimkina, and M. Yu. Kuchiev, "Wide emission bands in the radiation from atoms under the action of fast electrons," *Zh. Tekh. Fiz.*, **52**, No. 7, 1424–1426 (1982).

303. E. T. Verkhovtseva, E. V. Gnatenko, and P. S. Pogrebnjak, "Investigation of the connection between giant resonances and atomic bremsstrahlung," *J. Phys. B*, **16**, No. 20, L613–L616 (1983).

304. C. M. Lee and R. H. Pratt, "Comment on structure near the cutoff of the continuous x-ray spectrum of lanthanum," *Phys. Rev. A*, **12**, No. 2, 707–709 (1975).

305. M. B. Chamberlain, A. F. Burr, and R. J. Liefeld, "Excitation-energy dependent features in the continuum spectrum of cerium near the M_α and M_β x-ray emission lines," *Phys. Rev. A*, **9**, No. 2, 663–667 (1974).

306. R. H. Pratt, H. K. Tseng, C. M. Lee, et al., "Bremsstrahlung energy spectra from electrons of kinetic energy 1 keV $\leq T_1 \leq$ 2000 keV incident on neutral atoms 2 $\leq Z \leq$ 92," *At. Data Nucl. Data Tables*, **20**, No. 2, 175–209 (1977).

307. H. K. Tseng and R. H. Pratt, "Exact screened calculations of atomic-field bremsstrahlung," *Phys. Rev. A*, **3**, No. 1, 100–115 (1971).

308. M. Ya. Amus'ya, M. Yu. Kuchiev, A. V. Korol', and A. V. Solov'ev, "Bremsstrahlung of relativistic particles with regard for the dynamic polarizability of the atom-target," *Zh. Eksp. Teor. Fiz.*, **88**, No. 2, 383–389 (1985).

309. M. Gavrila, "Photon-atom elastic scattering," *op. cit.* [21], pp. 357–388.

310. T. Åberg and J. Tulkki, "Inelastic x-ray scattering including resonance phenomena," *op. cit.* [25], pp. 419–463.

311. V. I. Iveronova and G. P. Revkevich, *Theory of X-Ray Scattering*, 2nd edn. [in Russian], Izd. MGU, Moscow (1978).

312. L. Kissel, R. H. Pratt, and S. C. Roy, "Rayleigh scattering by neutral atoms 100 eV to 10 MeV," *Phys. Rev. A*, **22**, No. 5, 1970–2004 (1980).

313. B. L. Henke, P. Lee, T. J. Tanaka, et al., "Low-energy x-ray interaction coefficients: photoabsorption scattering and reflection. E = 100–2000 eV; Z = 1–94," *At. Data Nucl. Data Tables*, **27**, No. 1, 1–144 (1982).

314. J. C. Parker and R. H. Pratt, "Validity of common assumptions for anomalous scattering," *Phys. Rev. A*, **29**, No. 1, 152–158 (1984).

315. L. Kissel and R. H. Pratt, "Rayleigh scattering: elastic photon scattering by bound electrons," *op. cit.* [25], pp. 465–532.

316. C. Lin, K. T. Cheng, and W. R. Johnson, "Rayleigh scattering of photons by helium," *Phys. Rev. A*, **11**, No. 6, 1946–1956 (1975).

317. P. Eisenberger, P. M. Platzman, and H. Winick, "Resonant x-ray Raman scattering studies using synchrotron radiation," *Phys. Rev. B*, **13**, No. 6, 2377–2380 (1976).

318. G. S. Brown, M. H. Chen, B. Crasemann, and G. E. Ice, "Observation of the Auger resonant Raman effect," *Phys. Rev. Lett.*, **45**, No. 24, 1937–1940 (1980).

319. M. Cooper, "Compton scattering and electron momentum distributions," *Adv. Phys.*, **20**, No. 86, 453–491 (1971).

320. F. Briggs, L. B. Mendelsohn, and J. B. Mann, "Hartree–Fock Compton profiles for the elements," *At. Data Nucl. Data Tables*, **16**, No. 3, 201–309 (1975).

321. J. M. Jauch and F. Rohrlich, *Theory of Photons and Electrons*, Addison-Wesley, Reading, Massachusetts (1955).

322. G. Bhatt, H. Grotch, E. Kazes, and D. A. Owen, "Relativistic spin-dependent Compton scattering from electrons," *Phys. Rev. A*, **28**, No. 4, 2195–2200 (1983).

323. J. P. Briand, D. Girard, V. O. Kostroun, et al., "X-ray Raman and Compton scattering in the vicinity of a deep atomic level," *Phys. Rev. Lett.*, **46**, No. 25, 1625–1628 (1981).

324. A. A. Radtsig and B. M. Smirnov, *Parameters of Atoms and Atomic Ions*, 2nd edn. [in Russian], Énergoatomizdat, Moscow (1986).

325. W. L. Jolly, K. D. Bomben, and C. J. Eyermann, "Core-electron binding energies for gaseous atoms and molecules," *At. Data Nucl. Data Tables*, **31**, No. 3, 433–494 (1984).

326. K. D. Sevier, "Atomic electron binding energies," *At. Data Nucl. Data Tables*, **24**, No. 4, 323–371 (1979).

INDEX

Series Publications

Below is a chronological listing of all the published volumes in the *Physics of Atoms and Molecules* series.

ELECTRON AND PHOTON INTERACTIONS WITH ATOMS
Edited by H. Kleinpoppen and M. R. C. McDowell

ATOM–MOLECULE COLLISION THEORY: A Guide for the Experimentalist
Edited by Richard B. Bernstein

COHERENCE AND CORRELATION IN ATOMIC COLLISIONS
Edited by H. Kleinpoppen and J. F. Williams

VARIATIONAL METHODS IN ELECTRON–ATOM SCATTERING THEORY
R. K. Nesbet

DENSITY MATRIX THEORY AND APPLICATIONS
Karl Blum

INNER-SHELL AND X-RAY PHYSICS OF ATOMS AND SOLIDS
Edited by Derek J. Fabian, Hans Kleinpoppen, and Lewis M. Watson

INTRODUCTION TO THE THEORY OF LASER–ATOM INTERACTIONS
Marvin H. Mittleman

ATOMS IN ASTROPHYSICS
Edited by P. G. Burke, W. B. Eissner, D. G. Hummer, and I. C. Percival

ELECTRON–ATOM AND ELECTRON–MOLECULE COLLISIONS
Edited by Juergen Hinze

ELECTRON–MOLECULE COLLISIONS
Edited by Isao Shimamura and Kazuo Takayanagi

ISOTOPE SHIFTS IN ATOMIC SPECTRA
W. H. King

AUTOIONIZATION: Recent Developments and Applications
Edited by Aaron Temkin

ATOMIC INNER-SHELL PHYSICS
Edited by Bernd Crasemann

COLLISIONS OF ELECTRONS WITH ATOMS AND MOLECULES
G. F. Drukarev

THEORY OF MULTIPHOTON PROCESSES
Farhad H. M. Faisal

PROGRESS IN ATOMIC SPECTROSCOPY, Parts A, B, C, and D
Edited by W. Hanle, H. Kleinpoppen, and H. J. Beyer

RECENT STUDIES IN ATOMIC AND MOLECULAR PROCESSES
Edited by Arthur E. Kingston

QUANTUM MECHANICS VERSUS LOCAL REALISM: The Einstein-Podolsky-Rosen Paradox
Edited by Franco Selleri

ZERO-RANGE POTENTIALS AND THEIR APPLICATIONS IN ATOMIC PHYSICS
Yu. N. Demkov and V. N. Ostrovskii

COHERENCE IN ATOMIC COLLISION PHYSICS
Edited by H. J. Beyer, K. Blum, and R. Hippler

ELECTRON–MOLECULE SCATTERING AND PHOTOIONIZATION
Edited by P. G. Burke and J. B. West

ATOMIC SPECTRA AND COLLISIONS IN EXTERNAL FIELDS
Edited by K. T. Taylor, M. H. Nayfeh, and C. W. Clark

ATOMIC PHOTOEFFECT
M. Ya. Amusia

MOLECULAR PROCESSES IN SPACE
Edited by Tsutomu Watanabe, Isao Shimamura, Mikio Shimizu, and Yukikazu Itikawa

THE HANLE EFFECT AND LEVEL CROSSING SPECTROSCOPY
Edited by Giovanni Moruzzi and Franco Strumia

ATOMS AND LIGHT: INTERACTIONS
John N. Dodd

POLARIZATION BREMSSTRAHLUNG
Edited by V. N. Tsytovich and I. M. Ojringel

INTRODUCTION TO THE THEORY OF LASER–ATOM INTERACTIONS (Second Edition)
Marvin H. Mittleman

ELECTRON COLLISIONS WITH MOLECULES, CLUSTERS, AND SURFACES
Edited by H. Ehrhardt and L. A. Morgan

THEORY OF ELECTRON–ATOM COLLISIONS, Part 1: Potential Scattering
Philip G. Burke and Charles J. Joachain

POLARIZED ELECTRON/POLARIZED PHOTON PHYSICS
Edited by H. Kleinpoppen and W. R. Newell

INTRODUCTION TO THE THEORY OF X-RAY AND ELECTRONIC SPECTRA OF FREE ATOMS
Romas Karazija

VUV AND SOFT X-RAY PHOTOIONIZATION
Edited by Uwe Becker and David A. Shirley

GPSR Compliance

The European Union's (EU) General Product Safety Regulation (GPSR) is a set of rules that requires consumer products to be safe and our obligations to ensure this.

If you have any concerns about our products, you can contact us on ProductSafety@springernature.com

In case Publisher is established outside the EU, the EU authorized representative is:

Springer Nature Customer Service Center GmbH
Europaplatz 3
69115 Heidelberg, Germany

Batch number: 10761599

Printed by Printforce, the Netherlands